This book is dedicated to the students in my Nazi Germany seminar who encouraged me and to my husband, Wilhelm Schauer, who sustained me.

The Nazi Symbiosis

Human Genetics and Politics in the Third Reich

SHEILA FAITH WEISS

The University of Chicago Press   Chicago and London

SHEILA FAITH WEISS is professor of history at Clarkson University.

The University of Chicago Press, Chicago 60637
The University of Chicago Press, Ltd., London
© 2010 by The University of Chicago
All rights reserved. Published 2010
Printed in the United States of America
19 18 17 16 15 14 13 12 11 10   1 2 3 4 5

ISBN-13: 978-0-226-89176-7 (cloth)
ISBN-10: 0-226-89176-3 (cloth)

Library of Congress Cataloging-in-Publication Data
Weiss, Sheila Faith.
    The Nazi symbiosis : human genetics and politics in the Third
Reich / Sheila Faith Weiss.
       p. cm.
    Includes bibliographical references and index.
    ISBN-13: 978-0-226-89176-7 (cloth: alk. paper)
    ISBN-10: 0-226-89176-3 (cloth: alk. paper)
    1. Eugenics—Germany—History—20th century. 2. Human
genetics—Political aspects—Germany—20th century.
3. Human genetics—Government policy—Germany—20th
century. 4. Human genetics—Moral and ethical aspects—
Germany—20th century. 5. National socialism and science.
6. Germany—Politics and government—1933–1945. I. Title.
HQ755.5.G3W47 2010
363.9'20943—dc22   2010005549

♾ The paper used in this publication meets the minimum requirements
of the American National Standard for Information Sciences—
Permanence of Paper for Printed Library Materials, ANSI Z39.48-1992.

Contents

An Old Legend and a New Legacy

Concerns emanating from National Socialism will undoubtedly impact the direction of [biological] research such that the biological worldview of Nazism and the science of biology are united into one. We are merely standing at the beginning of such a symbiotic penetration and development. Only the future will shed light on the results of this symbiosis.[1] DR. WERNER SIEDENTOP, GERMAN HIGHER SECONDARY SCHOOL BIOLOGY TEACHER (1935)

In the southwestern German state of Baden-Württemberg, a short distance from the French border, lies the picturesque city of Staufen.[2] As one embarks from the train station and walks past the ruins of the medieval castle nestled among the vineyards, one immediately becomes aware that Staufen possesses a long and rich heritage. Yet just how rich that heritage is only really becomes clear when one enters the old town square or *Marktplatz*. Conspicuous among the old buildings of the *Marktplatz* is the one known to inhabitants of this quaint city as the Lion's Inn. On the façade of this old structure, visitors can still detect a story painted in a style reminiscent of a page from a medieval tome. This message relates the fate of Staufen's most famous resident, who, according to the inscription, found his demise in this building. The man was none other than Dr. Faustus—a scholar of the late Middle Ages—legendary for striking a deal with the devil in which the former agreed to sell his soul in exchange for twenty-four years of unparalleled knowledge in the black arts. As the message tells us, it was purportedly in

Staufen that, in 1539, Lucifer's right-hand demon, Mephistopheles, appeared to claim Faustus's soul for eternal damnation.

Staufen proudly bears the moniker "the City of Faust" for its association with this cultural icon of medieval Europe, and rightfully so. Hardly any other figure of Western European folklore has fueled the creative drives of so many intellectuals such as the German poet and playwright Johann Wolfgang von Goethe, the French composer Charles Gounod, and the German novelist Thomas Mann for so many centuries.[3] Arguably, however, it was the English dramatist Christopher Marlowe who created the first lasting image of the scholar in his 1604 play *The Tragical History of Dr. Faustus*.[4] In his tragedy, Marlowe portrays his title character as a scholar with an insatiable appetite for knowledge. The curtain rises on Faustus pacing the floor of his study in agitation; after mastering all scholarly pursuits, the Doctor still thirsts for knowledge and intends to turn to the black arts to gain omniscience. This he does by conjuring up the evil spirit Mephistopheles in a supernatural ceremony, in the hopes that Mephistopheles' master, the Devil, can give Faustus the knowledge he craves. But sure enough, there is a catch: Mephistopheles is willing to grant Faustus's wish only under the condition that the scholar pledge himself, body and soul, to the Devil at the end of his life. Dr. Faustus accepts these terms; he thus enters into what posterity has more recently dubbed the "Faustian bargain." Nevertheless, like all tragedies, Marlowe's story does not end well for the protagonist. As death draws near, Faustus realizes that his unlimited thirst for knowledge and its resultant fleeting thrill of power were not worth the price he would soon have to pay. And even though Faustus tries to repent his sins to attain salvation, his efforts are futile; evil spirits whisk away the pathetic frame of the once great scholar to suffer the miseries of hell. For Marlowe's audience, the lesson of the story is clear: people should beware of compromising their morals for personal gain, lest they, too, be eternally damned. The author also intended to question the hubris of modernity (Christopher Marlowe wrote at the dawn of the modern era): the belief that humans should and can control everything and that they could cross once impenetrable boundaries—both scientific and ethical—with impunity.

This book also examines a Faustian bargain, but one of recent origin whose legacy is infinitely more tragic than anything envisioned by Marlowe. The volume explores the deal struck between German specialists in human heredity (or, as it was later more frequently termed, human genetics) and representatives of the very epitome of evil in the twentieth century: Nazism. Both parties were relatively new to the world scene when they initiated the bargain in 1933. Human heredity owed its existence largely to the rediscovery of Gregor Mendel's (1822–84) laws of inheritance in 1900. For many practitioners in this fledgling discipline, eugenic concerns motivated their interest in questions of human genetics. Eugenics, according to Francis Galton (1822–1911), the British statistician, cousin of Charles Darwin, and researcher into the mathematical understanding of human heredity who coined the term in 1883, was "'the science' of improving human stock by giving 'the more suitable races of blood a better chance of prevailing over the less suitable.'"[5] Research into human genetics, especially into what would later be known as medical genetics, fueled a desire to apply this knowledge to improve what we today call the gene pool of a human population. These twin fields, human genetics and eugenics, were not separate entities during much of the first half of the twentieth century. As such, we should view them and their German practitioners as one unit and one party to the bargain. The other party to the deal was the political movement known as National Socialism (or Nazism, for short) and its key officials. Nazism, as is generally known, rose from the ashes of a humiliated, impoverished, and politically impotent Germany in the years immediately following the First World War (1914–18). Even though the origins of these future partners of the Faustian bargain were completely separate, historical circumstances would soon find the two courting each other when the Nazis became the most important political force during the last years of the ill-fated Weimar Republic (1918–33). By this time, the financial constraints put on many professionals by the worldwide depression as well as a sincere interest to advance their science—both theoretical and applied—motivated members of the German human genetics community to look seriously at the one political party that had "race" and "heredity" as its intellectual focal points: the National Socialist German Workers' Party (or NSDAP). The expectation that the Nazis would somehow further their professional interests existed both among individuals who produced human genetic knowledge and those who disseminated this science. Given the NSDAP's emphasis on biological politics—in particular, its desire to craft a genetically healthy and racially pure German national community, or *Volk*—it comes as little surprise that key Nazi bureaucrats would actively seek the help of experts in human heredity. These were the only scholars in Germany who could "scientifically" advance National Socialist racial goals both at home and abroad. For both sides about to seal the bargain, the German human geneticists and Nazi state and party officials, it must have appeared to be a marriage made in heaven.

Despite the fact that scholars have only recently referred to the relationship between German human geneticists and the political authorities during the Third Reich as a "Faustian bargain,"[6] historians have been examining the interaction between the two parties for decades. More than sixty years have passed since the liberation of Auschwitz—the time when the world saw the full horrors committed by the Nazis in the name of "racial purity" and "good heredity." However, it has only been about thirty years since researchers earnestly began their quest to understand the motives of those scientists largely responsible for these and other heinous crimes. German geneticist Benno Müller-Hill's *Tödliche Wissenschaft* (*Murderous Science*), first published in 1984, was the pioneering study.[7]

Since the publication of *Tödliche Wissenschaft,* more than two hundred books and articles have been written dealing with facets of the biomedical sciences and their practitioners under National Socialism. There is, to be sure, enough published material available on aspects of the topic to occupy an individual for a lifetime.[8] Even delineating a historiographical landscape of the current literature is difficult. Recognizing that what follows is, by necessity, a simplification, one can say that these studies differ greatly in respect to their scholarly intent, degree of historical contextualization, and interpretative framework.

Some of the oldest and pathbreaking studies have concentrated on laying bare the frightful particulars of Nazi medical crimes and their perpetrators.[9] Many of these publications focus on naming the medically trained individuals who undertook nonconsensual medical experiments in "euthanasia hospitals," as well as at slave labor or extermination camps and describing their grizzly deeds. As a rule, these works, like those of German journalist Ernst Klee, do not embed their subjects or their activities in their proper historical context.[10] While no one would wish to underestimate the importance of revealing the "witches' Sabbath"[11] of medical transgressions that form the content of books on medicine of this genre, publications dealing with these important topics do not adequately explain how such actions were possible in the first place. Indeed, they lack a viable framework in which to analyze them. Some books, like Edwin Black's *War against the Weak,* although not limited to medical crimes, is simply wrong-headed and sensationalistic. It assumes that the American eugenics movement is primarily responsible for racial policy, and ultimately, the Holocaust, in Germany during the Third Reich—an untenable position, the existence of a strong U.S.-German eugenic connection notwithstanding. It is purposely designed to shock readers (which it does) by mixing truth with half-truths and distortions. From reading Black's work, one would think that all American eugenicists were Nordic racists. We are also made to believe that all Rockefeller Foundation (RF) officers had no qualms about what was going on in the Kaiser Wilhelm

Institute (KWI) for Psychiatry during the early years of the Third Reich. No mention is made of Alan Gregg and Daniel O'Brien's critique. More to the point, the reader is never told that the RF's policy was to fund "good science," irrespective of politics. One can and should debate the appropriateness of adhering to a policy that supports seemingly good research under intolerable conditions, but it is important to understand the RF's position in order to explain why its funding went on as long as it did. Moreover, as historian of biology Garland Allen has rightfully claimed, Black's conclusions miss the point about the dangers of biological reductionism in an age of genomic medicine.[12] Indeed, Black's account is at odds with what this author is trying to achieve in the present study: contextualized history.

There are, however, notable examples of excellent, fully contextualized, studies on biomedical perpetrators—particularly those involved in the Nazi "euthanasia" project. Their purpose goes well beyond exposing the reader to the ill deeds of the medically trained "techno-bureaucratic intelligentsia." Indeed, the excellent work of Michael Burleigh tells us as much about the problematic profession of psychiatry in the twentieth century as it does about psychiatrists involved in the murder of "useless eaters." Henry Friedlander designed his monumental study of Nazi "euthanasia" to demonstrate its inextricable connection to the "Final Solution" as well as analyzing the actions of those involved in this murder project. In this process of intellectually multitasking, the authors have sparked debates surrounding the motivations of the "euthanasia" physicians—with Burleigh stressing the importance of economic and utilitarian factors in the arguments of the perpetrators and Friedlander emphasizing the role of ideology. Both, however, would agree that career opportunism—be it in order to rise in the professional hierarchy or to take advantage of unique outlets for vanguard research—played a significant role. Other scholars, such as Michael Kater, have aimed to provide a "group portrait" of medical professionals during the Third Reich—not just those involved in obvious medical crimes—employing the tools of prosopography.[13]

On the opposing methodological side of research, some investigators have adopted a biographical approach to the subject. By concentrating on an important German biomedical researcher, such as Eugen Fischer (1874–1967) or Ernst Rüdin (1874–1952), or a leading figure in the Nazi Party's "euthanasia" bureaucracy, such as Hitler's personal physician Karl Brandt (1904–48), authors such as Niels Lösch, Matthias Weber, and, most recently, Ulf Schmidt offer an in-depth intellectual and political profile of some of the key players in the construction, legitimization, and execution of the Nazi "racial state."[14] The expectation is that the worldviews and professional networks of these individuals will serve to illuminate their unholy career trajectories during the Third Reich. If done well, professional biographies are certainly excellent vehicles to reveal some of the unique as well as generational motivations for a biomedical scientist's decision to negotiate the "Faustian bargain."

Although the lay population may still believe that the research of German biomedical experts, especially in the field of human heredity, has been distorted or corrupted by the Nazi politics, serious scholars have thankfully never supported that apologia—at least since the influential studies of Gerhard Baader, Karl-Heinz Roth, Götz Aly, Robert Proctor, Hans-Walter Schmuhl, Paul Weindling, and Peter Weingart et al. in the 1980s.[15] That having been said, it is only fairly recently that we possess a large body of scholarship that outlines, in minute detail, the fruitful relationship between biomedical professionals and key organs of the Nazi state. These works investigated the institutes that were part of Germany's most important umbrella scientific organization, the Kaiser-Wilhelm-Gesellschaft (the Kaiser Wilhelm Society or KWS). In the context of a large five-year project (1999–2004) organized by the Max Planck Society, the postwar successor to the KWS, an international cadre of historians have published articles, anthologies, and books dealing with many of the forty-odd KWS institutes under the swastika. Many deal specifically with the six biology-related institutes of the Society during the Third Reich.[16] There were, of course, very important studies undertaken on the KWS and some of its institutes prior to this project. Kristie Macrakis provided the first English overview of the KWS during the Third Reich—no small undertaking. In it, she examined, among other things, the Kaiser Wilhelm Institute for Anthropology, Human Heredity and Eugenics (KWIA), a research center that is critical for the present investigation.[17] The pioneering work of historian of medicine Hans-Peter Kröner on the "posthistory" of the KWIA as well as its biomedical scientists in the immediate postwar period was indispensable for this study. It is a book that warrants a far larger audience than it has received in the United States.[18]

The rich and nuanced institutional studies stemming from the recent Max Planck Society project also deserve a wide readership. Among other things, they demonstrate that, in most cases, the researchers involved in the KWIs were engaged in vanguard science, thereby setting to rest the idea that the Nazis promoted "pseudo-science." Moreover, they also make clear that world-renowned figures such as plant geneticists Fritz von Wettstein (1895–1945) and Hans Stubbe (1902–89), biochemist and Nobel Prize–winner Adolf Butenandt (1903–95), as well as neuropathologist Hugo Spatz (1888–1969)—to mention only some of the biologically trained individuals not examined, or not examined in detail, in this present study—negotiated a highly productive relationship with relevant Nazi officials in order to continue their research.[19] They were not unwilling puppets of the National Socialist regime. Fortunately, at least some of these books and articles are already available in English translation. Just recently, Susanne Heim, Carola Sachse, and Mark Walker have done the English-speaking world a tremendous intellectual favor by publishing an anthology of some of the most important preprints stemming from the project.[20]

Although I was greatly enriched by having had the opportunity to participate in the Max Planck Society Project, the present volume is designed neither as an institutional history of the two KWS biomedical institutes (their different origins and methods of funding notwithstanding) that form chapters of this book nor as a biography of the three directors of these KWS research centers. I also do not aim to add to the collection of studies that are primarily preoccupied with exposing the variety and number of medical crimes perpetrated during this period, as important as this undertaking is for posterity.

This book has a different aim. First, it intentionally focuses on the "Faustian bargain" made between biomedical professional and officials of the Nazi state *itself*—be it located in KWS institutes, on the international stage, or in the college preparatory biology classroom. The study seeks to explain why and how this "deal" was negotiated as well as explore its ethical and professional consequences for the biomedical practitioners as well as the political ramifications for the institutionalization of Nazi racial policies. Second, the volume is designed primarily for the non-specialist, although it is hoped that scholars in the field can also benefit from this analysis.[21] It presents the nonexpert with the newest scholarship in the field of German human heredity under the Nazis, without assuming that the reader has previous detailed knowledge of the subject matter. In addition, it shows how important members of the German human genetics community functioned not only during the peak genocidal years of the regime, but also how they operated within the social, economic, and political contexts of the early years of the Third Reich. Even their activities and professional frustrations during the Weimar period are taken into account, since I believe that one cannot understand German human geneticists' willingness to enter into a deal with dignitaries of the Nazi regime without examining this earlier context. However, there is one important caveat I must make clear regarding the use of the "Faustian bargain" metaphor: unlike the deal made between Faust and the Devil in Marlowe's play, German human geneticists and officials of the Nazi state never sealed a once-for-all-time agreement. The relationship between them was ever-changing, if always useful to both parties. Third, although this is not the first study to place human heredity and racial hygiene in Germany, both before and during the Third Reich, in its larger international context, it does so much more systematically.[22] If one wishes to assess what, if anything, was different about the practice of eugenics under the swastika, one must be able to examine it as part of a larger international network.

Finally, this volume seeks to offer a nuanced, nonjudgmental assessment of the myriad ways in which human geneticists and German politics served to reinforce each other. The rich and wide-ranging archival source base used for this study (please see the list of archival sources) prevents any black-and-white interpretation of what I have termed the "Nazi symbiosis" between the science of human heredity and the racial policy aspirations of party officials during Third Reich. As such, I do not wish to "demonize" the biomedical scientists who actively took part in the most distasteful "fruits" of the "bargain." When one thinks of Josef Mengele (1911–79), perhaps the most notorious German human geneticist owing to the heinous medical crimes he perpetrated at Auschwitz, it is easy to believe that the scientists under consideration in this study were a different, indeed monstrous, strain of humanity. This was not the case. Tragically, the truth is that these researchers were all too human. The motivations for their actions were not intrinsically different from those of other professionals. Moreover, in discussing the actual science pursued by these human geneticists, the book will reinforce the efforts of other scholars who have tried to dispel a second myth common among the nonspecialist: that eugenics and racial anthropology, two essential subspecialties under the rubric of human heredity in the first half of the twentieth century, were "pseudoscientific" pursuits. Whatever one might think about them today, both were internationally respected and practiced by world-renowned human geneticists for most of the first half of the twentieth century. As the first chapter will make clear, neither eugenics nor racial anthropology was a Nazi invention. Both flourished inside and outside of Germany prior to the advent of the Third Reich.

If, indeed, this is the case, inescapable questions follow: What, if anything, was particularly "Nazi" about the way human heredity (broadly defined to include eugenics and racial anthropology) functioned under the auspices of the Third Reich? What induced so many trained German human geneticists to enter into this Faustian bargain in the first place, and what did the senior partner to this deal, the Nazi state and its functionaries, gain from these scientists? Why did these same individuals continue to work for the Nazis even as the truly evil nature of the regime began to rear its ugly head—long before, one might add, its foray into mass murder? And why, in the midst of a brutal and brutalizing war, did many of these same professionals leap into the moral abyss by engaging in research using victims from concentration, extermination, and slave labor camps, as well as "euthanasia" hospitals? To sum up these series of questions with the haunting query of eminent Holocaust scholar Omer Bartov, "what was it that induced Nobel Prize–winning scientists" and "physicians known throughout the world for their research . . . to become not merely opportunistic accomplices but in many ways the initiators and promoters of [the] attempt to subject the human race to a vast surgical operation by means of mass extermination of whole categories of human beings?"[23]

Providing historically satisfying answers to these questions is no easy task. It is, of course, relatively simple to single out a few of the anomalies of the human heredity-politics symbiosis under the swastika. For example, although mandatory sterilization laws existed in numerous countries, the number of Germans who were robbed of their fertility exceeded, by more than sixfold, that of the pioneer nation in this endeavor, the United States. In addition, although "positive eugenics" measures to increase the number of "valuable births" was echoed in practically every country boasting a eugenics movement, no nation went nearly as far as did Nazi Germany to take the combination of necessary steps to make this wish a seeming reality. Moreover, Germany was unique, even among fascist countries, in functioning as a "racial state"—a nation where the criterion for citizenship was determined by race and heredity. National Socialist Germany approached a biocracy, that is, a government where biomedical ideals and biomedical professionals were central for the regime in both word and deed. Finally, it goes without mentioning that none of the egregious medical crimes perpetrated by the biomedical professionals under discussion here—perhaps a potential logical result of a mindset that sought to separate the "valuable from the valueless"—have their counterparts elsewhere. We will examine these differences in the following chapters. But to facilitate an understanding of how the German biomedical community could have taken the path that eventually led to genocide, a theoretical framework is required—one that elucidates the interface between modern science and politics in general and applies it to this specific case.

This study employs such a paradigm. First proposed by historian of science Mitchell Ash in an article published in 2002 entitled "Science and Politics as Resources for Each Other,"[24]Ash's theory challenges the idea that governments, especially more tendentious ones such as the Nazi regime, tend to simply "misuse" or "mobilize" science to do the bidding of the political authorities. This rightfully undermines the notion that the scientists involved are little more than passive pawns instrumentalized to do the government's dirty work. Rather, Ash argues, we should view the relationship between politics and science in the modern world as mutually beneficial. Scientists and governmental authorities serve as "intellectual," "political," "rhetorical," and "financial resources" for each other. Their relationship is dynamic and symbiotic. And not only is this symbiosis mutually beneficial, it often changes the type of scientific questions of interest to researchers. In short, the symbiosis can change the very content and practice of science itself. Although Ash's paradigm is illuminating for many examples of science and technology during the Third Reich, it is, I believe, certainly the case for human heredity under the swastika, as my article of the history of the KWIA demonstrates.[25] Other authors have since recognized the explanatory power of Ash's model.[26]

Not only was the Faustian bargain advantageous for both parties; each served as a constellation of "resources" for the other. More importantly, this symbiosis functioned such that questions and practices in the field of human heredity that were irrelevant or ethically unthinkable during the early years of the Third Reich became vanguard science in Germany during the war. This, I argue, can help explain why human genetics in Germany took the tragic trajectory that it did. To give a concrete example, I believe that Ash's thesis can explain why, after years of constant exposure to Nazi rhetoric on "racial aliens"—rhetoric that German human geneticists both helped to fashion and systematically legitimize through their scientific investigations—some of these same biomedical practitioners became more willing than before to use these "racial outsiders" as guinea pigs in the interest of their research. And as we will see, this research was intricately connected—on all levels—to the political policies of the Nazi racial state. Moreover, as the quote at the outset of this introduction makes clear, even a relatively obscure secondary school biology teacher during the Third Reich recognized the symbiotic relationship between his science and the Nazi worldview. That he could not anticipate the results—scientific, political, and ethical—of this emerging symbiosis in 1935 suggests that few of his other academically trained colleagues in the German human genetics community had possessed this foresight as well.

This study of the Faustian bargain formed between Germany's human genetics community and government officials during the Third Reich certainly raises important questions regarding the ethical practice of science, especially regarding the proper use of human subjects in scientific experiments. But larger, more broadly applicable, lessons can be learned from this episode in history, especially for individuals concerned about potential ethical pitfalls of recent developments in genetic technologies. In particular, the volume reminds the reader of the danger of taking the first morally problematic steps in science, even if they are initially done for "pragmatic" reasons. Justifying a relatively harmless or "understand-able" ethically dubious action serves to morally desensitize scientists from making another, until, ultimately, they find themselves in an ethi-cal quagmire from which it is virtually impossible to escape. Returning, for a moment, to the theme of this study, we must keep in mind that our historical actors were not moral monsters in 1933, even if some were involved in actions that make them appear so in light of the Holocaust. Given this reality, it is imperative for individuals today to try to under-stand the professional and ethical dilemmas that these human geneti-cists faced during the Third Reich, if we are to explain the choices they made. If examined up close, even this most morally problematic field of human genetics under the swastika is not clear-cut. The professional dilemmas—often unpleasant—that these scientists faced under National Socialism were nonetheless real, their grave historical consequences not-withstanding.

Moreover, just as German human geneticists were not necessar-ily moral monsters from the outset, neither was the Nazi state initially viewed by most Germans as a regime that would plunge the country into a racial war and genocide. To the average non-Jewish German citizen in the beginning of 1933, the new government probably did not seem radically different from many of the short-lived Weimar coalitions that preceded it. The Nazis did not gain control of the government through a coup d'état; their rise to power was done within the legal contours of the very political system they were intent on destroying. And when Hitler was appointed chancellor on January 30, 1933, as head of a coali-tion government with the German Nationalists, the Nazis were in the numerical minority in the cabinet, and the Nationalists believed they had Hitler cornered. Hitler, many Germans thought, would help destroy the political left, eliminate the constraints of the hated Versailles Treaty, give them back their national pride, and reintroduce authoritarian, not introduce totalitarian, rule. The majority of Germans and foreign politicians alike did not read Hitler's Bible, *Mein Kampf* (*My Struggle*). Even if they knew Hitler's views in very general terms, most believed they were merely a rhetorical tool to become elected and that he would have to moderate them once entrusted with the reigns of power. We should also remember that ordinary Germans supported the NSDAP for a variety of reasons, but anti-Semitism was not high on the list. Hitler's anti-Marxism was far more important to middle-class and wealthy Germans than his view of the Jews.

Like most academics at the time, many, if certainly not all, of the human geneticists who remained at their posts after 1933 were politically conservative nationalists. Some were also *völkisch*—people who believed in the desirability of an ethnically pure Germany purged of all "non-Aryan" influence—while initially having reservations about the Nazis. A few were party members even prior to 1933. The point is that there was enough overlap between the aims of Nazis and nationalists to speak of a community of interests between them. If this was true of Nazis and nationalists (not just those who were members of the ultra-right-wing German Nationalist Party) generally, how much truer was it of nationalist human geneticists and Nazi Party officials who expected so much from one another? To make a Faustian bargain is to recognize, at least to some degree, the terms of the deal. That having been said, German human geneticist could not have predicted the end game of the Third Reich—the final price they would have to pay—in 1933. This is in part because the symbiotic relationship between their science and the Nazi state—not, of course, without their influence, but perhaps without their clear realization—took on an ever more radical form over the course of the regime. It ultimately led to the moral demise of human heredity (at least in Germany for a time) and the nihilistic destruction of the Nazi state. The bargain resulted in the untold suffering and death of millions.

Again, I must emphasize that my decision to refrain from delivering a moral verdict on the biomedical professionals in this study is not the result of any desire on my part to downplay their complicity in this large-scale murder—a complicity that I fully recognize and whose warnings for posterity I earnestly heed. Rather, I hope to give my readers an appreciation of the complexities surrounding the choices made by my historical protagonists. Since I, as author, do not have any special insight into the minds of the human geneticists under discussion, people can legitimately come to different conclusions about the same "facts" surrounding this most catastrophic chapter of the twentieth century. But if the nonexpert is to come away with anything more than knowledge of the crimes committed under the banner of eugenics and human genetics during the Third Reich, attempting to understand the historical contexts in which our biomedical professionals operated—including the conjunction of political, professional, and social circumstances that formed the parameters in which these individuals had to negotiate their choices—is a necessary prerequisite. In addition, to allow readers to "come to know" the main human geneticists under discussion well enough to understand their actions, I have purposely chosen to limit the number of scientists under investigation. With the exception of secondary school biology instructors, most German human geneticists in this story were KWS researchers. Since, however, the two human heredity-related Kaiser Wilhelm Institutes under discussion, despite their many differences, were the sites where most of the vanguard research in human genetics under the swastika was undertaken, I do not feel that this small sample of scientists leads to one-sided conclusions about the nature of the Nazi symbiosis.

This volume is in the tradition of much of the most recent work on the history of National Socialism. Obviously, by employing the theme of the "Faustian bargain" between human geneticists and the Nazi state, I am suggesting that what made the dictatorship function in the critical realm of racial policy was "complicity and consent," not, as was previously thought, merely "coercion and compulsion."[27] Indeed, the notion of a symbiosis goes beyond mere complicity and consent and suggests a meshing of interests on the part of the two parties to the bargain. And even the choice to withhold moral judgment is not new to historical analyses dealing with the Third Reich. Richard J. Evans, in his important study *The Coming of the Third Reich,* takes such an approach. As he states in his preface: "The purpose of this book is to understand: it is up to the reader to judge."[28] My own decision to withhold moral condemnation of any of the individuals studied in this work merely reflects the desire for readers to think critically about the issues raised; it should not be misinterpreted as an attempt to whitewash or legitimize the actions of these scientists. This last point cannot be stressed enough. I fully realize that, unlike Marlowe's Faustus, there was no deathbed recantation of past sins on the part of these biomedical practitioners. Rather like Marlowe's Faustus, even had there been, that would not have saved at least some of these human geneticists from being "eternally damned" in the eyes of many. Had these compromised biomedical professionals openly expressed second thoughts about their actions and displayed honest self-criticism, this might have altered the attitude that even a large number of contemporary Germans have about recent genetic technologies. It probably would have changed the assessment of the don of molecular biology, James Watson. In a speech held in Germany at the Max Delbrück Institute in Berlin-Buch in 1997, Watson concluded that had the former KWIA—one of the world's former premier research centers for human genetics, and the KWI with the notorious Josef Mengele connection—"been bulldozed to the ground immediately after the war," things would have been much easier for today's German geneticists.[29]

Hence despite the complexity of the circumstances under which these practitioners operated, German human geneticists were still ultimately complicit—to varying degrees, of course—in the crimes associated with the Holocaust. And for their culpability in the worst human tragedy in history, there can be no excuse. By writing such a book, I do not question the legitimacy of other scholars' desire to pass definitive ethical judgments on these human geneticists. It is simply my hope that through my particular kind of analysis I, too, can make a contribution to the history of biomedicine in the Third Reich that is meaningful to a readership not necessarily familiar with the rich historiography on this subject. In addition, I also want to challenge my audience to ponder the potentially grave implications resulting from making moral compromises in the scientific arena. Such reflection is demanded of all educated people as we embark upon genomic medicine and genetic technologies in the twenty-first century.

I trust that the main focus of this study—the examination of the symbiotic relationship between human heredity and Nazi politics—will be significant for scholars who are not experts in biomedicine in the Third Reich, especially researchers in the field of Holocaust studies. But beyond the central theme of the book, the analysis also highlights issues that are part of the historiography of National Socialism and the history of science, the domain of bioethicists and the concern of public school educators.

This work clearly supports the now commonplace understanding among historians of the Third Reich (but not necessarily of nonspecialists) that the Nazi regime was not monolithic but polycratic. We know today that there were numerous competing state and party organizations vying for power in Germany under the swastika, and it is instructive to view our historical protagonists as they were forced to negotiate their way through this political labyrinth—sometimes profiting, sometimes suffering from it. It also raises the significant issue of continuity and change before and after 1933, a central topic in much of the scholarship dealing with the Third Reich, for the scientific discipline of human heredity. The question of the nature of complicity under the swastika is also of concern to specialists in the field. In addition, this study also sheds light on the relationship between the KWS and the National Socialist state—an important area for historians of science that has been confirmed by numerous publications stemming from the Max Planck Society project.

The interface of human heredity and National Social racial policy at the heart of this work also raises numerous ethical questions regarding the guidelines and limits of acceptable biological research. This is the case not just in the extreme examples of human experimentation in concentration and death camps, "euthanasia," and mandatory sterilization, but in the more mundane cases of data abuse and the intrusion of other scientists' research material. Moreover, it also addresses the role of the function of public education in any given society. Are teachers obligated to prepare their charges for the society to which they belong, or are there universal values that should be taught regardless of the particular society in which one finds oneself?

Finally, and perhaps most importantly, we are all confronted with the thorny dilemma of how any person assesses the actions of individuals when numerous causal explanations for them are possible, a problem clearly transcending all disciplinary boundaries. This question—so obvious and yet so profound—necessitates posing numerous queries throughout the main section of this book. It should again remind the reader that history is about interpretation and that an author does not (nor should she pretend) to hold a monopoly on this mental exercise.

The book is organized as follows: I will attempt to give my readers some understanding of the complex situation in which this symbiotic relationship between human heredity and Nazi politics functioned by first exploring the science of human heredity[30] in its pre-Nazi, international context. Chapter 1 introduces some of the major internationally renowned human geneticists, especially the German ones, who laid the foundations of this new science since the turn of the twentieth century. It examines their intellectual and political concerns as expressed in their scientific publications, at professional conferences, and in personal correspondence with other researchers. What will become clear is that by the late 1920s and early 1930s, German human geneticists were part of an internationally respected scientific community of like-minded researchers. Even the First World War did not result in long-term isolation of German human geneticists from their colleagues in other countries. Yet despite their international recognition and acceptance, all was not well for human geneticists in Germany between the two world wars. The Weimar Republic, Germany's first, if ill-fated, attempt at democracy, simply could not make good on its promise to support the welfare state anchored in its constitution. This led, on the one hand, to an ever-greater call for human geneticists to apply their expertise in eugenics or *Rassenhygiene* (racial hygiene, as the applied science was frequently called in German)[31] as a scientific fix for Weimar's burgeoning social problems—something few had qualms about. On the other hand, German human geneticists, especially those employed by the KWS, faced budgetary cutbacks owing to the devastating impact of the Great Depression during the last years of the Weimar Republic. These same financial woes led the Weimar state to look for biotechnocratic solutions to the welfare problem in the first place. This lack of financial resources and German human geneticists' preoccupation with employing their scientific knowledge for the good of the state, made these members of the international community of human geneticists particularly receptive to a political party that promised them better times.

Indeed, it did not take long before the National Socialists became the rulers of Germany. The next four chapters are case studies examining the symbiotic relationship between human heredity and politics in the Third Reich—the main focus of the book. Chapter 2 will explore the nature of the Faustian bargain formed between certain Nazi bureaucrats and the scientific personnel at the KWIA in Berlin. We will see how the Institute's first director, Eugen Fischer, sold his Institute to important medical functionaries of the Nazi state. Having done so, both parties to the bargain advanced their interests. During the war years, Fischer's successor and protégé, Otmar Freiherr von Verschuer (1896–1969), carried out decisions initially suggested by his mentor that led both to the intensification of a scientific paradigm change at the KWIA and a brutalization of Nazi racial policy. The radical symbiosis between human genetics and Nazi politics helps account for the involvement of certain members of the KWIA in medical crimes.

A similar institutional study is the focus of chapter 3. Here, however, I turn my attention from Berlin to Munich to demonstrate how one of the most important international leaders in the field of psychiatric genetics, Ernst Rüdin, was not immune to striking a deal for himself and his Institute, the German Research Institute for Psychiatry (Kaiser Wilhelm Institute) (GRIP), with Nazi bureaucrats. I analyze this "Munich Pact" not only for what it can tell us about how Rüdin's field could serve the National Socialist state and vice versa, but because the GRIP received a substantial amount of money from the Rockefeller Foundation even after Hitler attained power. Although Rüdin, too, made a Faustian bargain with the Nazi regime, differences of temperament between Fischer and Rüdin, a conflict of opinions between the latter and his Institute colleagues, the various sources of funding for the two institutes, and Rüdin's unholy alliance with Heinrich Himmler's nefarious SS—the political organization in the Third Reich most complicit with the crimes associated with the Holocaust—resulted in as many differences as similarities between the fate of these two longtime rival institutes.

I move from the institutional production of human genetic knowledge during the Third Reich to its national and international dissemination in chapter 4. Since the most important function that German human geneticists could serve for the regime was providing scientific legitimation for the latter's racial policies, it comes as little surprise that many Nazi bureaucrats willingly provided these scientists with the means necessary to host and participate in conferences where they could publicly bestow their professional blessings on the "racial state." I investigate several national and international conferences where prominent scientists like Fischer, von Verschuer, and Rüdin would not only bear the Nazi racial banner, but also enhance the prestige of their employer, the KWS. This analysis provides us with a way of seeing just how mutually beneficial the relationship was—not only for the main parties of the Faustian bargain—but for the KWS as well.

Chapter 5 departs from my investigation of Nazi Germany's leading biomedical scientists to take a fascinating look at the way human heredity was taught in the college-preparatory schools prior to and during the Third Reich—another way of disseminating human genetic knowledge. Using never before analyzed "exit exams" written by actual students, we will see that eugenics was taught even before the Nazi "seizure of power." Moreover, employing similar exams for the Third Reich, one can demonstrate the differences and similarities between the racial science ideal and racial science practice in the biology classroom. This chapter demonstrates that human genetics instruction at the secondary school level was a professional and ethical gray zone in which the responsibility of biology teachers for inculcating their students with racial hygiene ideas comes into question. Given that they had a more direct influence on a larger number of Germans than did the human geneticists in the Kaiser Wilhelm Society, their moral culpability for the tragic trajectory of their science, if not as great as their more renowned academic colleagues, was nonetheless substantial.

I depart from these case studies on the Nazi symbiosis in chapter 6. Central to my examination here is the reaction of the international human genetics community to the use of its science by the National Socialist state. I analyze the critical position taken by so-called reform eugenicists, above all in Britain, the United States, and Sweden. Moreover, I pay special attention to the purported friendly connection between conservative and racist American and German practitioners of human heredity and eugenics. Finally, the chapter explores the international eugenic conferences during this period and how the German delegations increasingly attempted to control the course of events there. By the outbreak of war in 1939, the once highly publicized international eugenics movement, boasting members in over thirty countries, had all but ceased to exist. The term "eugenics" became taboo because of its trajectory in Nazi Germany, although it continued to be practiced in several countries largely under the guise of medical genetics.

Finally, in the conclusion, I come to my thesis: what explains the ethically reprehensible path taken by human heredity and eugenics under National Socialism was the unique manner in which human genetics and politics served as "resources" for each other. This deadly symbiosis radicalized both the science of human heredity as well as Nazi racial policy; it accounts for the heinous practices of all too many German human geneticists. The damage caused by this symbiosis, I might add, is not completely undone; its effects continue to cast a long shadow on humanity's collective memory of the twentieth century as well as the history of genetics. It reminds readers that the historically contingent nature of the symbiotic relationship between German human geneticists and the Nazi state notwithstanding, it is more important than ever to remain vigilant and avoid taking the first morally compromising steps in science—steps that can lead us in a direction that we surely would not wish to go.

In contemplating the design and substance of this book, I recognized, of course, that readers might be curious to know about the fate of several German human geneticists after the fall of the National Socialist state in 1945. It is an interesting and complex story—but one, I decided—that goes beyond the scope of this text. However, I am presently undertaking a full-length political and professional biography of the most controversial of these scientists, Otmar von Verschuer. Affording a prominent place to von Verschuer's largely unexplored postwar career, the book will demonstrate that his research association with Josef Mengele at Auschwitz notwithstanding, von Verschuer was able to "defeat the Devil" largely because of the "whitewashing culture" of the early postwar period. From quite humble beginnings in Münster where von Verschuer occupied the first university chair in human genetics beginning in 1951, he went on to become one of the leading medical geneticists in the early years of the Federal Republic of Germany. He did so by fashioning himself as a conservative academic and contributing to a new postwar synthesis between human heredity and politics during the tenure of the first West German Chancellor, Konrad Adenauer, (1876–1967). Von Verschuer's long and fascinating success in using the "sword of [his] science" to advance his research career in four distinct historical periods (Weimar Republic, the Third Reich, the Allied Occupation, and the Federal Republic of Germany) holds numerous ethical lessons for professionals and laypeople of the twenty-first century.

ONE

Human Heredity and Eugenics Make Their International Debut

By 1933, the existence of a lively international scientific community of human geneticists and eugenicists was one of the most important prerequisites for the symbiotic relationship forged between German biomedical scientists and functionaries of the Nazi state. Indeed, without focusing attention on the scientific status that human genetics and eugenics had attained as well as the professional prestige its practitioners enjoyed during the second and early third decades of the twentieth century, the critical role that German members of this international network of human geneticists played in the construction and legitimization of National Socialist racial policy remains inexplicable. As will become evident from our four case studies, the international *renommée* of German biomedical scientists was at least as much a "resource" for Nazi racial policy makers as the intellectual content of their research. German human geneticists, for their part, knew quite well how to exploit their status in the international arena to advance their own professional interests at home.

For this reason it is necessary to examine the origins and maturation of this international scientific community and the active involvement of key German human geneticists and eugenicists in it prior to the Third Reich. This is a long and complicated story with many nuances, and one whose contours, for our purposes, can only be sketched here. Its roots lie in the late nineteenth century: an era whose intellectual hall-mark was the belief in science as a tool to reform and advance society. This was a time still untouched by the brutalization of trench warfare, machine guns, and poison gas that would all too soon physically and psychologically scar an entire generation of men and radically alter the dominant European intellectual worldview. It also changed the nature of the German eugenics movement both at home and abroad. Although the divisive impact of the Great War on the international eugenics move-ment in the immediate years following the end of hostilities was largely overcome by the mid-1920s, deep-seated political resentments, especially on the part of conservative German geneticists and eugenicists, were not laid to rest. That having been said, by the beginning of the third de-cade of the twentieth century, eugenics worldwide was popularized and professionalized as never before. The development of a transnational eugenics movement in several regions around the globe was certainly spurred on by favorable international opportunities that went beyond a shared set of assumptions and values held by the nations involved.[1] Especially in Germany, new research institutes created to promote hu-man heredity and eugenics were established to keep up with interna-tional scientific trends in these fields. They would also contribute to the new, if fragile, Weimar Republic's vision of a healthy and efficient so-cial welfare state. With the waning of international tensions, Germany's established experts in the field once again played a major role in the world arena. As we will see, even the National Socialist regime in 1933 did not immediately affect Germany's position on the international stage.

The Specter of Degeneration, the Rise of Eugenics, and the Origins of Modern Genetics, 1890–1914

Although people always had been well aware that in the plant and ani-mal kingdom "like produces like," the modern science of genetics can be attributed to the efforts of a then relatively obscure Austrian monk, Gregor Mendel. Working in an experimental garden in his monastery, Mendel cultivated and tested thousands of pea plants between 1856 and 1863. The result of his work is his now famous laws of inheritance: the law of uniformity, the law of segregation, and the law of independent assortment. Based on his research, Mendel assumed that there were two factors (what we today call alleles) for each inherited trait. These fac-tors segregate during gamete production. Mendel also recognized that some hereditary traits are expressed only when both factors (one from each parent) are transmitted whereas other traits reveal themselves when merely one factor (from only one parent) is inherited. In the former case, the factor can be considered "recessive"; in the latter, "dominant." The transmission of one factor for a given trait will not affect the inheritance of the other. Moreover, in the second generation (F_2), dominant and recessive traits appear in a 3:1 ratio. Today we would say that the dominant and recessive phenotypes for a particular trait emerge in the F_2 in this proportion. In the case of a pure-bred dihybrid crossing, the F_2 reveals a phenotype ratio of 9:3:3:1.[2]

Interestingly, the significance of Mendel's work was not recognized during his own lifetime. There are numerous reasons for this, perhaps the most important being that the Austrian monk himself did not realize the general significance of his findings.[3] In 1900, thirty-five years after Mendel first published his experiments on pea hybridization, his work was simultaneously "rediscovered" by three scientists: the German botanist Carl Correns (1864–1933), the Austrian agronomist Erich von Tschermak (1871–1962), and the Dutch horticulturalist Hugo de Vries (1848–1935). Developments largely internal to the history of biology help explain why these three scientists were able to recognize the significance of Mendel's work in 1900 while earlier researchers did not. Within a decade of the rediscovery of Mendel's laws of inheritance, the modern study of genetics was becoming a distinct field in biology.[4] The term "genetics" itself was first introduced in 1906 by the British biologist and early supporter of Mendel, William Bateson (1861–1926).

The rediscovery and gradual acceptance of Mendelism during the first decade of the twentieth century served to legitimize and advance the incipient eugenics movements in the three countries where they first appeared: Great Britain, the United States, and Germany. As was mentioned in the introduction, the term "eugenics" was first coined by Francis Galton in 1883. His investigations convinced him that a broad range of human traits—moral and mental as well as physical—were passed down from generation to generation. Darwin's cousin firmly believed that some form of social control over the reproductive capacities of a population was necessary to elevate its hereditary substrate and halt what was assumed to be the threat of degeneration. As Galton himself remarked, "[if] the twentieth part of the cost and pains were spent in measures for the improvement of the human race that is spent on the improvement of the breed of horses and cattle, what a galaxy of genius might we not create!"[5] "Could not the undesirables be got rid of and the desirables multiplied," he mused?[6]

What did Galton and others mean by degeneration, and why were biologists, medically trained professionals, and social activists in these countries suddenly so concerned to adopt measures designed to improve the genetic endowment of their populations at the end of the nineteenth century? Why the desire to take evolution into their own hands?

Three contexts stand out as being particularly significant in addressing these queries and hence shaping the early history of eugenics in the above-mentioned three countries: the social problems resulting from industrialization and urbanization; the intellectual currency of social Darwinism, especially its "selectionist" variety that denied the importance of environmental influences;[7] and a state interventionist policy in the fields of health and welfare based on scientific expertise.

During the nineteenth century, most Western countries were transformed from agricultural to industrial societies. This transformation resulted not only in profound structural changes but social and economic strife, class conflicts, the rise of socialist movements that appeared to threaten the reigning capitalist order, the increase of criminality, pauperism, alcoholism, prostitution, and the heightened awareness of the existence of a large number of mentally ill and so-called feebleminded individuals. This latter group, the so-called mental defectives, was singled out by scientists, physicians, and lay observers as posing a grave social and financial liability for the state.[8] Given the general faith in science to solve all of humankind's problems, numerous attempts to flesh out the larger philosophical and social meaning of Charles Darwin's (1809–82) now generally accepted theory of evolutionary change as well as state interventionist policies in many countries designed to manage their welfare problems, it is hardly surprising that those concerned with explaining the disconcerting outgrowths of the social transformations taking place employed the rhetoric of biologists. In hindsight we can say that such individuals "scientized" or "biologized" what we would today consider social and economic problems.

Future eugenicists in the Anglo-Saxon countries could take their lead from Darwin himself who, after reading Galton's writings on the threat of the decline of hereditary talent and the resulting degeneration of the human race that inevitably follows, commented on the problem in his second most important book, *The Decent of Man:*

If the various checks . . . do not prevent the reckless, the vicious and the otherwise inferior members of society from increasing at a quicker rate than the better classes of men, the nation will retrograde, as has too often occurred in the history of the world.

We must remember that progress is no invariable rule. It is very difficult to say why one civilized nation rises, becomes more powerful, and spreads more widely than another; or why the same nation progresses more quickly at one time than another. We can only say that it depends on an increase in the actual number of the population, on the number of men endowed with high intellectual and moral faculties, as well as on their standard of excellence.[9]

In his same work, Darwin suggested a solution to the problem of degeneration, though much less vigorously than his cousin: "Man scans with scrupulous care the character and pedigree of his horses, cattle and dogs before he matches them, but when it comes to his own marriage he rarely, or never takes any care. . . . Yet might selection do something not only for the bodily constitution and frame of his offspring, but for their intellectual and moral qualities. Both sexes should refrain from marriage if they are in any marked degree inferior in body or mind. . . . Everyone does a good service who aids towards this end."[10]

Darwin and many "social Darwinists," individuals who believed that Darwin's laws could also explain social and political changes, did not themselves disavow the impact of environmental influences in the origin and development of species. At a time when the laws of inheritance were not understood (remember that Mendel's laws were not generally known until after 1900), most scientists who supported evolution had little choice but to accept a role for "Lamarckism." This was the emphasis on the inheritance of acquired characteristics as a mechanism of evolutionary change put forth by the turn-of-the-nineteenth-century French naturalist Jean-Baptiste Lamarck (1744–1829). By the end of the nineteenth century, however, the inheritance of acquired characteristics was challenged both by Galton himself and, even more thoroughly, by the German embryologist August Weismann (1834–1914). As a result of his research, Weismann totally rejected Lamarckism and afforded Darwin's principle of natural selection an even greater role in organic and social evolution than the author of the *Origin of Species* himself. His notion of the "continuity of the germ-plasm," which presupposed that the hereditary substance was distinct from and unaffected by somatic cells, suggested that it was impossible to improve a human being's condition by means of mental or physical training. As one later German eugenicist noted, "only selection can preserve and improve the race." Indeed for those who accepted Weismann's views with respect to heredity and the "all-supremacy" of selection, eugenics was the only practical strategy to improve the human species and avert its degeneration.[11]

In Britain, America, and Germany, the first stirrings of eugenics began before the rediscovery of Mendel's laws at the turn of the twentieth century. Even after 1900, Galton and some of his British eugenic-minded colleagues like Karl Pearson (1857–1936) first rejected the universality of Mendelism; they attempted to demonstrate the laws of human inheritance statistically through what was then known as biometry, but they could not provide a scientific mechanism for the transmission of the hereditary material. By the end of the first decade of the twentieth century, however, Mendelism won the day even in Britain. It was used by those interested in the social question to attack the problem of the "pauper class," or "residuum," a large group of unskilled laborers at the margins of society whose alleged low intelligence and high fecundity was perceived as a danger to the British state. Earlier "unscientific" reform strategies to control these individuals were replaced by the modern applied science of eugenics.[12] In the United States, Mendelism was the hereditary theory of choice after it became known in scientific circles. Charles B. Davenport (1866–1944), one of the most important American human geneticists at the time (known for his work on the inheritance of human skin and eye color as well as Huntington's chorea), used it to advocate eugenics, "the science of the improvement of the human race by better breeding." Davenport first became interested in heredity after spending a sabbatical year with Galton and Pearson. By 1908, however, he had become an adherent of Mendelism and began to apply his principles to the study of human traits. The link Davenport made between Mendelism and eugenics can best be seen in his popular book *Heredity in Relationship to Eugenics* (1911). Like their counterparts in Britain, American enthusiasts for the cause such as Davenport viewed eugenics as a way to apply science to the problems of a class-ridden and socially heterogeneous industrial society.[13]

Germany also became preoccupied with eugenics at the end of the nineteenth century. Two physicians, Wilhelm Schallmayer (1857–1919) and Alfred Ploetz (1860–1940), the cofounders of the nation's incipient eugenics movement, wrote treatises arguing for the need to take action against degeneration. It was no accident that medically trained professionals were in the vanguard of German eugenics, and three future directors of the two Kaiser Wilhelm Institutes we will examine, Eugen Fischer, Ernst Rüdin, and Otmar von Verschuer, were all medical doctors by training and early supporters of the gospel of Galton. Physicians enjoyed extraordinary prestige in Germany because of the medical breakthroughs, particularly in bacteriology, of the nineteenth century. This reinforced their view of themselves as the one professional group possessing the expertise to safeguard the health and welfare of the young nation. Eugenics in Germany was an attempt to apply a biotechnocratic approach to the hotly debated "social question"—how to keep the large and militant

American Philosophical Society. Noncommercial, educational use only.

1   The undisputed leader of the American eugenics movement, Charles B. Davenport, ca. 1929.
     As director of the Eugenic Record Office at Cold Spring Harbor, Davenport corresponded with
     most European eugenicists, including German racial hygienists such as Eugen Fischer and
     Ernst Rüdin. Photo courtesy of the American Philosophical Library.

working class true to the state—and boost national efficiency.[14] Whereas
medical professionals would be strongly represented later in eugenics
movements in other countries, this was not the case in Great Britain and
the United States.

Although Galton first coined the word "eugenics" and had been interested in the inheritance of human traits since the 1860s, it was Germany,
not Britain, that established the first eugenics society and journal. In

1900 a prize competition funded by the Krupp munitions family to answer the question, "what can we learn from the theory of evolution about internal political development and state legislation?" provided the impetus for the incipient German eugenics movement. The munitions baron hoped, through support of this contest, to take the wind out of the sails of Germany's left-wing Social Democratic Party (SPD) that threatened his political interests. Supported largely by skilled workers, the SPD used Darwin's theories to argue for democratic political change. Schallmayer's 1903 award-winning work *Vererbung und Auslese im Lebenslauf der Völker* (*Heredity and Selection in the Life Process of Nations*) did support a change in governmental policy, but not that advocated by the SPD. Rather, Schallmayer insisted that long-term national power depended upon the biological vitality of its citizens and neglect of hereditary fitness would lead to the downfall of the state.

More important than Schallmayer, especially for the subsequent contours of the German movement, was Ploetz. His 1895 book *Die Tüchtigkeit unserer Rasse und der Schutz der Schwachen* (*The Fitness of Our Race and the Protection of the Weak*) employed the term "*Rassenhygiene*" (racial hygiene) as a German synonym for eugenics. His ambiguous definition of the term "*Rasse*" as any interbreeding human population that, over the course of generations, demonstrated similar physical and mental traits could be understood to mean the entire human race or a portion of humanity such as an anthropological or ethnic group. As we will see, this double meaning of the term racial hygiene would be especially welcome by those German eugenic enthusiasts who accepted ideologies of Aryan or Nordic supremacy. With the help of his numerous influential connections, Ploetz founded the *Archiv für Rassen- und Gesellschaftsbiologie* (*Journal of Racial and Social Biology*) in 1904, the first eugenics journal in the world. In the early years, the *Archiv* carried articles dealing with genetics, human heredity, and population policy as well as eugenics. One year later Ploetz and some of his close intellectual friends, including the psychiatrist and future brother-in-law Rüdin, established the Society for Racial Hygiene, the first professional eugenics organization. Initially it was designed to be international and attract members from all "nations of culture." Rüdin was sent to Scandinavia to help recruit members; he won over the Danish geneticist Wilhelm Johannsen, the scientist who, in 1909, first explained the difference between genotype and phenotype (an organism's actual genetic makeup and its outward expression, respectively).[15] Despite Ploetz's and Rüdin's best efforts, the Society would ultimately fall short of its goals to increase its international membership. By 1910 the several individual German chapters united under the title

German Society for Racial Hygiene, by far the most numerous section of the Society. The largest percentage of the German Society's four hundred or so members by 1913 comprised physicians and medical students.[16] As we will see, the Society was totally nationalized during World War I.

During the first decade of the twentieth century, Britain and the United States also founded national professional societies and organizations that embraced not only eugenics narrowly defined, but more theoretical work on human heredity. In 1907 the British Eugenics Education Society was founded; its first president was a friend of Galton. Although its early membership was relatively small (approximately 1,200), what it lacked in numbers it made up for in the importance of its members. Approximately 80 percent of them were eminent enough to be included in the *Dictionary of National Biography*. Those who joined belonged to the "modern professions," especially those based on the biological sciences. Moreover, many members of the Society were also involved in other related organizations like the Society for the Study of Inebriety and the Moral Education League. This argues for the prominence and visibility of the early British Eugenics Society members. After Galton's death in 1911, he bequeathed a healthy amount of money for a Laboratory for Eugenics at the University of London to fund research on human inheritance.

In the United States, there were several organizations with an interest in human heredity and eugenics. The American Breeders Association, with its Section of Eugenics, and the Race Betterment Foundation spread the eugenic gospel early in the twentieth century. The most important organization for eugenics in the United States, however, was the Eugenics Record Office (ERO) established in Cold Spring Harbor in 1910. It was founded by Davenport with the financial support of Mary Harriman, the widow of the railroad magnate E. H. Harriman. Davenport convinced the wealthy widow that money donated to a research institute devoted to human heredity and eugenics was a wise investment. After all, Davenport claimed, much of the social violence and crime in America was due to hereditarily determined "social inadequacy." Such arguments made good sense during America's Progressive Era—when science and technology were employed to advance "the cult of efficiency." The day-to-day running of the ERO was left to Davenport's superintendent, Harry H. Laughlin (1880–1943), a man with lackluster scientific qualifications but enormous energy for the cause of eugenics.

The ERO had two main functions: to pursue scientific research into the inheritance of human traits; and to popularize eugenic ideas and lobby for eugenics-related legislation.[17] Interestingly, an American Eugenics Society was not founded until 1922, long after eugenic measures, especially

American Philosophical Society. Noncommercial, educational use only

2　Harry H. Laughlin. Laughlin's lobbying efforts were instrumental in helping to pass the Immigration Act of 1924, which adversely affected the number of individuals allowed entry to the United States from Eastern and Southern Europe. According to the testimony Laughlin gave at the House of Representatives Committee on Immigration and Naturalization, such people were more likely than those from Northern and Western Europe to pollute the American gene pool by harboring a wide variety of "defective traits." Photo courtesy of the American Philosophical Society.

sterilization laws, were established in many American states. Nonetheless, the majority of American geneticists supported eugenics in the early years. Indeed, all the members of the first editorial board of the American journal *Genetics,* established in 1916, endorsed eugenics at the time.[18]

Despite differences in emphasis, all Anglo-Saxon human geneticists and eugenicists stressed the need for so-called positive eugenics, those measures designed to increase the number of the "fitter" members of the population, as well as negative eugenics, legal and educational measures to reduce the number of the "unfit" or "defective" members of the state. In all three countries "fitness" was equated with old-fashioned educated middle-class values, whereby in Germany and the United States "race" would also become a component of "fitness," especially after the Great War.

Somewhat later, Canada also embraced eugenics on the model of other Anglo-Saxon countries. Here the Anglo-Canadian fear of French-Canadian fertility was an important motivating factor. Perhaps surprisingly, Canadian eugenics enjoyed its greatest impact in the 1930s, at a time when supporters in the United States and Britain were becoming more critical.[19] But eugenics and human heredity were by no means limited to the countries already mentioned. Scandinavian biologists, physicians, and social activists were also interested in genetics and its new applied science, and incipient eugenic movements sprouted in Norway, Denmark, Sweden, and Finland. By 1900, the social and economic changes associated with industrialization and urbanization affected the Scandinavian countries just as they had affected Britain, the United States, and Germany in the nineteenth century. However, unlike the three Anglo-Saxon countries, Scandinavia ultimately made peace with its labor movement.[20] In the context of their social democratic governments, eugenics in the Scandinavian countries appeared part of scientific social reform. We will very briefly examine two countries: Norway and Sweden.

Norwegian eugenics, under its two early leaders, Jon Alfred Mjøen (1860–1939) and Ragner Vogt (1870–1943), won some popular support by the First World War. A pharmacist and organic chemist by training, Mjøen's interest in racial hygiene, like that of numerous German practitioners, was stimulated by his acquaintance with Alfred Ploetz. In 1906 he founded a private research institute for his professional passion, and only two years later, he was on the lecture circuit spreading the eugenics gospel in his homeland. Vogt, a practicing psychiatrist, was a trained human geneticist. His work on manic-depressive mental illness supported the generally accepted view that mental disorders were hereditary. Far more scientifically respectable than Mjøen, he argued that the "socially most important hereditary diseases like deaf-muteness, feeblemindedness and mental illness" were probably recessive.[21] As will become clear, Mjøen was a "Nordic enthusiast" who viewed racial hygiene similarly to his *völkisch* colleagues in Germany, many of whom rose to prominence during the Third Reich. Although Mjøen was later scorned by so-called reform eugenicists who were critical of the racism and class prejudices of mainliners like himself as well as reputable human geneticists in Scandinavia and elsewhere, he played a vital role on the international stage

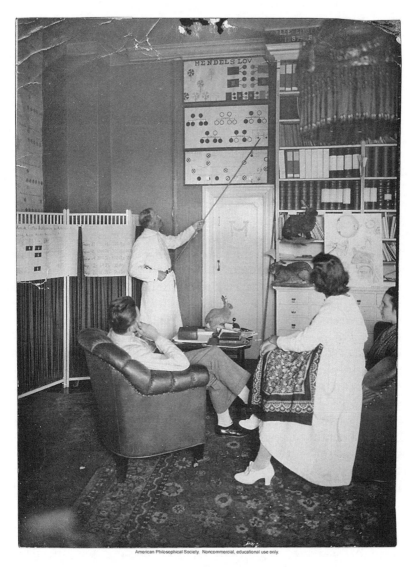

3   Jon Alfred Mjøen, director of the Vindern Biological Laboratory in Oslo, Norway,
    demonstrating the inheritance of traits through a pedigree chart. In his eugenic outlook,
    Mjøen closely resembled his American friend and colleague Davenport. Photo courtesy of the
    American Philosophical Society.

throughout his career. Like many eugenicists with an outward racist bent, he had excellent contacts with Charles Davenport and the ERO.[22]

Of all the Scandinavian countries, the popularization of human genetics and eugenics had the greatest success in Sweden. Swedish eugenics paralleled the German racial hygiene movement. Both movements harbored members concerned with the decline of the "Nordic population," and both had input from anthropology as well as boasting a large percentage of medically trained professionals. In 1909, the Swedish Society for Racial Hygiene was established in Stockholm. Prior to that, almost all its members were part of Ploetz's International Society for Racial Hygiene that was stripped of its international status in 1916. In the early years, there were strong ties between eugenicists in both countries. The future Kaiser Wilhelm Institute for Anthropology, Human Heredity and Eugenics was modeled on the Swedish Institute for Race Biology established in 1922, the first research center of its kind in Europe. The first director of the Kaiser Wilhelm Institute for Anthropology, Eugen Fischer, was personally enamored by the work undertaken at the Swedish Institute during his visit there. Other German practitioners like the plant geneticist and eugenicist Erwin Baur (1875–1933) and Fischer's future colleague Hermann Muckermann (1877–1962) spent time at the Institute in Uppsala. Sweden's most important human geneticist and eugenicist during the early years was undoubtedly the psychiatrist Herman Lundborg (1868–1943), an expert on the genetics of epilepsy. For him, "heredity was everything."[23]

The meaning of eugenics and the uses made of human genetic knowledge was not the same in every country. We would say today that eugenics—perhaps more so than any other applied science—is socially constructed. Broadly speaking, this means that practitioners' particular understanding of eugenics was influenced by the social, economic, and political circumstances that gave rise to it. Its meaning is not universal; it has always been a "protean concept."[24] As there were eugenics movements in over thirty countries by the 1920s, it seems reasonable to assume that supporters in different nations understood and interpreted eugenics differently. This can be observed most clearly when we examine what can be described as "Latin eugenics"—eugenics practiced in countries where a Romance language is spoken.

The differences in style and understanding of eugenics become apparent if we examine two Latin countries, France and Brazil. More so than in Anglo-Saxon countries, the French movement owes its origins to perceptions of catastrophic population decline. Perhaps most surprisingly, at least for those not considering the country's intellectual heritage, French eugenic practitioners were Lamarckians; like their older compatriot La-marck, they believed that environmental influences had a significant in-fluence on heredity. There were relatively few eugenicists in France who embraced Mendelism. Moreover, in France, negative eugenic measures like sterilization were unpopular, and few French eugenic supporters ad-vocated them. Again, there are social circumstances that account for this, above all the influence of the Catholic Church. Instead, positive eugenic measures, especially those that would reduce the mortality rate of in-fants, were the heart of the French eugenic project. The idea that bound the largely medically trained French eugenic practitioners together was coined by one of the founders of the movement, the obstetrician Adol-phe Pinard (1844–1934): *puericulture.* Pinard defined the term as "knowl-edge relative to the reproduction, the conservation and the amelioration of the human species." Pinard founded the French Eugenic Society; it boasted over a hundred other members in 1912.[25]

Brazilian eugenics established its organizational and institutional base just after World War I. The Brazilian Eugenics Society dates from 1918. Like its counterpart in France, it was largely supported by medically trained individuals who believed in the power of science to solve social problems. The social issues in Brazil, however, were not quite the same. Although both countries shared a concern with health issues, Brazilian eugenics was linked to the racially heterogeneous nature of the popu-lation and concerns about Brazil's racial identity. More so than in France, eugenics in Brazil, as in the Anglo-Saxon countries, was fueled by class tensions owing to industrialization and urbanization. But like French and unlike Anglo-Saxon and Scandinavian eugenics, Brazilian practitioners were largely attracted to positive eugenics and identified their science with what we might call social or public hygiene mea-sures. For the eugenic practitioners in Latin America's largest country, "to sanitize is to eugenize." Like France, the movement's early adherents in Brazil adopted a Lamarckian view of heredity. Later, however, Brazil-ian Lamarckian eugenicists advocated the sterilization of "degenerates." Such views demonstrate that not all Lamarckian eugenicists automati-cally rejected so-called negative eugenic measures.[26]

Viewed from the trajectory that racial hygiene took under the Nazis, it would be easy to believe that eugenics was always and everywhere a weapon of right-wing forces trying to suppress the poor and eliminate ethnic minorities. Nothing could be further from the truth. It is certainly the case that many eugenicists were both conservative and racist, and they used their influence to further their prejudices. As we will see, how-ever, not only were there left-leaning, socialist, and Jewish supporters of eugenics in Germany prior to the Third Reich,[27] but the former Soviet Union, the home of the Bolshevist revolution, had its own eugenics movement until Stalin undermined it in the late 1920s. Even then, it continued until the mid to late 1930s as a stealth operation of sorts under the protection of the state-supported Maxim Gorky Institute for Medical Genetics in Moscow. Its director, Solomon Levit (1895–1938), was not only a Bolshevist, but a world-renowned human geneticist who worked in the laboratory of the famous left-wing American geneticist, eugenicist, and Nobel Prize winner H. J. Muller (1890–1967).[28] Muller is justly renowned for his work on the impact of radiation on the mutation rate of genes. Indeed, Muller and several important British Marxist geneticists and eugenicists such as J. B. S. Haldane (1892–1964) demonstrate the interest that socialists and communists with excellent scientific credentials had in the new applied science.[29] It goes without saying that these individuals rejected the crude racism and class biases of right-wing eugenicists in their own and other countries. Eugenics was even embraced by anarchists in eastern Spain in their discussions on sexuality and by those seeking to modernize interwar Romania.[30] Nor was it limited to the West. In Asia, both Japan and China had eugenics movements during the early twentieth century. Even Africa was not immune from these developments. Kenya was also the site of a eugenics project.[31]

Given the diversity of the uses of human heredity in the opening decades of the last century, is there anything that all eugenics supporters and national eugenics movements had in common? At first glance, it would seem improbable that conservatives and communists, Mendelians and neo-Lamarckians, male chauvinists and feminists as well as anthropologists and public health officials could find common ground. To be sure, eugenic advocates of different political and professional persuasions frequently bickered about the means and ends of their cause. But all eugenicists viewed human beings, consciously or unconsciously, as *human resources* whose numbers could be manipulated for some transindividual purpose. By indirectly controlling the genetic quality of these human resources through positive and negative eugenic measures, the long-term efficiency of the nation, "the race," "civilized countries," or humanity could be achieved. The power of science, in this case genetics, to solve social problems in a state interventionist framework was accepted in capitalist, communist, and non-Western developing countries alike. Adherents of ideologies of Nordic supremacy as well as advocates of socialist utopias could embrace the technocratic, managerial logic inherent in eugenics. Herein lays the common bond uniting all those who lobbied, worldwide, under its banner.

Early on, human geneticists interested in the research of their international colleagues began to forge personal and professional contacts with their peers. Davenport, the leading American human geneticist and eugenicist, began corresponding with the future first director of the KWIA, Eugen Fischer, as early as 1908.[32] At that time the medically trained nationalist and political conservative Fischer was a *Privatdozent* (lecturer), not yet a professor, at the University in Freiburg for anatomy. He was beginning, however, to turn his attention away from traditional anatomy toward racial anthropology. In 1909, for example, we see this intellectual transition in a course he offered in Freiburg entitled "The Anatomy of Human Races with Special Emphasis on the People Living in Our Colonies." Fischer quickly became one of the pioneers in the introduction of Mendelism into his newly adopted field. He never tired of reminding all who would listen that anthropology and heredity were inseparable. Fischer was also a convinced racial hygienist; like many German enthusiasts of the new applied science, Ploetz won him over to the cause. Indeed, in 1910 Fischer established a local chapter of the German Society for Racial Hygiene in Freiburg. The young nationalist medical student and Fischer's later head for the Division of Eugenics at the KWIA during the Third Reich, Fritz Lenz (1887–1976), served as its secretary.[33]

In his earliest correspondence with Davenport, Fischer praised the American's work on the inheritance of eye color and asked whether his colleague would be interested in his own study on the subject. Fischer politely excused himself for not answering Davenport's letter sooner, but he had just returned from German Southwest Africa (current day Namibia) where he was undertaking studies that would lead to his major early work on the so-called Rehobother Bastards. Fischer's research was an investigation of the biological effects of racial intermarriage between Dutch settlers and native Hottentots in the region—published in 1913. His study was billed as the first successful demonstration of Mendelism in a human population, although that claim has since been rightfully called into question.[34] As we will see, the two colleagues would continue to share their research results and their obsession with eugenics for many years to come.

Davenport also established early contact with the future director of the Munich-based German Research Institute/Kaiser Wilhelm Institute for Psychiatry, Ernst Rüdin. The first direct correspondence between the two goes back to 1910. The topic: the important upcoming eugenics exhibit at the prestigious German Hygiene Museum in Dresden in 1911—an event designed to popularize racial hygiene among physicians and educated laypeople. A year after the exhibit Davenport and Rüdin exchanged ideas about the role of genetics in nervous diseases; Rüdin bemoaned the limitations on genealogical studies in Germany.[35]

Rüdin corresponded frequently with Davenport in both the pre-WWI and post-WWI period. Born in Switzerland, the medical student Rüdin was attracted to eugenic ideas by the end of the nineteenth century. Once again, Ploetz proved the decisive motivator. Rüdin's sister Pauline, a medical student at the University of Zurich, met Ploetz while he was studying economics there. They soon married, and Ploetz became well acquainted with his brother-in-law. Ploetz intensified an interest in eugenics that was also nurtured by Rüdin's acquaintance with the alcohol prohibition advocate and Zurich psychiatrist August Forel (1848–1931). During Rüdin's studies in medicine and human heredity—a journey that took him to numerous universities in Switzerland and Germany—the future psychiatric geneticist became more and more involved in the circle of intellectual friends surrounding his brother-in-law Ploetz. By 1910, Rüdin was one of Ploetz's closest organizational collaborators in both the German Society for Racial Hygiene and the journal, the *Archiv,* this despite the fact that Ploetz and Pauline Rüdin divorced owing to their inability to produce children. Although Rüdin did not join openly anti-Semitic, pro-Nordic organizations like his brother-in-law, his book reviews in the *Archiv* and a negative letter he wrote regarding Schallmayer's critique of Aryan ideology suggest that, at the very least, he was sympathetic to Ploetz's views. His membership in the *völkisch* Munich chapter of the German Society of Race Hygiene (along with Ploetz and Lenz) further confirms this. As will become clear in chapter 3, Rüdin was appointed head of a "coordinated" German Society for Racial Hygiene in 1933, as he was viewed to be totally loyal to the new Nazi regime. There can be little doubt that the psychiatric geneticist left an indelible mark on the history of German eugenics from its inception until its demise in 1945.[36]

By the beginning of the second decade of the twentieth century, individual human geneticists and eugenicists were making contact with their colleagues in other countries, but there was no official forum for intellectual exchange and professional enhancement. This would change with the decision to hold the First International Eugenics Conference in 1912. It would be hosted in London and concern itself with what was considered the serious hereditary degeneration of the white population of "Western cultured nations."

A broad spectrum of individuals interested in eugenics—physicians, biologists, statisticians, sociologists, feminists, social reformers, clergy, and anthropologists—were among the more than seven hundred participants who gathered to inform themselves about the rising tide of the "unfit" who posed a "grave danger for the future of the entire human race." The president of the Conference was none other than Major Leonard Darwin (1850–1943), the chairman of the British Eugenics Society and son of Charles Darwin. In his opening address he expressed the hope that the conference would lead those in attendance to recognize the "dangers of the present social situation," exchange views on the matter, and discuss "concerted schemes for action."[37] Representatives from Italy, France, the United States, Germany, Norway, Denmark and, of course, Great Britain spoke on a wide range of eugenics-related topics. Delegates came from as far away as Toronto and Sydney. Given their view of eugenics, the French experts discussed measures that would lead to the decline of infant mortality. Pinard stressed the need for reproductive education before procreation.[38] The British and Americans attempted to convince their audience that the cause of degeneration was the result of hereditary diseases and social conditions that allegedly could be understood in terms of Mendel's laws and Weismann's theory of the continuity of the germ-plasm. Raymond Pearl, a prominent American geneticist and eugenicist, suggested that the results of his research on chickens be applied to humans. Davenport and his colleagues from the Eugenics Record Office provided evidence that most mental illnesses were passed on to future generations through recessive genes. Reginald Punnet, the Cambridge geneticist (known for his "Punnet square" to demonstrate the possible number and combination of genetic combinations), argued that feeblemindedness was transmitted according to Mendel's laws.[39] Although neither Fischer nor Rüdin gave speeches, both belonged to the German Consultative Committee for the International Conference. They left it to Agnes Bluhm (1862–1943), a good friend of Ploetz and the leading female racial hygienist in Germany, to present a talk on eugenics and obstetrics.[40]

The lack of a consensus regarding the aims and seriousness of eugenics within the Conference's press reports notwithstanding, the emphasis that the new applied science combined science and politics had a positive impact on the development of the international eugenics movement and the importance of human genetics. As research into human heredity was a prerequisite for halting the ostensible degeneration of humankind, money was made available for this young science.[41] Likewise, the scientific grounding of eugenics gave it a level of respectability in the eyes of important state officials that it otherwise might not have achieved. The symbiosis between human heredity and politics, although in its infancy, was already visible. This interface became even clearer owing to the establishment of a Permanent International Eugenics Committee a year after the Conference. Its members, representing the older and newer national eugenics movements, held its first meeting in Paris. In order to cast a programmatical net large enough to encompass the myriad interpretations of "eugenics," the Committee adopted the Norwegian Mjøen's platform, since it did not clearly commit to Mendelism and Weismannism. Its hallmark: the difference between the right to life and the right to give life. The first, Mjøen proclaimed, was an inalienable right; the second, a privilege that only healthy and "fit" couples should enjoy. Apparently, all delegates could agree to this. In addition, those in Paris discussed means and ways to professionalize eugenics as an applied science. They concluded that the best method was through meetings, congresses, and research initiatives on the international stage.[42] A year earlier, in 1911, the Fourth International Congress for Problems in Heredity met in Paris under the leadership of the British Mendelian geneticist, William Bateson (1861–1926). It was a testimony to the growing international network in the broad field of genetics. As Erwin Baur was to report later, "all of us knew the available literature in genetics and everyone knew what we were all working on."[43] Unfortunately, very soon such international cooperation in the name of heredity and eugenics was seriously threatened.

The Great War, German Racial Hygiene, and the International Eugenics Movement

Even before the first shots were fired in 1914, eugenicists in most Western countries were fearful that war would initiate disaster for the hereditary substance of their various nations. This concern intensified with the beginning of hostilities, especially after it became clear that the fighting would last longer than anticipated. With a slow birth rate already in effect, war would mean further population problems, especially among the "fittest" elements in society. Leonard Darwin and Oxford eugenicist Edward Poulton argued in the British *Eugenics Review* that "war will undoubtedly kill the better variations and, as a result, is highly dysgenic." Several months after America's entry into the maelstrom in 1917, Irving Fisher, economist at Yale University, held a talk at the Unitarian Church in Portland, Oregon. He informed his listeners that "as someone who himself studied eugenics, the outbreak of the war nearly broke his heart." Other American eugenicists complained about the negative genetic impact the fighting would have in such publications as the *Journal of Heredity*. Italian eugenicists also lamented that the war would allow the physically and mentally weak—those exempt from conscription—the chance to reproduce unhampered by any restrictions.[44]

In Germany eugenicists not only bemoaned the lost superior germ plasm that would follow from the Great War. They were also concerned about its political repercussions. For example, Schallmayer was particularly worried that Russia, not yet a member of the international eugenics community, would outreproduce his fatherland. The Slavs, according to Schallmayer, did not yet employ birth control measures and were hence more fecund than their western neighbors. Not long after the fighting began, he predicted that "[a] decision concerning the existence or non-existence of an independent Germany in the face of the Russian threat will hardly be more than a few decades forthcoming."[45]

Their efforts notwithstanding, the leaders of the German racial hygiene movement—an elite social network largely centered on Ploetz's literary and scientific acquaintances—were unsuccessful in driving home their racial hygiene message to government officials prior to 1914, although the state was not completely oblivious to eugenic developments elsewhere. As early as 1901, the German Ministry of the Interior and the Foreign Ministry took note of an 1899 law in Michigan that forbade marriage among the "feeble-minded and idiots" as well as those suffering from venereal disease.[46] Perhaps not surprisingly, the German government's lack of interest in the views of its own eugenics enthusiasts changed during the war. If there was anything positive about these otherwise tragic developments, at least for most German racial hygienists, it was that the state finally took notice of eugenic issues.

In the context of a situation where welfare and health were informally state centralized during the war and population policy became an integral part of plans for postwar reconstruction, several state-sponsored eugenics-related organizations and committees were established.[47] For example, the military was concerned about the scourge of venereal disease—a problem lamented by German racial hygienists. The German Society took up this issue, as well as the task of altering existing "material conditions" so that larger families became feasible. Moreover, the Deutsche Gesellschaft für Bevölkerungspolitik (German Society for Population Policy) was founded in 1915. It counted high-level civil servants among its members. So significant did population policy appear to the government by this time, that even German Chancellor Bethmann-Hollweg sent a representative to the Society's first meeting. During a several day conference in November 1916 held in Darmstadt, the Society wrestled with the high-priority issue, "The New Beginning of German Family Life after the War." The Society also stressed the "national importance of the occupation of housewife" and the education of girls for a career as mothers. By 1917, the government was officially represented at an important conference held at the Royal Agriculture College in Berlin. Other organizations with state ties included the Bund zur Erhaltung und Mehrung der deutschen Volkskraft (League for the Preservation and Increase of German National Strength) and the Deutsche Gesellschaft zur Bekämpfung der Geschlechtskrankheiten (German Society for the Protection against Venereal Disease).[48]

In mid-1917, the German Society for Racial Hygiene attempted to capitalize on the disastrous impact of war to make a plea for money from the German government. It would be used to ensure that the Reich did not lag behind its enemies in Britain and the United States. With nationalistic rhetoric and appropriate pathos, Erwin Baur, a leading plant geneticist and eugenicist, sought to make his case:

The decision in the national struggle has arrived and the German people are faced with the difficult question of how they will defend themselves from the actions of their open and secret enemies during the World War. Everything that is connected to this—every hope for the future—depends upon whether the German people will have a sufficiently large and [genetically] fit number of progeny.

America and England have already recognized how important the number and quality of the population is in the national struggle for survival. . . .

The times demand that racial hygiene, so necessary for Germany's future, be given the means to continue and expand its promising research and put it into practice. . . . What is planned is a popular-scientific journal . . . and, if possible, a research institute, such as one that already exists in London and near New York.[49]

With obvious reference to the Galton Laboratory and the Eugenic Record Office, Baur hoped to awaken German scientific pride and stir national political interest in demanding similar support in his own country. As we will see, however, whereas a new, more popular journal would appear soon after the creation of the Republic, it would take an additional ten years before Germany had an equivalent institute to those of her wartime adversaries.

Although Baur clearly employed his patriotic language as a rhetorical resource in order to further his own scientific interests and those of his fellow eugenic practitioners, the nationalist turn within the German racial hygiene movement was real. We will recall that in 1916 the Society for Racial Hygiene gave up all pretensions to internationalism and became a thoroughly German organization. During the war, German racial hygienists turned their attention from the "race" in general to the eugenic problems confronting their own country—a trend that would continue in the future. More importantly for the future of racial hygiene in Germany, eugenic practitioners were radicalized by the continuing hostilities. Important figures like Ploetz, Lenz, and Rüdin moved to the political right. Especially in Bavaria, future home of the Nazi movement, several members of the Munich section of the German Society for Racial Hygiene joined secret and not-so-secret right-wing, pro-Nordic, and anti-Semitic organizations. Max von Gruber (1853–1927), Munich Professor of Hygiene and eugenics supporter, championed the extreme war aims of the Pan-Germanists, a right-wing, militarist group dedicated to the annexation of territories outside the border of the Reich. Even left-leaning eugenics supporter Alfred Grotjahn (1869–1931), like most of his fellow Social Democrats, did not condemn the war but rather supported a peace settlement without annexations.[50] Although eugenicists in all countries involved in the fighting supported their respective nations and their alliances, racial hygienists in Germany were more affected by the war and its aftermath then their colleagues elsewhere.

There are perhaps good reasons for this. The war, which most Germans, including German racial hygienists, initially believed was purely defensive, led to a humiliating defeat and significant loss of territories and population. It toppled the monarchy, initiated a German revolution (which in Bavaria resulted in a short-lived Bolshevik-style republic), and gave birth to a largely unloved Weimar Republic (1919–33)—Germany's first attempt at democratic government. Exacerbating the already tense situation was what virtually every German viewed as the "dictate of Versailles," the Versailles Peace Treaty signed in 1919. Especially incomprehensible from their perspective was the "war guilt" clause assigning to Germany and her allies sole responsibility for initiating hostilities. In addition, catastrophic inflation, right-wing (including the infamous 1923 Nazi Beer Hall Putsch) and left-wing political attempts to destroy the government, and the occupation by the French of the resource-rich Ruhr region darkened the German political and cultural landscape in the early postwar years. It should perhaps come as no surprise that divisions among members of the German racial hygiene community that simmered beneath the surface during the relatively untroubled prewar period ultimately erupted in the face of the chaos that marked the first four years of the Weimar Republic.[51]

The aftermath of the Great War also had a profound impact on eugenics in the international arena, especially with regard to Germany's participation in future international conferences. Concerns regarding the dysgenic impact of the war led to a meeting of the Permanent International Eugenics Committee in London in October 1919. Representatives from several countries decided to push for a second International

Eugenics Conference in New York as quickly as possible. The newly formed League of Nations sent a hopeful signal that a new spirit of internationalism would support such meetings. The Conference was set for 1921 at the American Museum of Natural History in New York. As one future participant at the meeting would express it, one outcome of the war "has been to develop what may be called the international sense among the peoples of the world," resulting in the "evolution of international organizations."[52]

In addition to the eugenic catastrophe caused by the war, organizers felt that more eugenics education was needed for state interventionist purposes. Henry Fairfield Osborn, the American host of the 1921 Conference, never doubted the right of governments to lead the eugenics charge. As he emphasized in his welcome address: "The right of the state to safeguard the character and integrity of the race or races on which its future depends is, to my mind, as incontestable as the right of the state to safeguard the health and morals of its people. As science has enlightened government in the prevention and spread of disease, it must also enlighten government to the prevention and spread and multiplication of worthless members of society, the spread of feeble-mindedness, of idiocy, and of all moral and intellectual as well as physical diseases."[53] Indeed the main theme at the Conference dealt with eugenics' theoretical underpinning: genetics. In particular, the focus was on chromosomes and mutations. Leonard Darwin emphasized that human heredity is a "pure science" and, as such, it is the "guiding star" of eugenics. But it was more than a mere basis for eugenics: genetics and eugenics were partners in one large enterprise. Among the numerous prominent geneticists at the International Conference were the Americans T. H. Morgan, H. J. Muller, H. S. Jennings as well as the British R. A. Fischer and R. Ruggles Gates. About four hundred eugenicists from the United States, Britain, France, Belgium, Norway, Sweden, Denmark, Mexico, Venezuela, India, Australia, New Zealand, and Uruguay attended the meeting. Perhaps most important from the standpoint of the professionalization of eugenics worldwide, a resolution was prepared that laid the groundwork for the first international eugenics organization, the International Federation of Eugenics Organizations. Such an organization appeared politically opportune. It was finally called into existence in 1925.[54]

Conspicuously absent in New York were representatives from the war's vanquished nations, all discussion of internationalism notwithstanding. Prior to the Conference there was a heated controversy over whether or not eugenicists from Germany, Austria, and Hungary should be invited. The French and Belgian delegates threatened to boycott the meeting should

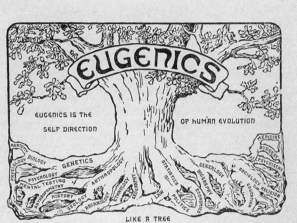

EUGENICS

EUGENICS IS THE
SELF DIRECTION

OF HUMAN EVOLUTION

LIKE A TREE
EUGENICS DRAWS ITS MATERIALS FROM MANY SOURCES AND ORGANIZES
THEM INTO AN HARMONIOUS ENTITY.

The Second International Congress of Eugenics, devoted to researches in all fields of science and practice which bear upon the improvement of racial qualities in man, convey this expression of appreciation of the generous gift of

Mrs E. H. Harriman

of New York, which made possible the exhibition of eugenical materials which were assembled and displayed, in connection with the Congress, at the American Museum of Natural History.

New York, September 1921

President of the Congress.   Chairman of the Committee on Exhibits

Fig. 3. Copy of Certificate Awarded for Meritorious Exhibits

15

4   Certificate awarded to "meritorious" exhibits at the Second International Congress of
Eugenics held in New York in 1921. The certificate pictures the "eugenics tree"—a symbol
that was supposed to demonstrate that the new "applied science" has its roots in numerous
disciplines. "Eugenics," it states "is self-directed evolution." Photo courtesy of the Cold Spring
Harbor Laboratory Archives.

this happen; even Darwin's compromise to allow these countries to send representatives after they joined the League of Nations came to naught. In general, the Anglo-Saxon and Scandinavian eugenicists were in favor of bringing the one-time enemy nations into the eugenics fold. After all, eugenics was an international science. As the Swede Herman Lundborg wrote to Davenport, an international eugenics conference to which not all "civilized nations" are invited is not international and only harms the cause of their project. In order not to offend their sensitive French and Belgian colleagues who hardly viewed the Germans as "civilized," Darwin and Davenport negotiated a strategy that would effectively keep Germany and her former allies from attending the Second International Conference but simultaneously left the door open for their participation in future meetings.[55] Scientific cooperation and contact between Anglo-Saxon and Scandinavian eugenicists continued unabated.[56]

German eugenicists did not make things easier for the moderates. Immediately after the terms of the Versailles Treaty were publicized, racial hygienists refused to involve themselves in international activities. For example, in a sarcastic letter to representatives of the Smithsonian Institution who asked to receive the *Archiv,* Ploetz replied that since the "Americans" charged the "Germans" with being solely responsible for the war, it naturally followed that the Second International Eugenics Congress would view them as "mangy barbarians" and keep them away. Since the *Archiv* is a project undertaken by individuals "who consciously adhere to the German *Volk,*" scientific exchange with Americans is no longer possible.[57] Baur touched on the social problems that made work as usual impossible. In a formal letter to American colleagues written in 1920, the plant geneticist expressed his frustration:

The entire work of eugenics is very difficult with us, all children in the cities are entirely insufficiently nourished. Everywhere milk and fat are lacking, and this matter will become yet greater if we now shall give up to France and Belgium the milch [*sic*] cows which they have requisitioned. The entirely unnecessary huge arm of occupation eats us poor, but eugenically the worst is what we call the Black Shame, the French negro regiments, which are placed all over Germany and which in the most shameful fashion give free rein to their impulses toward women and children. By force and by money they secure their victims—each French negro soldier has, at our expense, a greater income than a German professor—and the consequences is a frightful increase of syphilis and the mass of mulatto children. Even if all French-Belgian tales of mishandling by German soldiers were true they have been ten times exceeded by what—in peace!—happens on German soil.[58]

As strained as relations were between the Germans and her former enemies, the situation would soon worsen. Although Davenport was proud
that even the French had agreed to accept members of the German Society for Racial Hygiene at a 1922 meeting in Brussels, the Germans were
so incensed by the French occupation of the Ruhr in 1923 that they were
even less willing than before to take up their leading role again in the
international eugenics movement. "The German delegates" the American eugenicist was informed, "would not sit on the same Commission
with the French and Belgians." Davenport pleaded with Baur to use "his
influence to prevent such a backward step." "The only way we can heal
the wounds caused by the late war," the director of the Eugenic Record
Office continued, "is to suppress these sad memories from our scientific
activities." Baur appreciated the good will of the American eugenicists,
but claimed "the truth of what is happening in the Rhine and Ruhr cannot be known from the American press." "There is hardly a day," Baur asserted, "that a German is not murdered, and the number of German civil
servants and workers who are incarcerated or driven from their homes is
more than several thousand." Baur assured Davenport that if these forms
of "cultural shame" were ended and real "peace" secured, the Germans
would attend the next international conference. Unfortunately, the present actions of the French and Belgians is "war and nothing else—war
against a completely defenseless population."[59]

The Norwegian eugenicist Mjøen, a man whose sympathies were
closely in line with the Nordic enthusiasts among the German racial hygienists, informed Davenport that given the politics of the French and
Belgians, it would be unlikely that the Germans would change their position. The optimistic American ERO Director, however, still hoped that
"it would be possible for German eugenicists and French eugenicists to
sit in the same room—as eugenicists primarily and not as Germans and
Frenchmen." In his desire to find some way that "politics will not be
permitted to interfere with the progress of eugenics in various countries,"
he appealed to the scientific responsibility of the Munich eugenicist and
human geneticist Fritz Lenz. "There is no country," Davenport told his
colleague, "which has higher ideals [in respect to eugenics] than Germany and we assume that she will assume a leading position at the next
congress."[60]

Given the role that Lenz would play in the German eugenics movement during the Third Reich, his answer to Davenport is worth quoting
in full. It reveals the worldview of this conservative German nationalist as well as that of many of his contemporaries radicalized toward the

right because of the country's upheavals. It also demonstrates that many intellectuals of Lenz's political ilk could not imagine that the Great War would be the last one. In his Germanized English, he summarized his position to his American colleague.

A cooperative work between the Germans and French seems to be impossible so long as the Ruhr invasion lasts. If in America a foreign power had entered and held in its grasp the chief industrial area surely no American man of science would sit with a representative to that other nation at a table. . . . When you write that each injustice it is to be hoped may soon be removed, I know that it is certainly an intimation of a sense of justice and well wishing toward all nations but I believe that you see the situation in Europe in too hopeful a light. Europe goes with rapid steps toward a new frightful war, in which Germany will chiefly participate as an object; if America does not once again participate in the war it will end in the world domination of the British empire. This view is not only general in Germany, but it is also present already in leading political circles in England. America has, therefore, a great task to fulfill were it only to recognize it at the proper time. . . . I do not believe that the time for international congresses have arrived so long as France occupies the Ruhr, that is not before the second World War. I do not wish this certainly; I know that our race in it would suffer more heavily than in the past World War but it cannot be avoided.

In the same response to Davenport, Lenz mentioned that the Americans only fought out of "Anglo-Saxon national feeling" toward England, and that Britain had no choice but to enter the war. Had she not taken that step, Lenz continued, France would have been lost and England would have had to depend on the mercy of France's "conqueror." The letter hence indicates the interconnection between scientific conferences and politics; it also lays bare bitterness among German right-wing intellectuals that helped doom the fragile Republic from the outset.[61]

Owing to the more relaxed international political situation following the appointment of Gustav Stresemann as foreign minister of Germany in 1924, the French retreated from the Ruhr. Germany breathed easier in the relative calm before the storm. Once again, it took up its former position in the international arena. Indeed, between 1925 and 1933 human heredity and eugenics were professionalized and popularized on the international stage as never before. In addition, new institutes and organizations for these sciences were created in countries all over the world, including Germany. International meetings and scientific correspondence between human geneticists in various countries, in particular between the United States and Germany, promised a fruitful development of the gospel of Galton.

The Maturation of Human Heredity and Eugenics in the International Arena

Even during the years that German human geneticists were absent from the international stage, Baur, Lenz, and the racial anthropologist Fischer did much to advance the role of Germany's prestige in the international scientific community through the so-called standard work, *Grundriss der menschlichen Erblichkeitslehre und Rassenhygiene* (*Principles of Human Heredity and Racial Hygiene*). The *Grundriss* was first published in 1921 with the financial backing of the Munich-based conservative-*völkisch* J. H. Lehmann publishing house, the same publisher that supported Ploetz's *Archiv*. "Baur-Fischer-Lenz," as the *Grundriss* was frequently dubbed by insiders, went through four editions in German; its third edition was translated into English in 1931. Apparently, Hitler himself read portions of the *Grundriss* while he was serving time in Landsberg prison in Bavaria following his attempt to overthrow the Weimar government during the infamous 1923 Beer Hall Putsch.

Its importance for Nazi racial hygiene notwithstanding, the *Grundriss* was widely read and reviewed internationally prior to the Third Reich. It was a two-volume work. The first volume had a theoretical orientation and contained chapters by Baur on the principles of heredity, Fischer on the world's racial groups, and Lenz on human inheritance. The second, composed entirely by Lenz, dealt exclusively with racial hygiene. Such respected American geneticists as Raymond Pearl and H. J. Muller considered the section written by Baur to be a clear and state-of-the art summary of classical genetics. Even Fischer's contribution, his tendentious racial portraits aside, was largely evaluated as "good science" by the more than three hundred reviews of the *Grundriss,* including twenty-seven written by individuals in non-German-speaking countries. Indeed, the scientific acclaim he enjoyed as coauthor of the treatise would play an important role in his appointment as director of the KWIA in 1927. The international human genetics and eugenics community was most critical of the portion written by Lenz—especially his treatment of the inheritance of mental traits, his penchant for projecting his class prejudices on various "races," and his enthusiasm for the virtues of the "Nordic race." His sympathy for the well-being of Nordics was also shared by some non-German eugenicists, especially in American and Scandinavian quarters. For example, in the late 1920s, the American Eugenics Research Association offered a prize for the best eugenic studies dealing with the "relative fecundity" of the Nordic race. Surprisingly, only one review, by the sexologist Max Marcuse, stressed the potential political danger of Lenz's views on the "Jewish question" prior to 1933. Marcuse was just one of the numerous German Jewish scientists who would become targets of the Nazi regime's state-sponsored anti-Semitism partly legitimized by the "objective" portrayal of so-called racial character traits found in the *Grundriss*. But this is all in the future. During the 1920s and early 1930s, more than 70 percent of those who reviewed the *Grundrisss* were overwhelmingly positive about the treatise as a whole. This reminds us that what we may today view as "pseudoscience" and ideology was not necessarily understood as such by respectable scientists in the not-too-distant past.[62]

In addition to the contribution of the "Baur-Fischer-Lenz," German human geneticists and eugenicists left their mark on the international arena through their participation, beginning in 1925, in the newly formed International Federation of Eugenics Organizations (IFEO). Originating out of the need to incorporate the ever-increasing number of national eugenic movements worldwide into an organization promoting the professionalization of the applied science as well as the political agendas of its leading members, the IFEO, in practice at least, became the arena of operation for conservative eugenic practitioners. There were men who, like Davenport, Mjøen, and Ploetz, were always sympathetic toward the interests of the "white race," or the "Nordics." As we have seen, this group was especially dedicated to improving the hereditary quality of Northern and Western European stock included many first-generation leaders of the Anglo-Saxon and Scandinavian movements. Indeed the IFEO was originally founded to promote the genetic betterment of the "white race." The growing eugenics movements in Asia and nonwhite Africa were initially not represented.

Davenport, the chairman of the IFEO from 1927 to 1932, made the intellectual and physical differences between races and the crossing of so-called main races the nerve point of the IFEO. Older forms of racial research had become discredited, and he as well as many of his international colleagues believed that such investigations into the nature of "race" could only remain respectable if practitioners agreed on standardized methods to evaluate racial differences. The effort to coordinate investigations into racial mixing worldwide and to ensure uniformity in its research techniques became Davenport's top priority. Given Davenport's professional interests, it is little wonder that the chairman of the IFEO had much to discuss in his correspondence with Eugen Fischer and Ernst Rüdin.[63] In 1929, for example, Davenport wrote to the latter asking if he could head a committee on racial psychology for the IFEO. The Munich

5   A picture taken of prominent members of International Federation of Eugenics Organizations
    when it met in Stonehenge, England in 1930. Visible on the extreme left is Laughlin. The man
    with the long white beard is one of the cofounders of the German racial hygiene movement,
    Alfred Ploetz. Next to him, on his right, is the Swiss-born German psychiatric geneticist, Ernst
    Rüdin. Photo courtesy of the Cold Spring Harbor Laboratory Archives.

psychiatrist agreed.[64] As we will see, Fischer was nominated to succeed
Davenport as IFEO chairman in 1932.

   Although some leading IFEO members were already being criticized
by so-called reform eugenicists, who questioned the scientific accuracy of
the racial prejudices embedded in their work, there were enough reputable researchers among the traditional "mainline" practitioners such that
the activists in the organization could not be accused of being a mere
assemblage of quacks. Whatever one might think of Laughlin's scientific
credentials, those of the statistician and geneticist R. A. Fischer and his
compatriot, the plant geneticist R. Ruggles Gates, were impeccable. Both
men were members of the British delegation to the IFEO. Nor could one
doubt the importance of the investigations of the Swedish psychiatrist
Torsten Sjögren (his later pro-Nazi sympathies notwithstanding) or those
of the head of the first academic department of genetics in Finland, Harry
Federley. They, too, were members of their respective national delegations to the IFEO. And the research of Fischer and Rüdin, both future
directors of prestigious Kaiser Wilhelm Institutes in the 1920s because of
their international scientific reputations, could also hardly be dismissed
as amateurish.[65]

That having been said, it would be hard to deny that the IFEO, even as early as the late 1920s, increasingly provided the international professional meeting grounds for eugenicists who made no apologies for their preoccupation with the genetic worth of alleged "races" and strategies to improve the most valuable ones. Although at the beginning of the 1930s the IFEO had changed its position on representatives from "nonwhite" countries to further professionalize the gospel of Galton (by this time twenty-two nations boasted eugenics movements and were incorporated into the international organization), the Anglo-Saxon, German, and Scandinavian representatives continued to set the tone. Slowly but surely, Lamarckians, especially from France, socialists, feminists, and neo-Malthusians advocating birth control technologies were marginalized within the IFEO.[66] This trend toward the one-sided advancement of the interests of "mainline" eugenicists at the expense of the multifaceted concerns of the worldwide community became further exacerbated during the Third Reich. As will become clear in chapter 6, in the years just prior to the outbreak of war, the IFEO was deliberately manipulated by Germany. Only those representatives of countries who shared the racist prejudices of Nazi racial hygienists had any influence in the organization. Reform eugenicists, as we will see, wanted nothing to do with what they considered unscientific and even fascist applications of their program for human betterment.

Among the strongest intellectual and institutional ties within the international eugenics community were the ones between mainline American and German practitioners. Even prior to the Great War, German racial hygienists praised the accomplishments of their American brethren. After the First International Congress in London in 1912, Ploetz remarked to a German newspaper that the "United States was a bold leader" in advancing the cause of eugenics.[67] During the early and mid-Weimar years, German racial hygienists looked across the Atlantic in envy of the practical successes of their New World colleagues, even if most would not themselves have embraced all such measures, especially mandatory sterilization. As is generally known, American eugenicists were instrumental in the passage of sterilization laws in numerous states; they were also one of the driving forces behind the notorious 1924 Johnson-Reed anti-immigration bill greatly limiting the number of "genetically less valuable" immigrants from Eastern and Southern Europe. Indeed American eugenicists possessed such a powerful lobby that even the U.S. Supreme Court upheld the decision to deny allegedly hereditarily tainted individuals from procreating. "We have seen more than once that the public

"THREE GENERATIONS ENOUGH"

Cold Spring Harbor Laboratory. Noncommercial, educational use only.

6   Oliver Wendell Holmes, associate justice of the American Supreme Court from 1902 to 1932. Holmes upheld the constitutionality of mandatory sterilization in the infamous 1927 case of *Buck vs. Bell.* He went on record as saying "three generations of imbeciles are enough." Photo courtesy of the Cold Spring Harbor Laboratory Archives.

welfare may call upon the best citizens for their lives," Chief Justice Oliver Wendell Holmes proclaimed in 1927 in the infamous Buck vs. Bell case. "It would be strange if it could not call upon those who already sap the strength of the State for these lesser sacrifices . . . in order to prevent our being swamped with incompetence. . . . The principle that sustains compulsory vaccination is broad enough to cover cutting the Fallopian tubes." In short, Holmes concluded, "three generations of imbeciles are enough."[68] This judgment from the American high bench was well known in the international human genetics community.[69]

As we have mentioned, German and American human geneticists regularly corresponded about professional matters. In 1924 Lenz wrote Davenport to ask whether he thought Baur-Fischer-Lenz would sell in the United States. In 1931 the two-volume work was translated into English. In 1926, the Munich eugenicist's article "Are the Gifted Families in

America Maintaining Themselves?" appeared in the American *Eugenical News*. Lenz also established good relations with Laughlin and Davenport at the Eugenics Record Office in Cold Spring Harbor as well as with Paul Popenoe (1888–1979), a leading mainline American eugenicist on the West Coast. Davenport returned Lenz the honor he bestowed upon his American colleagues in 1927. In that year he traveled to Germany and met the Munich professor for racial hygiene. In a follow-up letter to Lenz, he suggested that the IFEO might hold its next meeting in the Bavarian capital. It was obvious, Davenport remarked, that "Munich is at the center of eugenic research in Europe."[70] Davenport communicated with other German human geneticists and eugenicists such as Hermann Muckermann, Günther Just (1892–1950), and Hans Nachtsheim (1890–1979). Nachtsheim even went to Cold Spring Harbor to visit the ERO. In a letter preceding his visit, the American eugenicist wrote to Nachtsheim, "I learn that you have arrived in America and are now with Professor [T. H.] Morgan. . . . A warm invitation is extended to you to visit us. . . . We should be glad if you could come out on Thursday and speak to our group here. . . . It would be a pleasure to show you the work going on here." Naturally, Davenport was also in frequent touch with Rüdin and Fischer.[71]

In 1932 Davenport became a corresponding member of the Berlin Society for Anthropology, Ethnology, and Pre-History, a professional organization headed by Fischer. The racial anthropologist sent Davenport a twelve-page questionnaire to be used when scientists undertook research on racial crossing. Perhaps Fischer believed that his research methodology would be adopted by all members of the IFEO interested in this kind of investigation. His questionnaire left no aspect of the physical, intellectual, and psychological traits of the "racial bastards" unmentioned. Among the questions: "Are the [racial bastards] morally more degenerate than both their racially pure parents?" In addition, the form asked researchers to comment on the "intelligence, energy, industry, patience, thrift, temperament, irascibility, pride, generosity, cruelty, musicality, and dexterity" of their subjects. Fischer also thought the observers should know whether their racially mixed subjects suffer from "insanity, alcoholism, criminality, prostitution, and vagabondism." Finally, the questionnaire requested that hair samples and pictures of the "bastards" as well as those of both their parents be attached to the report.[72]

During the Weimar Republic the German government took a great interest in human hereditary and eugenics as scientific tools to construct a modern welfare state. Because it was generally accepted that the health of a population was linked to national efficiency, the former was a central state concern. Fortunately, government officials believed, health was a "good" that could be manipulated according to scientific principles. If individuals were a form of human capital, healthy people were especially valuable resources. Hence the preoccupation demonstrated by the number of related museum exhibits, education and propaganda projects (especially for youth), and the creation of genetic counseling centers in major cities throughout the Reich.[73] As such, it should come as no surprise that government officials kept abreast of what other counties were implementing in eugenics and health management. For example, in 1923 the German Foreign Office informed the Ministry of the Interior about the establishment of a Danish commission to discuss possible measures to tackle the problem of the unfit, including mandatory sterilization. Governmental representatives in Berlin received foreign eugenics journals, such as the British *Annals of Eugenics*. As we have seen, the Reich government was long informed about American eugenic measures. By 1923 America's sterilization procedures and laws were common knowledge in the highest governmental levels in Germany.[74]

Only a few years after the establishment of the first German Republic, voices clamored for the creation of an institute of human genetics and eugenics, both to keep up with international developments in this field as well as to serve as a think tank for German researchers whose investigations would provide the state with useful biological knowledge.

In fact, by this time an institute already existed in Munich that concentrated on psychiatric illnesses from a modern standpoint. Although what would become the German Research Institute for Psychiatry in 1917 did not contain the words "eugenics" or "human genetics" in its name, calls for such a research center prior to World War I used racial hygienic arguments for its establishment. As one physician argued as early as 1910, "[t]he many thousands of sick that form a burden to the state—the huge sums that the common weal must pay for the mentally ill and the mental institutions—cry out for . . . [biological-psychiatric] research. Our provincial and communal mental hospitals cannot afford this research. . . . It requires an especially large biological and experimental-therapeutic research center, one located, if at all possible, in a large city."[75] The most renowned German psychiatrist of his day, Emil Kraepelin (1856–1926), the mentor of Ernst Rüdin, pushed for the realization of such an institute. His success is intimately connected to his acquaintance with the American Jewish philanthropist James Loeb (1867–1933). The Loeb family emigrated from Germany in the nineteenth century and acquired their fortune through banking. Loeb became interested in Kraepelin's mission and agreed to provide much of the funding needed to realize the German psychiatrist's dream. Exactly why he did so is unclear; it is reported, how-ever, that Kraepelin helped Loeb fight his own mental illness.[76]

In addition to Loeb's generous contribution, the Ruhr-based steel magnate Krupp von Bohlen und Halbach as well as other wealthy patrons offered needed financial support for the Munich institute. As will become clear in chapter 3, the Research Institute for Psychiatry eventually be-came a Kaiser Wilhelm Institute in 1924, largely as a result of the efforts of its future director, Rüdin. Long before the Swiss psychiatrist became head of the entire complex in 1931, he had made a name for himself well beyond his adopted German homeland. The research that made him world renowned in the field of psychiatric genetics was his creation of the "empirical hereditary prognosis" for nervous diseases. This statistically based methodology replaced the traditional emphasis on pedigrees, such as existed at the ERO. In an important article written in 1916, the Swiss human geneticist applied his new statistical technique to schizophrenia for which he received international acclaim.[77] As we mentioned, Rüdin was also one of Ploetz's closest associates in the publication of the *Archiv* and a very active member of the Munich section of the German Society of Racial Hygiene. It should be noted that his research interests in psychia-try were part and parcel of his eugenic concerns. As was typical for the first third of the twentieth century, eugenics and human genetics were mutually reinforcing.

Rüdin's international reputation ensured that when the Munich In-stitute was finally a reality Kraepelin would ask him to head one of its departments, the Demographic-Genealogical Division. Here he worked with several assistants, the psychiatrist Hans Luxenburger (1894–1976) as well as the researchers Adele Juda (1888–1949) and Bruno Schulz (1890–1958) on the genetic basis of psychiatric disorders.[78] This work, expensive and time consuming because of the necessity to travel to their patients, won them international prestige. Luxenburger and Juda remained in Mu-nich until Rüdin, unhappy with the poor working conditions, left with them to take a position in Zurich. As we will see, however, they returned to the Bavarian capital in 1928 after a new, state-of-the-art building was completed for the Munich-based KWI.

Plant geneticist Erwin Baur, we will recall, was eager to see an insti-tute devoted to genetics and eugenics in Berlin as early as 1917. Unfor-tunately, the costs of funding a war and, from the state's perspective, more serious financial concerns, prevented its realization at that time. In the immediate postwar years, however, social conditions were more favorable to a serious consideration of genetics and eugenics. In the larg-est state, Prussia, health issues were now part of a Ministry for Public

Welfare. The director of the Department of Health was an individual who had a broad interest in family history, genetics, and eugenics. In 1920, owing to the initiative of Otto Krohne (1868–1928), an important medical bureaucrat in Prussia as well president of the German Society for Racial Hygiene, a "Racial Hygiene Advisory Board" was established within the Department of Health to examine the severity of the eugenic consequences of the Great War on the German population as well as to take steps to remedy the problem. Among the eight members of the Board, five were geneticists, including Baur. Interestingly, at this time, three of them were members of the Kaiser Wilhelm Institute for Biology in Dahlem. Perhaps related to this Advisor Board, Baur, Carl Correns (one of the three "co-discovers" of Mendel's laws), and Richard Goldschmidt (1878–1958), a Jewish geneticist and eugenicist who would eventually be forced to flee Nazi Germany, founded the German Society of Genetics in 1921. To Baur, the time appeared to be ripe to push for an institute dealing specifically with human genetics and eugenics under the umbrella of the Kaiser Wilhelm Society. He might have also been professionally self-motivated: his sincere interest in eugenics notwithstanding, he was also anxious for the Society to oversee an experimental breeding institute with himself at the helm. Baur viewed the two projects, plant breeding and human genetics, as symbiotic.[79]

As early as the war years, Baur tirelessly lobbied the KWS to establish such an institution—unfortunately in vain.[80] His first mark of success occurred during the early Weimar years; he was able to win over the first president of the Society, theologian Adolf von Harnack (1851–1931), to his idea. It was necessary, however, to find an appropriate future director for the institute. Using his contacts with Eugen Fischer, Baur did all in his power to convince his friend and coeditor of Baur-Fischer-Lenz that he ought to consider leaving his Freiburg homeland and take up residence in Berlin. Only Fischer, von Harnack was convinced, was sufficiently prestigious to occupy the position. Moreover, with Fischer, the KWS would have an expert on the science of "race," a controversial topic during the mid-Weimar years. As we will see in the next chapter, Fischer was indeed persuaded to move to the capital and take up the directorship. Recognizing that the term "racial hygiene" would be an inappropriate intellectual resource in the Catholic–Social Democratic political milieu of Prussia at the time, Fischer decided to use the term "*Eugenik*" as part of the name of his new institute, regardless of his Nordic sympathies.[81]

The ceremonial opening of the Kaiser Wilhelm Institute for Anthropology, Human Heredity and Eugenics (KWIA) was held on September 15, 1927. This date was hardly accidental; the specific day and the gala events surrounding the opening were designed to coincide with the Fifth International Congress of Heredity—the first international conference on any subject held in Germany after the war. At the previous Congress held in Paris, Baur was charged with organizing its successor in Berlin in 1916. Obviously, the war prevented such an activity. Given animosities on the part of Belgium and France as well as Germany, the date for the next conference was delayed. At the 1925 meeting of the German Society for Genetics, Baur decided that instead of holding the next German meeting in Berlin, he would suggest that the Fifth International Congress for Genetics would be held in the Reich capital. As it turned out, it was an intellectually exciting and successful meeting with over nine hundred geneticists in attendance.[82] Some of those attending the International Congress took the opportunity to officiate the opening of the KWIA. Certainly these two events—the opening of the KWIA and the convening of the Fifth International Congress of Genetics—appeared to usher in an era of international cooperation in genetics and eugenics.

The meeting of the International Federation of Eugenics Organizations in Amsterdam that same month further served to reconcile the worldwide human genetics community.[83] The IFEO kept conservative eugenicists in the organization in touch with each other. In a 1929 letter written by Fischer to then president of the Federation, Charles Davenport, he emphasized the special nature of the upcoming IFEO meeting in Rome. The Berlin director remarked that Benito Mussolini would almost definitely attend a session of the meeting. *Il duce,* Fischer continued, was sincerely interested in "our questions," and assured his American colleague that "there will never again be such an opportunity to make our thoughts clear to a national leader." Moreover, the German anthropologist continued, "he [Mussolini] is the only politician who really can, and perhaps will, carry out eugenic measures." Fischer suggested that Davenport, as president, compose a memorandum listing the Organization's position on eugenics in Italian, English, and German. Mussolini should receive a special greeting at the outset of the meeting.[84] At its 1930 meeting in Great Britain, the IFEO decided to establish a "Committee on Human Heredity" that would bring together international researchers in the field, stimulate more investigations on the subject, and serve as a springboard for discussion.

Further east, the opening of the third decade of the twentieth century witnessed some starling new developments in the institutionalization of human genetics and eugenics in the Soviet Union. Although Soviet authorities had frowned upon eugenics as a bourgeois science that was not compatible with Marxism-Leninism, Baltic Jew and Bolshevik Levit, who

7 Soviet eugenicists. Few nonexperts realize that the Soviet Union had a short-lived eugenics movement. In the 1930s, Solomon Levit headed it. The Jewish Soviet medical geneticist (*extreme left*) is seated next to the renowned American geneticist, H. J. Muller, at his lab in Austin, Texas, in 1931. The Argentine geneticist Carlos Offermann and the Soviet geneticist Israel Agol are pictured on Muller's right. Levit and Agol were executed during Stalin's purges. Photo courtesy of the Lilly Library, Indiana University, Bloomington, IN.

had been instrumental in cultivating it during the late 1920s, was offered a chance to establish a world-class institute for the study of human and medical genetics in 1932. Between 1930 and 1932 he had worked in H. J. Muller's laboratory at the University of Texas.

Upon his return to Moscow, Levit found himself with the support and financial resources to study a whole host of problems from a genetic point of view. He and his coworkers also secretly pursued eugenic issues—albeit divorced from the kind of racism and class prejudice that was the hallmark of most mainline practitioners, especially in the United States and Germany—in what came to be known as Maxim Gorky Medical Genetics Institute. During its short life, the Moscow Institute was probably the most advanced research center of its kind. Its director, like his German colleague von Verschuer, specialized in twin studies.[85]

Although new eugenic institutions were established worldwide and the international human genetics network was thriving, differences in the understanding of eugenics among individual practitioners notwithstanding, the onset of the Great Depression in 1929 did not fail to leave its mark on the discipline. It certainly accounts for the inability of many researchers from continental Europe to attend the Third International Eugenics Conference held in New York in 1932. Most of the papers dealing with the theme of the Conference—"A Decade of Progress in Eugenics"—came from Americans. The financial and political situation in Germany was so grave that none of the important leaders in the field were able to attend; indeed only four researchers were able to make the journey.[86] As Davenport expressed the situation in a letter to Ploetz, "the prospect of many visitors from Germany is very small." The director of the ERO commented to his German friend that Fischer regretted his absence, but was told "that directors of institutions must not leave their posts in this crisis."[87] The most that Fischer could do was to send population policy and racial charts as posters representing the work undertaken at the Kaiser Wilhelm Institute for Anthropology. Rüdin, now director of the German Research Institute/Kaiser Wilhelm Institute for Psychiatry in Munich, would have to skip the meeting as well.[88]

The worldwide depression also impacted the emphasis placed on sterilization as an allegedly efficient cost-cutting measure. Long before the late 1920s, the United States led the way in the adoption of mandatory sterilization measures on a state-by-state basis. Beginning with the first law in Indiana passed in 1907, twenty-three American states had sterilization laws on the books by 1929. California, which enacted a manda-

8   Several sets of twins together with a caretaker outside the Institute for Medico-Genetics in 1932. Like the KWIA in Berlin, Levit's Institute in Moscow employed twins studies to assess the role of heredity in medical disorders. Photo courtesy of the Lilly Library, Indiana University, Bloomington, IN.

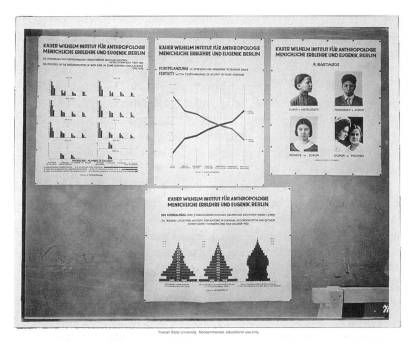

Truman State University. Noncommercial, educational use only.

9   An exhibit dealing with population studies and racial crossing from the KWIA. Unable to
attend the 1932 Third International Eugenics Conference in New York owing to the financial
impact of the Great Depression, the German Society for Racial Hygiene had to be satisfied
with simply sending exhibits without a delegation. Photo courtesy of the Prickler Memorial
Library, Truman State University.

tory sterilization law only two years after Indiana, boasted the largest
number of operations. By 1921 over two thousand had been performed.[89]
Eventually, the number of mandatory sterilizations would reach some
sixty thousand. In the United States, the number of sterilizations rose as
budgets for institutions housing the "defective" declined following the
disastrous stock market crash.[90] Although Britain was never able to enact
sterilization legislation, the Great Depression increased the number of
voices clamoring for such a measure.[91]

The Great Depression and the Radical Turn in
German Eugenics

Prior to the last years of the Weimar Republic, most German eugenicists generally did not believe that their country was ripe for American
measures, even if some personally believed in their eugenic desirability.

A few attempts, however, were made to change the law that penalized eugenic-related sterilizations, even when undertaken with consent of the patient.[92] At the time, they lacked sufficient political support. This all changed with the desire to reexamine the continued expansion of the Weimar welfare state in the wake of the depression. For a variety of reasons, the stock market crash hit Germany especially hard. By 1933 there were over 6 million unemployed in Germany, approximately one-third of the working population. For comparative purposes, the depression in the United States, as catastrophic as it was, left one-quarter of the prior employed out of a job.[93]

Numerous articles lamenting the burgeoning financial costs of housing the "unfit" could be found in Germany's leading eugenic journals. Hermann Muckermann, a former Jesuit priest with close ties to the Catholic Center Party in Prussia and an outspoken advocate of eugenics, complained in an article that institutionalized mental defectives were costing the state over 185 million marks a year at a time when there was barely enough money to keep the healthy from starving. His solution: differential welfare. A distinction should be made between those who could be "brought back to work and life" and individuals who would always remain nonproductive owing to bad heredity. In hard economic times, the latter group could only expect to receive the minimum required to maintain its existence. Naturally, preventive care—i.e., ensuring that the so-called defectives were never born—was the most cost-effective measure.[94]

The urgent call for racial hygienic action was naturally championed by right-wing, overtly racist eugenicists as well as the leaders of the Nazis, now the largest political party in the German Parliament. But as we have seen in the case of the Catholic prelate Muckermann, even those loyal to the Weimar Republic were becoming more and more radical in their rhetoric. Indeed the preface to *Eugenik*—hardly a publication of the radical right—appeared far less humane and sympathetic to those considered a financial and genetic liability to the commonweal than had previously been the case. "A crushing and ever-growing burden of useless individuals unworthy of life," the preface proclaimed, "are maintained and taken care of in institutions at the expense of the healthy. . . . Does not today's predicament cry out strongly enough for a 'planned economy,' i.e., eugenics, in health policy?"[95]

The term "useless individuals," would, in the not-too-distant future, become a synonym for a large number of genetic and racial undesirables who would ultimately pay for their alleged taint with their lives. This is not, of course, to suggest, that the editors of *Eugenik* had anything like this in mind. In fact, many of the regular contributors to the journal would themselves later be forced out of their positions owing to their political persuasion or racial background. Yet their harsh language reminds us that if those who still backed a democracy over a dictatorship could view the handicapped in such a manner prior to the Third Reich, how easy would it be for the vanguard of the rapidly approaching "racial state" to eventually put such rhetoric into deadly practice? Although admittedly impressionistic, it appears that none of the other eugenic movements in the myriad countries where they were flourishing at this time could claim such a quick and radical rhetorical hostile turn toward the "unfit" as that which occurred in Germany during the waning Weimar years.

Calls for action against the welfare problem and the growing cost of the defective in the Reich were made known in another, ostensibly moderate, quarter: the German Protestant Church. To be sure, eugenics was preached from the pulpit elsewhere, especially in the United States.[96] As such, this interest in theological circles was not completely unique to Germany. There were, however, differences. Enjoying close ties to the Reich since the founding of the Empire in 1871, the German Protestant Church was, in essence, a state church. Moreover, it and its members had a long history in involvement with social and welfare questions that potentially undermined the nation.[97]

The social-welfare arm of the German Protestant Church was known as the Innere Mission (roughly translated as National Missionary Work). Throughout the Weimar period there was a growing interest within the Mission in eugenic questions. After all, it was as concerned about limiting the money spent on the unfit as the state. Indeed, within the Mission the desire to take some kind of action in the form of eugenic measures in the late Weimar years was so strong that special conferences on the subject were held in 1931 and 1932. They were organized and headed by population policy expert and eugenicist Hans Harmsen (1899–1989), an important spokesperson within the Mission. Among the numerous individuals in attendance at the first conference held in the city of Treysa from May 18 to May 20, 1931, were several key heads of leading Protestant hospitals and mental institutions, as well as influential pastors and human geneticists. The director of the Division of Human Heredity of the Kaiser Wilhelm Institute for Anthropology, Otmar Freiherr von Verschuer—about whom we will hear much more in the following chapter—was invited as an expert witness. His strong Protestant convictions made him the obvious choice to represent Fischer's institute. The main topic of discussion at the meeting was the pressing need and theological justification for a voluntary sterilization law.[98] By this time, the United States was

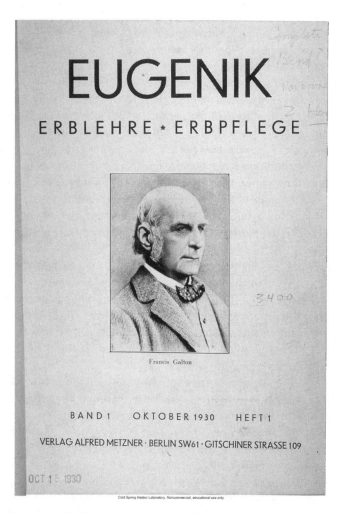

EUGENIK

ERBLEHRE * ERBPFLEGE

Francis Galton

BAND 1   OKTOBER 1930   HEFT 1

VERLAG ALFRED METZNER · BERLIN SW61 · GITSCHINER STRASSE 109

OCT 15 1930

10 The front cover of the German eugenics journal *Eugenics: Heredity and Hereditary Care,* featuring Francis Galton, the "father" of the discipline. This publication was advanced by the "moderate" eugenicists in Germany; those, on the other hand, who accepted ideologies of Aryan supremacy tended to use the word *Rassenhygiene* instead of *Eugenik.* The competing German journal, the *Archive for Racial and Social Biology,* although not filled with overtly racist articles during the Weimar years, certainly aligned itself with the dictates of the new order after 1933. *Eugenics* disappeared shortly after this last issue in 1932. Photo courtesy of the Cold Spring Harbor Laboratory Archives.

the country with the largest number of sterilization laws on the books. The results of these measures were circulated freely, especially the coauthored book, by Paul Popenoe and Ezra Gosney (1855–1942), *Sterilization for Human Betterment: A Summary of the Results of 6,000 Operations in California*—a work that would be influential for and legitimize the future Nazi sterilization law.[99] As we know, across the Atlantic almost thirty states made such procedures mandatory. But the United States was no longer the only nation that prevented the unfit from reproducing. Closer to Germany, the small state of Denmark had legalized voluntary sterilization in 1928—whereby its voluntary nature could seriously be questioned.[100] In addition, by the time of the conference in Treysa, two cantons in Switzerland also permitted individuals to be rendered infertile.

It would lead us too far a field to outline the details of the first conference. For our purposes, it is important to learn what von Verschuer, at the time one of Germany's two or three experts on human heredity, had to say. Even prior to the Great Depression, von Verschuer was a strong adherent of the form of "differential care" discussed at the conferences. He spoke in favor of voluntary sterilization in cases of individuals suffering from feeblemindedness, schizophrenia, manic-depressive illness, epilepsy, and Huntington's chorea. Psychopaths and the hereditary blind and deaf should also choose to come under the knife.[101] Von Verschuer argued that sterilization could be a condition for release from one of Germany's hospitals or mental institutions; in a paper von Verschuer delivered at the conference entitled "The Contemporary Biological Foundation for Assessing Sterilization," he argued, along the same lines, that a "feeble-minded, completely adrift girl who year after year brings an illegitimate child to the world" might continue to receive support for herself and her children if she agreed to be sterilized.[102]

In this same talk von Verschuer also sought to legitimize sterilization theologically. As the devout Protestant human geneticist viewed it, "Christians should adopt the model of their Lord who demands that we sacrifice in the name of loving one's neighbor." This love, he continued, should be extended to children not yet born. "I believe it is appropriate," von Verschuer stated, "to demand a smaller sacrifice from individuals than the sacrifice of life, in other words, to refrain from having progeny out of love for children who are expected to be [hereditarily] ill." At other Protestant venues, the Dahlem human geneticist offered similar arguments.[103]

As far as the two Treysa conferences were concerned, there was a consensus among the participants of the need for a "new eugenic orientation for welfare" and the introduction of "differential care" for individuals on the basis of cost-benefit analysis. And although killing the hereditary ill was clearly rejected by all in attendance, the Treysa report argued that the "artificial dragging out of dwindling life can just as easily be seen as an interference in God's will of creation as is euthanasia."[104] Clearly such statements—in particular in the language in which they are couched— would have been unthinkable from members of the German Protestant Church even ten years earlier.

Parallel to the first Treysa conference in 1931, an initiative was brought to the Prussian Upper House from physicians who had been involved with population policy and eugenic issues on the state level. Stressing the enormous strain to the public purse, the petition suggested that all physicians, pedagogues, and theologians be trained in human heredity and eugenics; future doctors should be tested in these areas before leveling medical school. Moreover, eugenics and human heredity should be taught in the upper classes of the secondary schools. In addition, eugenic health certificates should be exchanged by couples planning to get engaged. Finally, everything possible needed to be done to make the general population aware of the fundamentals of human genetics and eugenics. The hope was that other German states besides Prussia would strive for the same. The author of the petition once again reminded Prussian state officials of the unfavorable situation in the distribution of welfare and demanded a "differentiated" form of care that reflected considerations of efficiency. In this he was very influenced by the writings of Muckermann. Although the Upper Prussian House agreed with these proposals, it was not ready to take a positive stand on voluntary sterilization—Muckermann and Verschuer's outspoken support for the measure not withstanding. The division heads of the KWIA, including the director Fischer, decided to lobby for action on the sterilization issue at an upcoming meeting of the Prussian Health Council. The time for change, they believed, was now.[105]

Indeed, the strongest lobby group for a voluntary sterilization law at the Prussian Health Council's July 2, 1932, meeting, "Eugenik im Dienste der Wohlfahrt" (Eugenics in the Service of Welfare), was Fischer's institute. Among the twenty-three members of the Council and thirty-six scientific experts who attended the meeting, all three division heads (including Fischer, who was simultaneously director and head of the Division of Anthropology) of the KWIA were not merely present but active. In addition, the Jewish geneticist Goldschmidt, co-director of the Dahlem Kaiser Wilhelm Institute for Biology, was also engaged at the meeting, as were other members of the KWIA's board of trustees. Baur, friend and early supporter of the construction of Fischer's institute, also served as a scientific expert. He now headed his own KWI for Breeding Research in northern Bavaria. Although most of the coworkers at Rüdin's German Research Institute for Psychiatry in Munich were not present, they, too, supported the idea of voluntary sterilization.[106] Their absence and that of Fritz Lenz, however, is significant. It demonstrates that at this time, only months before the final curtain would fall on the Weimar Republic, the moderate Berlin group of eugenicists, not the openly right-wing racist Munich contingent, called the political shots at this important political gathering.

Perhaps the most interesting aspect of the meeting was the role played by Muckermann. We will recall that he was a former Jesuit, a devout Catholic, and an enthusiast for eugenics. Whereas the earlier role of the Catholic Center Party in eugenic matters concentrated on so-called positive measures to increase the number of the fit, with the coming of the depression both Muckermann and his party embraced voluntary sterilization. This is even more surprising when one realizes that Pope Pius XI issued the papal encyclical *Casti Connubii* on December 30, 1930. Among other things, it specifically forbade eugenic considerations in contemplating marriage and procreation.[107] That Muckermann and many other Center Party activists deliberately ignored this edict in their deliberations in Berlin demonstrates just how strong eugenic ideas had permeated the Catholic Church in Germany. This contrasts with the situation in Britain and America where Catholics were frequently outspoken critics of any attempt to apply scientific principles to procreation.[108] We have already seen how German Protestants supported the eugenic outlook. In other words, both major churches in the Reich were convinced that without some negative eugenic intervention, the nation would not be able to shoulder the financial burden of the "degenerates." Unlike the situation in other countries, there was no open critique of sterilization among German geneticists.

Fischer, the KWIA director, played a crucial role at the meeting with his deft handling of conflicts and differences of opinion on acceptable eugenic measures and the role of sterilization. As will become evident in the next chapter, by playing the "science card"—by appearing to represent a totally objective scientific outlook of the human geneticist—Fischer was able to smooth over potential difficulties. By the time he was finished with his talk, he received applause from all in the room. Those at the meeting set up a commission to write a draft of a voluntary sterilization law for Prussia. On July 30, 1932, the task was completed. Interestingly, it bore the stamp of Muckermann and members of the Catholic Center Party—hardly a group of individuals one might imagine would lead the charge of eugenic sterilization.

By the time the draft was written, however, the independence of the state it was supposed to serve no longer existed; the so-called Prussian Coup that followed an emergency degree from the aging German President Hindenburg dismissed the Cabinet of Prussia, Germany's largest state, only ten days earlier. The bastion of Weimar democracy, Prussia, was placed under leadership of the right-wing aristocrat and Reich Chancellor Franz von Papen. This emergency degree measure was pronounced allegedly owing to the center-left cabinet's inability to maintain law and order as street fighting between Nazis and Communists intensified. The true reason was so that the authoritarian von Papen, as Reich Commissioner of Prussia, could control the largest police force in Germany. The Prussian Coup spelled the death of Weimar government. Indeed, only six months later, Hitler was appointed Chancellor of Germany. In the midst of all the political intrigue and confusion, the draft sterilization law never attained legal standing.[109] As we will see, it did serve as the basis for the Nazi mandatory sterilization law of July 1933, the Gesetz zur Verhütung erbkranken Nachwuchses (the Law for the Prevention of Genetically Diseased Offspring).

———

On the eve of the Third Reich human genetics and eugenics had a strong presence on the international stage. From its modest beginnings at the turn of the century, human genetics and eugenics—at the time the theoretical and applied sides of the same science—quickly became the most socially significant field of biology. The synergistic relationship between the two in conjunction with political, social, and economic changes quickly led to incipient eugenics movements in the Anglo-Saxon and Scandinavian countries. Yet as we have seen, the meeting of the First International Eugenics Conference in London in 1912 suggested that a new worldwide scientific community dedicated to the "gospel of Galton" was beginning to emerge.

The Great War and its aftermath tested the ideals and the tenacity of the international eugenics movement. Believing, nearly to a person, that war was dysgenic and that the impact on the gene pool of "civilized nations" would suffer, human geneticists from all belligerent countries nonetheless supported their political leaders in what was supposed to be a short, patriotic war. As we have seen, the outcome of the bloodbath that emerged held special challenges; after first suffering pariah status in the international scientific community of human geneticists, Germany later exacerbated her exclusion by refusing to cooperate in international conferences owing to her feeling of humiliation at the hands of the former Entente powers. These tensions notwithstanding, communication among the human geneticists on both sides of the fighting continued after the war. By the mid-1920s, the international scientific community in this field was stronger than ever. It was politically opportune in the new interwar global arena. As we have seen, new organizations like the IEFO had been formed in which Germany, as well as other countries, played a central role. The number of nations boasted eugenics movements increased. In addition, new eugenic research institutions such as the Institute for Race Biology in Sweden and the German Research Institute for Psychiatry as well as the KWIA in Dahlem quickly became established sites for investigations into human genetics; they joined older such institutes like the American Eugenics Record Office and the British Galton Laboratory. The leaders of the new German research centers, Fischer and Rüdin, were lauded as world-renowned experts in their fields—a critical factor, as we will see, in the Nazi symbiosis between human heredity and politics to come. And it was during the 1920s, the Weimar years, that the German government embraced human genetics and eugenics as a tool to solve a variety of social and welfare problems.

Although German human geneticists had bemoaned the lack of eugenic legislation during the Weimar Republic, almost none wished for or expected the kind of negative eugenic measures like mandatory sterilization such as existed in the United States—that is until the coming of the Great Depression. As we have seen, the impact of the U.S. stock market crash radicalized German racial hygienists—both the openly racist, conservative practitioners based largely in Munich, as well as those who worked with the Catholic–Social Democratic coalition in Prussia. Eugenics would become a weapon in the desperate campaign to reduce welfare expenditures. No leading German geneticist protested the increasingly condescending, indeed inhumane, manner in which the so-called degenerates were described. The end result, as we know, was the failed attempt at a voluntary sterilization law.

The appeal of eugenics in the context of the cost-cutting mentality of the final years of the Weimar Republic was certainly central in the general acceptance of the future draconian Nazi sterilization law. It should be noted, however, that no international eugenicist condemned Germany in the waning months of its first attempt in democracy, although as we will see in chapter 6, voices were raised by "reform eugenicists" outside of Germany that the "mainline" outlook prevalent in the Reich and elsewhere was scientifically outdated and misguided. However, we should remember, it was not clear that it would be the openly conservative, racist elements in the Reich that stood poised to move eugenics forward in 1932. Although many German practitioners simply hid their true feelings about the issue of "race" owing to political pressure, Fischer, the president of the German Society for Racial Hygiene, for example, argued with Ploetz, the man who coined the term "racial hygiene," for a compromise in the name of the Society. In 1931 the Society's name was officially changed to the German Society for Racial Hygiene-Eugenics to please its Berlin moderate wing and allegedly attract more members. That Ploetz, representing the overtly *völkisch* Munich wing of the Society, argued that "laws and administration was not [as Fischer believed] dependent on Jews and Social Democrats," admittedly demonstrated the tension among Germany's eugenicists; it certainly did not bode well for the future.[110] But it was neither the suspect Jews nor the Social Democrats, but rather members of the Catholic Center Party, particularly under the auspices of Muckermann, who hammered out the foiled draft voluntary sterilization law. Just prior to the "new order," Jews and those on the political left and center were still among Germany's eminent eugenicists—individuals just as anxious as the openly racist conservatives to apply the advances of human genetics in the name of better breeding. Needless to say, much would change in the Third Reich.

Thus far we have examined the vicissitudes of human genetics and eugenics in the international arena. This discussion was designed to outline their trajectory prior to the victory of the National Socialists in Germany on January 30, 1933. At this point, the focus will change. The next four chapters will serve as venues of sorts for the main theme of our study: the symbiotic relationship between human heredity and politics during the Third Reich. In each of these chapters a short story will serve as an introduction to the particular site for the Faustian bargain in question. Before turning to the relationship between human heredity and politics under the Nazis, the chapters examine the way in which the two served as mutually beneficial resources during the Weimar Republic, raising the issue of continuity and change as Germany's political system was transformed. Although the focus is on the Third Reich, we should not forget that science and politics frequently serve as mutually beneficial resources—even if the resulting symbiosis was not as deadly as under Hitler. After the four case studies, the study will return, in chapter 6, where it leaves off here: to a discussion of the interaction between human genetics and eugenics on the international stage and the practice of human heredity in Nazi Germany.

TWO

The Devil's Directors
at Dahlem

The first meeting of the Kaiser Wilhelm Institute for Anthropology's *Kuratorium* (or board of directors) in the "new state" was held on July 5, 1933. With the recent death or negotiated removal of numerous former members, the new composition of the *Kuratorium* was decidedly national conservative.[1] The only relatively unfamiliar face at the meeting—at least to most in attendance—was that of Dr. Arthur Gütt, SS officer and high-ranking medical bureaucrat in the Reich Ministry of the Interior. An early advocate of eugenics during the Weimar years, Gütt quickly emerged as one of the major "coordinators" of National Socialist racial policy in the early years of the Third Reich, whereby research in the field of eugenics and human genetics would now be "coordinated" or brought into line with the political interests of the new regime. Through propaganda, intimidation, or "bargaining" methods, National Socialism would eventually leave its mark on the biomedical sciences, its practitioners, and its institutions, just as it was doing in all other areas of German society. However, for the various biomedical sciences this "coordination" was greatly aided by the willing participation of the researchers themselves. Although Gütt appeared at the *Kuratorium* meeting as a "guest," his agenda took precedence over all others: to "suggest" that the Kaiser Wilhelm Society "systematically place itself in the service of the Reich," and to formally ask the Dahlem Institute's help to carry out the draconian Law for the Prevention of

Genetically Diseased Offspring (the official name of the Nazi Sterilization Law) along with an intended Reich citizenship law. Although appearing to be little more than self-serving flattery, Gütt's appreciation
of the "important work" already accomplished in the Kaiser Wilhelm
Institute for Anthropology as well as his assurance that "the Reich Government placed a lot of value in the Institute's counsel" turns out to have
been quite genuine.[2] The new racial policy czar in the Reichsministerium
des Innern (Reich Ministry of the Interior) had every reason to believe
that the Dahlem director would be eager to help advance this national
mission.

This July event marks a key turning point in the history of the "Eugen
Fischer Institute."[3] The *Kuratorium* meeting sealed the Faustian bargain
between the Dahlem director and the Nazi state medical bureaucracy.
Yet the smooth-functioning and mutually beneficial deal that would last
throughout the twelve-year history of the Third Reich—even beyond
Fischer's own tenure as director—belies its rocky origins. As we will see,
the agreement made by the Dahlem racial anthropologist on behalf of
his Institute to serve the Nazi regime was made partly under duress. The
haughty, world-renowned national conservative human geneticist could
have hardly anticipated that he would be subjected to a vicious denunciation campaign levied against him by scientific rivals and Nazi Party
bureaucrats. This threat to Fischer's professional prestige at precisely the
moment when his long-nurtured research project in racial genetics at
Dahlem seemed more relevant and worthy of state funding than ever
certainly came as a shock. It took a short time for him to regain his bear-
ings. Yet ultimately, the Dahlem director would triumph over this obstacle: after all, when his science was at stake, Fischer was the political
chameleon par excellence. He was an expert at professional and political
self-refashioning.

As we will see in the course of this chapter, Fischer was no stranger to the
dictates of politics in science policy. Although sympathetic to *völkisch* racial views, this former member of the ultraconservative and anti-Semitic
German National People's Party was astute enough to realize that these
leanings were best left publicly unexpressed within the Centrist and Social Democratic–run Prussian government during the Weimar Republic.
After all, he would have to find an arrangement with these pro-Weimar
parties in order to win financial support for his new Kaiser Wilhelm Institute. His *völkisch* sentiments would hardly have been a valuable resource
at this time. Indeed, Fischer would go out of his way to demonstrate
that his vanguard Institute would not be used as an intellectual weapon
by political right-wing forces to destroy the fragile democratic govern-

ment, as some people charged. As we have seen, even after he was firmly in the saddle in Dahlem, Fischer helped orchestrate a name change for the long-established German Society for Racial Hygiene. The ambiguous term *Rassenhygiene* was offensive to many Jewish and left-leaning politicians whom he had to court during the Republic to keep his Institute up and running. His political misgivings notwithstanding, the Dahlem director used the international reputation of the Institute as well as his own *renommée* to serve the Republic; whether he approved of Weimar democracy or not, it was in the interest of his science and the *Volk* to find biomedical solutions to Germany's myriad social problems. The lessons that Fischer learned in his attempts to win government support for his line of research would be put to good use under more favorable political circumstances (for Fischer, at least) with the "national revolution" that placed Adolf Hitler behind the Chancellor's desk on January 30, 1933.

With this "national revolution" and the subsequent establishment of the "racial state," it became increasingly clear that government officials would need scientific experts in the broad area of racial science to put its dystopian, and later, genocidal, vision into practice. The Dahlem Institute was too vital a research center to be easily overlooked as a site where Nazi racial policy could be scientifically legitimized. A prerequisite, however, was the absolute collusion of its director, Fischer. Although known to be nationalistic, the Dahlem director was not a Nazi Party member; perhaps more important, one of his coworkers was a "political Catholic" whom the new regime did not trust. And finally, Fischer's own views on racial science did not exactly square with the main party line.

In the course of his denunciation campaign, it became clear to the Dahlem director that he could only remain an authoritative academic spokesperson for his science and expect to receive funding for it if he aligned himself totally with the new order. Conditions—some of them unpleasant—would have to be met. It appears that the politically savvy Fischer soon came to realize that the current Nazi-Nationalist government was different from the various coalitions preceding it. Having learned the nature of the new Nazi masters, the Dahlem director "sold" his Institute to the National Socialist state. Although he did not join the party immediately, he and his research center nonetheless moved with the times. With the exception of the Rüdin Institute in Munich (discussed in the next chapter), there was no center for human heredity and eugenics in Germany that did as much to provide the intellectual underpinnings for the Nazi regime's racial policy than did the Kaiser Wilhelm Institute for Anthropology. The Dahlem director and his coworkers reaped rich rewards for their part in this Faustian bargain. But as we will see, the price that the senior partner in this bargain levied upon him would be high indeed. What is tragically ironic, however, is that the costs of this deal were ultimately borne upon the backs of those who either suffered from or were killed by the very racial policies Fischer and his Institute's personnel promoted throughout the Nazi years. Indeed, there is very little evidence that Fischer or his colleagues thought long and hard about the consequences of placing their science at the service of the Third Reich.

The political chameleon Fischer viewed himself first and foremost as a scientist, and he wanted to advance a research project in racial genetics second to none. When it became clear to him that traditional approaches to understanding the inheritance of normal and pathological traits were wanting, the Dahlem director initiated a far-reaching paradigm shift at the Institute.[4] By 1940, he hoped to pursue *Phänogenetik,* or developmental genetics, as a way to understand what we would today loosely term the process of gene expression. During the war years—at a time when many other Kaiser Wilhelm Institutes were experiencing budget cuts—Fischer's new important party patron, SS-Brigade Leader Leonardo Conti (1900–45), helped him secure the funds necessary for his lavish project. As we will see, the racial anthropologist's decision to pursue the vanguard science of developmental genetics would have enormous moral consequences for the Devil's second disciple in Dahlem and Fischer's devoted protégé: the medical geneticist Otmar von Verschuer.

Handpicked by Fischer to serve as head of the Department of Human Heredity and ultimately as his successor as Institute director, von Verschuer was known internationally for his use of twins to study the inheritance of disease. When he inherited the directorship in late 1942, von Verschuer continued to advance this new methodology in Dahlem. As we will see, human heredity and politics had a radicalizing symbiotic impact during the brutal and brutalizing conditions of Germany's "racial war." When it became evident that investigations as part of the Institute's new research agenda had advantageous political implications as well as scientific ones, a series of events unfolded that ultimately led the second director and at least one of his coworkers toward the moral abyss associated with the biomedical crimes committed at Auschwitz. Another member of the Institute, Hans Nachtsheim undertook research made possible by Nazi Germany's "euthanasia" project. He, too, was hired by Fischer.

In assessing the extent of Fischer's and his colleagues' moral culpability, we must consider the role that the director played in establishing a research program at Dahlem that made such crimes possible in the first place. We may well wonder whether the Devil's first disciple is really any less ethically compromised than von Verschuer, even though he never got his hands as dirty as that of his protégé. In order to contemplate such issues intelligently, we must begin on a seemingly more benign note: with the decision to establish the Dahlem Institute and the preparations for its gala opening under its first director, Fischer.

"True Racial Science . . . Will Bring Segments of the Nation Closer Together"

As we have noted, the tireless behind-the-scenes efforts of German plant geneticist and eugenicist Baur to convince KWS President von Harnack to establish an institute devoted to human heredity in the Reich capital ultimately paid off. Of critical importance was his ability to persuade his colleague and friend, the world-renowned anthropologist and coauthor of the German human genetics and eugenics standard work, *Principles of Human Heredity and Racial Hygiene,* to leave his beloved Freiburg and assume the directorship. It was to be located in the idyllic southwest section of Berlin, Dahlem. Two other Kaiser Wilhelm Institutes devoted to the biological sciences were within easy walking distance of what became the twenty-ninth member of a growing family of research centers under the Society's umbrella.[5] Their close proximity to the Fischer Institute would come in handy over time.

Yet negotiating a commitment for such an institute from von Harnack was only the first step toward bringing the plans to fruition. The Society president had to convince the Social Democratic–Catholic Center Party led coalition government in Prussia of this "great national task."[6] More particularly, he had to demonstrate that the Institute's scientific work would not serve as an intellectual resource in the hands of rabid anti-Republican forces and anti-Semites. The terms "race" and "heredity" were already circulating in national conservative and *völkisch* circles. On the national level, there was a well-justified concern about the potential usefulness of the sciences to be housed in the proposed Dahlem research center for the political right wing. This can be gleaned by a comment made by a Social Democratic delegate to the Reichstag, the German lower house of parliament. During a meeting to discuss the Institute's financing, Dr. Julius Moses, a Jewish social hygienist, reminded the delegates as well as von Harnack that anthropology, human heredity, and eugenics had indeed supported nationalism both before and during the Great War. Fortunately for the future of the Dahlem Institute, Moses became convinced that the planned research center would examine these disciplines "from a strictly scientific viewpoint." As a result, the Social

Democrats, although leery, lent their support to the establishment of the Institute.[7] Catholic Center Party members were won over by the lobbying of Muckermann. His propaganda campaign on behalf of eugenics had already convinced political leaders and health professionals with a clear allegiance to the Weimar Republic that the new applied science and its allied disciplines were a state imperative compatible with the democratic order. His close ties to Prussian Center political circles together with von Harnack's solemn pledge that "the Kaiser Wilhelm Society [would take] full responsibility" to guarantee the Institute's scientific objectivity undoubtedly helped shore up funding for the new center.[8]

Von Harnack knew, however, that the best way to secure state money for the Dahlem Institute was to sell it as a national health and welfare resource. In a letter asking for additional support for the new building, the Society president adopted the language of cost-benefit analysis that, as we have seen, colored much of Weimar eugenic discourse: "If I ask for 100,000 RM more than was planned . . . ," von Harnack pleaded, "I do so with the complete conviction that if the ideas of Herr Prof. Fischer and his close scientific associates are able to penetrate relevant public health circles, enormous savings of Germany's human, as well as financial, assets could be achieved."[9] In the end, more than 85 percent of the money for the new Institute would be paid for by public funds. As a state-sponsored Kaiser Wilhelm Institute, it would naturally lend itself to confronting pressing national biomedical tasks, regardless of all the future director's disclaimers that the new institution was "purely theoretical."[10] Given the context in which it was founded, the chance that the Institute would remain above the fray of politics was remote from the outset.

The president realized that he would have to legitimize the establishment of the proposed Institute to the Kaiser Wilhelm Society Senate in a somewhat different manner than he did to the politicians. Knowing that the senators were particularly sensitive to Germany's international stature in the world of science, the president played up the need to retain its competitive edge in the biomedical sphere. "Now that Sweden, the United States, France and England have taken the lead in founding their own research institutions," von Harnack informed the senators, "it is absolutely imperative for Germany to also establish a scientific center for anthropology, human heredity and eugenics, especially as the insufficient and dilettantish approaches in this field must be met head on."[11] But von Harnack's persuasiveness notwithstanding, no individual was able to master rhetorical resources better than the future director himself. When the president asked Fischer to deliver a programmatic address on the nature and tasks of the proposed Institute before the Society's senators as well as dignitaries from state and industry on June 19, 1926, Germany's premier anthropologist tailored it perfectly to his immediate and wider audience. Had the senators not already given the new research center "the green light" that very afternoon (without Fischer's knowledge), they certainly would not have hesitated long after hearing the Freiburg anthropologist's showcase speech.[12]

Recognizing the prejudices of his listeners on the topics of anthropology and "race," Fischer began his talk by attempting to set his audience straight. Anthropology was no longer merely preoccupied with measuring skulls and studying the shapes of noses. It was intricately linked to the new science of genetics. Indeed, "anthropology and human heredity," the future director affirmed, "cannot be separated." Racial science or "racial biology," as Fischer called it in his address, was a misunderstood concept. It was not limited to the study of traits within any of humanity's major "racial" groups; it also investigated normal and pathological characteristics of any "hereditary line" or interbreeding population.[13] More particularly, it studied both from a genetic perspective. There could be no doubt that race, in the first sense of the term, played at least some role in history, Fischer informed his listeners. Past civilizations perished partly because of "racial chaos," the speaker's description for unfavorable racial mixtures among populations. The crossing of two closely related "appropriate races," however, was viewed by Fischer as positive.[14] Indeed he deliberately separated his understanding of "race" from that of Hans F. K. Günther (1891–1968), author of the extremely popular *völkisch* book *Racial Science of the German People*. Günther's treatise was precisely the kind of work that alarmed the Jewish and liberal circles Fischer needed to reassure. The fact that Fischer would soon (or had already) reviewed the book favorably in a journal he edited was left unmentioned that evening.[15] What he did stress favorably was the research of the German-Jewish American cultural anthropologist Franz Boas—allegedly to show how complicated the subject matter of "race" really was. In an important study, Boas demonstrated that traits formally viewed as fixed racial characteristics were actually affected by environmental factors. Fischer also did not neglect to mention that the politically liberal Boas, long established at Columbia University, had been a "benefactor" of Germany in its hour of need after the First World War.[16]

From here the future director moved easily to a discussion of human heredity as a scientific and political resource. "We need a thorough investigation of the normal and pathological hereditary lineages of our population," Fischer maintained, if we are to have an adequate scientific estimate of the number of "cretin, criminal, idiot and other constitutionally abnormal hereditary lines." Naturally, the state had an interest in ascertaining the number of talented lineages as well. In former times, the Freiburg anthropologist pointed out, physicians studied disease. Today people, not illnesses, are once again the rightful object of medical research. Race and illness are not separate entities, Fischer emphasized. It may turn out that the one-time racial components of Europe's present population have different levels of resistance to disease. It would be important to understand, for example, what, if any, impact racial crossing might have on an individual's predisposition to serious illnesses such as cancer, tuberculosis, and epilepsy. A person of racially mixed origin might appear healthy on the outside; however, on the inside, he or she may be subject to "chemical-physical" imbalances making the individual more prone to disease. "We know nothing about all these things," Fischer emphasized. The speaker reminded the dignitaries assembled that "even as non-physicians" they could appreciate how such information would be well worth knowing.[17]

The guest of honor then turned his attention to eugenics—perhaps the most socially relevant topic discussed that evening. Here again he felt the political necessity to enlighten his listeners. In Germany, Fischer continued, the applied science frequently went under the name "*Rassenhygiene.*" The speaker conveniently forgot to mention that right-wing adherents had adopted the multivaried German word for their own advantage. Racial hygiene, Fischer insisted, did not imply the breeding of a better "race" in the usual anthropological sense. Nor did it denote the superiority of one particular racial group over another. Rather, it was synonymous with all actions aimed at improving the hereditary substrate of any population. Since the German word has led to such misunderstanding, Fischer added, "we gladly avoid '*Rassenhygiene*' and "select the English term *Eugenik.*" Coined by the British "hereditary theorist" Galton, he continued, eugenics involved nothing more than the recognition and care of the "favorable" hereditary lines."[18]

Fischer naturally failed to mention that he himself had a weakness for the so-called Nordic race; were it not politically incorrect in the Reich capital during the middle years of the Weimar Republic, he would have openly embraced the term "racial hygiene." Fischer did, however, inform his audience that as early as 1910, he was preaching the eugenics gospel. Even at that early date the Freiburg anthropologist knew that controlling the reproductive capacities of a population was a "fundamental question of existence for states." Unfortunately, at that time, his words fell on deaf ears. "We allowed our culture [to develop] as it wished," Fischer continued. "There was . . . a completely incomprehensible indifference in the entire civilized world regarding . . . [how cultural amenities useful for the welfare of the individual] . . . could impact the long-range future." Today, an unmistakable series of "warning signs" has appeared on the landscape. We have to march forward, Fischer informed his audience, armed with "positive eugenic measures."[19]

The future Dahlem director closed his performance with an appeal to his listeners—scientists, politicians, and leaders of industry—to harness biology in the service of the nation.

We so often claim: we have made ourselves masters of nature. With [our] expansive technology we control an infinite number of things. . . . The space on Earth, at least, has practically dwindled to nothing, when we think of modern planes, wireless telegraphs and similar things. What we have *not yet*, however, even *begun* to master and get a handle on are those biological phenomena that have harmed our culture. To work on the preservation of hereditary lineages—to study them and affect them in a positive manner . . . has not yet been undertaken! This is the essential and final task inherent in all [the research I have talked about]. This task—and this is something everyone will admit—is a matter of life and death for the well-being of all our people. It cannot suffer any delay; it demands our complete concern and energy.[20]

Fischer's impassioned plea had its intended effect. He would have his vanguard Institute for the human sciences.

Although there were financial problems that threatened to slow progress of its completion, the new building was erected in record time—in just eleven months. It was important to finish the work on schedule, since the Society planned that the new Dahlem Institute should open its doors on September 15, 1927, a day carefully chosen, as we have seen, to coincide with the Fifth International Congress on Genetics hosted in Berlin during that week. The festivities surrounding the inauguration were meant to attract national and international attention. In front of an audience of Prussian politicians, university dignitaries, and over nine hundred conference members, including Charles Davenport from the Eugenics Record Office, von Harnack held his inauguration speech.[21] The press was on hand to report this important event.

The Society president used his talk at the Institute's dedication ceremony to once again reassure those who worried about the possible negative political fallout of a research center devoted to anthropology, human genetics, and eugenics. "True racial science," the aging von Harnack argued, "will bring segments of a nation closer together, not divide them [in a hostile manner]."[22] When he turned over the Institute keys to Fischer, the new director instinctively recognized the need to formulate a politically appropriate statement. The new research center, Fischer proclaimed, "will naturally deal with all questions surrounding the problematic concept of 'racial science' from a purely scientific basis."[23] With these benign and seemingly promising words, the KWIA entered upon Germany's, and the world's, genetics and eugenics scene.

"Is the Patient a Twin?"

After winning the long battle to get the Kaiser Wilhelm Institute for Anthropology established, Fischer turned his attention to creating an organizational framework for the Institute based upon the tenets of the "Harnack Principle." According to this guiding philosophy allegedly operative since the inception of the Kaiser Wilhelm Society during the Empire, Kaiser Wilhelm Institutes are built up around a director chosen by the Society for his research excellence who is then given a free hand to use the available financial resources to shape both the organizational structure and research orientation of his institute.[24] Given Fischer's academic background outlined in the previous chapter, it comes as no surprise that he established a research program that focused on racial genetics, the inheritance of normal (i.e., racial) and pathological traits. As Fischer explained it, these two research foci would find their culmination in a third, eugenics.[25] This research concentration, with its strong emphasis on the genetics of race, did not change appreciably over the course of his directorship, regardless of its later "modernization" through the application of the newest trends in developmental genetics. Fischer organized these research foci in the Institute into three separate departments whose boundaries he always viewed as somewhat arbitrary and fluid. It would not be an exaggeration to say that the Dahlem research center was the first leading interdisciplinary institute in Germany for the study of human beings from a genetic perspective. As such, Fischer established an Institute that already extended beyond the boundaries of the more traditional German university departments. These scientific boundaries would not be the only barriers he and his successor, von Verschuer, would eventually cross.[26]

Fischer himself was Institute director and head of the Department for Anthropology—positions he combined with his professorship of anthropology at the University of Berlin. He had the good fortune of heading the Institute at a time when one of America's premier philanthropic organizations, the Rockefeller Foundation (RF), was interested in supporting population studies in Germany. Fischer was instrumental in organizing a large-scale, national anthropological survey of the German population that promised to generate information on virtually all the meaningful racial, biological, and medical particulars of the *Volk* (this time meaning all people living in Germany). This carefully crafted interdisciplinary project involving a host of different institutes and different disciplines (but with an emphasis on anthropology) enabled Fischer to acquire a share of $125,000 distributed over a five-year period (beginning in 1930) for his Institute. The project, although funded by the Social Science Division of the RF, would supplement the support he already received for this research from the Emergency Association of German Science, the German equivalent of the American National Science Foundation. It would also be managed by that same German organization. The project was sold to the RF on eugenic grounds. As Friedrich Schmidt-Ott, president of the Emergency Association explained to an RF officer, "one hopes that through [this survey] new insights will be acquired regarding the origin of possible degenerative phenomena," especially the distribution of hereditary pathological character traits. "Questions regarding the biological conditions of families, the numbers of births and miscarriages . . . and lastly, the issue of birth rate decline and birth control [will be analyzed] from a eugenic perspective."

Fischer surveyed university professors, high school teachers, and elementary school teachers as part of his research. All such individuals were members of the German civil service. He also received permission from the Prussian Minister of Interior to request information from another source—the Prussian police. According to a report, thirty-nine thousand forms were distributed to members of the Prussian police force in 1931. In other words, the Dahlem director demanded and received access to a significant amount of private data for his work. As we will see, Fischer was not the only biomedical scientist who would require such information. Although his Institute did not receive the largest share of this money (in fact, Rüdin's Institute did), Fischer's ability to organize this significant project demonstrated his talents as a research manager. The Dahlem anthropologist was also able to attract other outside money for two other research projects on the genetic component of tuberculosis and germ plasma toxins undertaken by members of his Institute. This support was obtained from the Medical Division of the RF. Indeed, it appeared that the RF thought extremely highly of the Dahlem director. As RF Officer Robert Lambert reported to a Foundation colleague of his in 1932, "[as] far as I know everyone had been well impressed by F[ischer] both as a scientist and as an administrator, and by the sound work coming out of his institute."[27]

The Dahlem director appointed Muckermann to oversee the institute's Eugenics Department. This proved especially useful owing to the Catholic eugenicist's talent for popularizing his subject, especially in the all-important Catholic political milieu. We have already noted the large role he played on the Prussian State Health Council in 1932 in advancing racial hygiene measures. Indeed, by his own account, Muckermann gave approximately six hundred eugenics-related talks between 1927 and 1933—at least some of which were sponsored by the Kaiser Wilhelm Society.

Fischer's deliberate selection of his medically trained former student,[28] von Verschuer, to head up the Department of Human Heredity, was a professional choice that would have far-reaching consequences for the future of his Institute. Born into a Protestant noble family with a military heritage, von Verschuer became interested in the racial worth of the members of his privileged class long before he turned to medicine.[29] He ultimately decided to make medicine a career rather than becoming a military officer because of his political unwillingness to serve in the new Republican army following the German Revolution of 1918–19.[30] Once, however, he settled on mastering the biomedical sciences, he never wavered. We can observe the drive with which he would later embrace his science while he was still a medical student completing his dissertation on the effects of caffeine on the swelling of blood serum proteins. Using language eerily similar to Mary Shelley's description of Victor Frankenstein's single-minded pursuit to solve the riddle of reanimation, von Verschuer related his obsession with his doctoral research. "I plunged myself into my task with all the fire of my scientific zeal; I worked for months intensively in the laboratory of the clinic, first testing subjects then myself. . . . [I] allowed venal and capillary blood to be drawn from me every two hours from early morning until afternoon for two [consecutive] days while [not under the influence of caffeine]. The second time [I repeated the test] with caffeine. I analyzed the results of the tests until late into the night."[31] As we will see, von Verschuer's fervor for biochemical research would not diminish throughout his career.

Upon completing his dissertation, the ambitious aristocrat first volunteered at the Psychiatric Clinic at the University of Munich, having developed an interest in mental disorders as a result of attending Emil Kraepelin's lectures there. However, von Verschuer, unlike Rüdin, would not follow in the footsteps of the renowned psychiatrist. His first paid medical position was acquired through the help of Lenz, the Munich human geneticist and racial hygienist who would later replace him in Dahlem. Through Lenz's family connections, von Verschuer secured a position as a medical assistant at the University of Tübingen in 1923. Together with Wilhelm Weitz (1881–1969), his senior colleague, the young clinician began using twins as part of his medical research. It was at this time that von Verschuer became more and more interested in what we today would call medical genetics.[32]

Although von Verschuer had already developed a scientific reputation in the field of twin studies by the mid-1920s, he never forgot his earlier preoccupation with "race" and eugenics. His interest in these subjects induced him to attend Fischer's classes in anthropology during his medical studies in Freiburg. Fischer's "force of personality" managed to have a large impact on von Verschuer during his brief stay in the Baden college town; the reverence he would show Fischer throughout his scientific life probably has its origins in this period. Earlier, when he began his preclinical medical training in Marburg, von Verschuer involved himself with various right-wing organizations and activities in the aftermath of the German Revolution. He never shed his deeply felt *völkisch*-nationalist sentiments, although when he heard the future *Führer* speak years later, he allegedly found Hitler too brutal, anti-intellectual, and fanatical.[33] While in Tübingen, he became secretary of the city's newly formed section of the German Society for Racial Hygiene; indeed, the budding young human geneticist took every opportunity to lecture and write on eugenics. Perhaps a similar *völkisch* worldview along with a shared penchant to investigate the newest questions in human heredity accounted for the interest that Lenz took in von Verschuer.

At any rate, in 1926, Lenz suggested that he send Fischer one of von Verschuer's early professional manuscripts; the Munich human geneticist apparently knew that the renowned Freiburg racial anthropologist would soon be heading up a research institute in Dahlem. Fischer was obviously impressed with von Verschuer's scientific work. He offered his former student the post as director of the Human Heredity Department, although it was contingent upon von Verschuer's completion of his *Habilitation* (second dissertation)—the necessary qualification to teach at a university.[34] This would ensure the scientific respectability demanded by the Kaiser Wilhelm Society and help compensate for Muckermann's lackluster research portfolio. Von Verschuer held his obligatory *Habilitation* defense at the University of Berlin in May 1927; it was the first time anyone had completed an advanced degree in the new field of human genetics anywhere in Germany.[35] Von Verschuer, like his mentor Fischer, hid his *völkisch* outlook during the Weimar years. Prior to 1933, von Verschuer was also active in the Berlin section of the German Society for Racial Hygiene; again, like Fischer, he seemed to have a good working relationship with officials of the Centrist/Social Democratic Prussian state.

Almost immediately, von Verschuer used his new position to establish the Dahlem Institute as one of the world centers for twin research, at the time the most innovative way to study human genetics. Prior to his arrival in Dahlem, von Verschuer won scientific acclaim for this theoretical work in twin research. He was instrumental in improving upon the so-called polysymptomatic similarity method—a means of using numerous characteristics to determine whether twins were identical or fraternal. Yet as we will see, for all the promise of twin studies, by the late 1930s it was clear that this method alone could not answer all the important questions posed by its German practitioners. A new paradigm, whose development lay several years in the future, would be necessary to account for such shortfalls. When von Verschuer arrived at the Fischer Institute in 1927, his scientific reputation was linked to the efficacy of twin studies as a way of determining the roles of nature and nurture in the origin of diseases. The RF was quite impressed by his innovative research.[36]

Proving himself especially adept at finding ways to locate subjects for his research at hospitals, schools, and through newspapers, von Verschuer filled his files at Dahlem with data from hundreds of twins. The human geneticist stamped the query "is the patient a twin?" on all his inquiries to medical facilities to increase his database. By the time he left the Dahlem Institute in 1935, he had amassed information on over four thousand twins—albeit not without the help of party and state institutions in the years after the National Socialist "seizure of power."[37] Many of von Verschuer's subjects, especially school children, were examined in the Institute itself. Both alone and working with his assistants, he used twin studies to demonstrate the inheritance of mental traits and criminality—research that was quite generously supported by the Reich and Prussian governments as well as by the RF. While at the Kaiser Wilhelm Institute for Anthropology, he finished his massive study, coauthored with his old school friend and medical colleague Karl Diehl (1896–1969), on tubercular twins, also with outside funding. Perhaps armed with the results of von Verschuer and Diehl's research, Muckermann allegedly gave public talks in which he advised against spending the country's ever-dwindling financial resources on the prevention of tuberculosis. Individuals with TB should be encouraged not to reproduce.[38] Other investigations by von Verschuer, such as his work on blood groups, promised to have direct applications in paternity cases facing the courts.[39]

At this point, one may wonder whether von Verschuer was unwittingly laying the necessary scientific foundations for the Nazi racial state

11 Otmar von Verschuer, head of the Division of Human Heredity, standing in front of his large twin file at the KWIA during the late Weimar years. Photo courtesy of the Archiv der Max-Planck-Gesellschaft, Berlin Dahlem.

through his research in the Weimar years. We should not forget that at this time, even before the Third Reich, institutions such as schools and hospitals gave von Verschuer the names of potential subjects for his research on twins—an activity whose legality would be questionable today. According to medical guidelines that were passed during the Republic

12 Von Verschuer assessing the exactness of eye color in a pair of twin school boys with the aid of an eye chart. Photo courtesy of the Archiv der Max-Planck-Gesellschaft, Berlin Dahlem.

and remained unaltered with the coming of the "national awakening" in 1933, people under the age of eighteen could not be used for human experiments designed to advance science if such tests "could in any way hurt them." Some of von Verschuer's subjects were as young as fifteen. They did not undergo these tests for any therapeutic advantages but for research purposes. As the experiments that these youngsters were encouraged to endure were sometimes painful if not actually dangerous, von Verschuer and the Dahlem Institute were already operating in an ethical gray zone prior to the Third Reich.[40]

As we have noted, the Dahlem director initially promised that a major function of his Institute would be to inform and legitimize eugenics. It accomplished this goal—both at home and abroad—in several ways. Fischer, Muckermann, and von Verschuer headed, took part in, or otherwise had close ties to numerous national and international eugenics organizations and societies, such as the German Society for Racial Hygiene and the International Federation of Eugenics Organizations. They published their research in the journals of these organizations, often in popularized form. In addition, Institute personnel maintained close relationships to members of the health and welfare bureaucracy of Germany's largest state, Prussia. As we have seen in the last chapter, their expertise was sought when these bureaucrats contemplated any kind of eugenics-related legislation. The Dahlem scientists demanded a more differentiated, eugenically grounded social welfare policy during the financially troubled final years of the Republic. Imbuing his eugenic outlook with a healthy dose of class prejudice, von Verschuer, like Lenz, argued for the need to increase the hereditary substrate of those who did not have to toil with their hands. "The traits that make mental work possible . . . ," the twin specialist argued, "are necessary for the security and advancement of the state and for the life of the *Volk*."

As we have noted, von Verschuer and his colleagues were also intricately involved in the 1932 draft sterilization law. The devout Protestant defended sterilization as an act of Christian charity on the part of individuals toward their progeny who, in all likelihood, would be born defective.[41] Moreover, the Dahlem Institute soon began to distinguish itself from other Kaiser Wilhelm Institutes through its teaching and service functions. Beginning in 1929, it offered a host of human genetics and eugenics related courses for state-appointed medical and welfare officials, a function that secured it additional funding and won the KWIA the goodwill of grateful Prussian bureaucrats. What is particularly important to note is that much of the instructional material used to train these officials was based on research currently carried out at the Institute. And finally, the research undertaken in Dahlem could be used by Institute members in their admittedly limited role as scientific experts in paternity cases. The existence of this increasingly intertwined research-propaganda-instructional complex at the KWIA during the final years of the Republic can perhaps help to explain the relative ease with which its members could respond to the new realities following the "national revolution," at least once its director recovered from the intrigues surrounding its "coordination."[42] By 1933, serving the state and *Volk* was indeed nothing new for Fischer or his Institute.

The Faustian Bargain in Berlin: Act I

Political Pressure, Professional Self-Fashioning, and the Power of Human Heredity

Even prior to the Dahlem Institute's first *Kuratorium* meeting under the Third Reich, SS doctor Gütt had the opportunity to become acquainted with Fischer in connection with a series of denunciation campaigns directed toward the director from the end of May 1933 to late summer 1934.[43] To understand how such a denunciation campaign functioned within Nazi Germany, it is important to realize that contrary to popular belief, the Third Reich was not a monolithic regime in which every aspect of life was personally controlled by the *Führer*. It was a polycratic system where power was divided between various state and party institutions; even the military and big industry could affect the interplay between the government and society. There was no such thing as a prototypical "Nazi." Individual National Socialist Party officials, big and small, had different ideas on important ideological and governmental matters and often exploited the nature of the system to gain power for themselves—competing with other Nazi Party members for prestige in the eyes of their superiors or even, for the biggest of Nazi bigwigs, for Hitler's approval. As such, political denunciations, serving as a form of intrasocietal competition for favor, were commonplace in many institutions under National Socialism; despite its international prestige, the Kaiser Wilhelm Society was not immune to the sometimes petty viciousness that characterized these campaigns.[44] But in their intensity, in the number of party and state offices involved, and in the importance of the issue at stake, the attacks against Fischer are probably unique among denunciations within Germany's most elite research organization.[45]

The details of this denunciation campaign are complicated and the exact motivations of the parties involved are not always clear. Nonetheless, it appears that the diatribes directed against Fischer were unleashed owing to concern expressed by National Socialist medical bureaucrats, party officials, and potential competitors in the field of racial science, perhaps including Rüdin, that this "non-Nazi" academic (Fischer was not yet a party member) was poising himself to assume a leading position to help shape racial policy under the swastika. By late May 1933 when the battle began, Fischer was simultaneously director of the internationally respected Kaiser Wilhelm Institute for Anthropology, senator of the Kaiser Wilhelm Society, president of the German Society for Racial Hygiene, president of the Berlin Society for Anthropology, Ethnology and Pre-History, and Rector of the University of Berlin, the latter position having been achieved against the wishes of senior members of the Nazi Party. He was also a world-renowned authority in a field highly visible in numerous international organizations. Indeed, it was only his expressed wish not to stand for election as president of the IFEO that enabled his rival Rüdin to acquire that post. For those both inside and outside the Dahlem Institute who found Fischer's power base a career threat, the director's ability to work well with politicians and bureaucrats in the Weimar Republic[46] and his somewhat nonconformist views on the race question could be used as a political weapon against him.

The event that instigated the denunciation campaign against Fischer occurred on February 1, 1933, several months before the *Kuratorium* meeting that would seal the deal between the Dahlem Institute and the Nazi regime, when Fischer delivered his first public lecture in the new *"völkisch* state." It was held in a prominent place: the capital of the new Reich, Berlin. His well-attended and much anticipated talk was entitled "Racial Crossing and Intellectual Achievement." In his lecture, Fischer not only contradicted mainstream National Socialist doctrine on the desirability of racial crossing within the so-called European races, but he also clearly took a "soft" stand on the "Jewish question." Whereas the official National Socialist view on racial mixtures between Jews and "Aryans" was uncompromising, Fischer found the "judgment of the results of such a crossing" to be "very difficult." Although he would not rule out the possibility of a "psychic disharmony" arising from such a racial mixture, he argued that "[it] undoubtedly makes a huge difference whether the offspring of long-standing cultivated German Jewish families or whether the progeny of newly arrived Eastern European Jewish families mate [with non-Jews]." In this statement, Fischer's prejudices regarding the difference between allegedly acceptable assimilated German Jews and the politically and racially dangerous Eastern European Jews, or *"Ostjuden"*—prejudices common among nationalist conservatives like himself—demonstrate that he had not yet fully adopted the National Socialist Party line on "race." The director did, however, admit that the result of such a crossing would introduce a (racially) different element into the German population. The long-range effect of such a crossing, Fischer insisted, could not be known. Hence, there can be little doubt that this highly visible professional talk gave his political opponents ammunition; it probably set the stage for all subsequent denunciations against him.[47]

Gütt's role in the "Fischer case" wavered between that of the heavy-handed arbiter of intraparty disputes, on the one hand, and the Dahlem director's political advocate, on the other. Clearly, former sins could not be left unpunished. Gütt did not attempt to reverse Fischer's dethronement as president of the German Society for Racial Hygiene, a position now held by Rüdin.[48] After all, as former leader of the moderate Berlin faction of the Society—the group that had close alliances with the old Centrist/Social Democratic coalition in Prussia—Fischer was no longer as ideal for this post as the psychiatric geneticist Rüdin, a member of the unabashedly *völkisch* wing of the Society in Munich. Moreover, unlike Lenz or even Muckermann, Fischer had declined to lend the Nazi Party any open support until it was in power. Allegedly, this was to guarantee the independence of his Institute.[49] More likely, however, Fischer wanted to hedge his bets and avoid throwing his cards into the brown box until the new political constellation was clear. What the nationalist Fischer may have thought about the Nazis privately prior to this is, of course, open to speculation. Owing to lack of information, however, the issue cannot be resolved. At any rate, Gütt had sufficient reasons not to bend over backward to make life too easy for the Dahlem director. As we will see in the next chapter, he was politically partial to Fischer's rival, Rüdin.

During the course of a series of meetings, Gütt also convinced Fischer of the need to "coordinate" his Institute. Since there were no "racial enemies" employed at the Dahlem research center, what he meant was the removal of the politically unreliable elements, first and foremost the Catholic Center Party supporter and Eugenics Department head Muckermann. Given its ties to the Vatican, the Center Party was distrusted as a potential pillar of the new state. Muckermann, as a "Catholic eugenicist," could hence not be counted on to support all National Socialist racial policy measures, despite his attempt to find common eugenic ground with certain Nazi health officials and his large role in garnering support for the draft sterilization law prior to 1933. As such, his tenure at the Fischer Institute had to come to an end, and this was made clear to the director during these meetings. "Any cooperation with the [Kaiser Wilhelm Institute for Anthropology] is impossible as long as Herr Muckermann is a member," Gütt insisted.[50] And although Fischer could never be made to state that Jews were genetically inferior to "Nordics" or that racial mixtures between "Nordics" and other so-called European races were intrinsically bad, Gütt impressed upon him the need to be more cautious with public pronouncements that "could be interpreted as being opposed to state policies."

Despite Gütt's warning, however, it appears that throughout most of the Third Reich, Fischer consciously attempted to find a compromise on the "Jewish question" that would simultaneously reflect his own national conservative prejudices (such as his distinction between the racial crossing of "Aryans" with so-called respectable German Jews and such biological mixing with "Ostjuden"), be marginally acceptable to important officials in the government, and most importantly, not undermine his scientific credibility in foreign countries. When he was forced or felt the need to deal with the "Jewish question," Fischer employed the phrase he first formulated in his controversial February 1933 talk. He claimed that Jews were racially "andersartig" (different) from people who were allegedly a composite of the so-called European races. He insisted that this said nothing about the relative genetic worth of any "race."[51] It will be interesting to observe whether his allegedly "scientific" attitude toward the

Jews would hold up in the more radical atmosphere of the war years. That having been said, Fischer certainly did not protest when his renowned German Jewish colleagues at other KWIs were sooner or later ousted from their positions. As we have mentioned, he did not have to worry about this issue in his own Institute.

Despite his own qualms about Fischer, Gütt felt obligated to communicate a central truth to those ready to go to extremes and politically crush the Dahlem director: Fischer and his Institute were irreplaceable resources for the new state. Although Gütt had recognized this reality since his earliest encounters with Fischer, he articulated the point most clearly in a 1934 letter to Richard Walther Darré, Reich Minister of Food and Agriculture and architect of the notorious SS Race and Settlement Office:

Cooperation with Professor Fischer appears unavoidable, since, at present, there is no other equally valuable institution; also, his institute and staff are absolutely necessary for the appropriate training of medical officers and physicians, and especially for carrying out the Law for the Prevention of Genetically Diseased Offspring. Furthermore, Professor Fischer is a nationally and internationally recognized authority in the field of genetics and racial science. For that reason he was also, without my knowledge, appointed as the first "coordinated" Rector of the University of Berlin. A split between him and official quarters could easily give the impression both nationally and internationally that Professor Fischer doesn't approve of the path that the government has taken in the field of racial care—that [our] governmental policies contradict science.[52]

Whether Gütt was ever this blunt with Fischer about the latter's importance for the new state is unlikely, but also irrelevant. Almost a year earlier, only weeks before the fateful *Kuratorium* meeting, Gütt helped Fischer find the most "generous" way of carrying out the necessary "coordination" of his Institute.[53] Perhaps most important to the Dahlem director, the SS doctor had not forced the expulsion of von Verschuer. Interestingly, especially in light of the human geneticist's later career trajectory, he, too, was politically suspect at this time. Von Verschuer's prior participation in national conservative organizations, his early publications in right-wing *völkisch* journals, as well as his anti-Semitism seemed to count for little.[54] Gütt's actions showed the Dahlem director that, within the limits of his position, the SS doctor tried to be of help. Fischer had always recognized his worth as a scientist and the value of his international scientific reputation; what he quickly learned was that Gütt did, as well. Gütt clearly belonged to those Nazi medical bureaucrats who valued human genetics for its potential to provide a respectable scientific foundation for National Socialist racial policies—as a key intellectual resource for the Third Reich. It is sometimes argued that all National Socialists were anti-intellectual and opposed to scientific research, but this is a myth. Although the infamous Franconian District Party leader Julius Streicher's (1885–1946) maniacal Jew-baiting undoubtedly served an important propaganda function for the Nazis, his vile, if highly profitable,[55] journal, *Der Stürmer* (approximately translated as *The Attacker*), was not taken seriously among German biomedical professionals; it certainly could not be used to win international acclaim for Nazi racial policy—something important to the regime, at least before the outbreak of war. As such, Gütt could be expected to support Germany's community of human geneticists—not just in word, but in deed. After all, National Socialist policy in this key area must not appear to the world to contradict internationally accepted views on human heredity.

Fischer capitalized on this knowledge. Anticipating a visit by Gütt at the *Kuratorium* meeting, Fischer assembled a special report, "Research Institutes for the Scientific Support of German-*Völkisch*, Racial-Hygienic Population Policy." It apparently was produced on his own initiative. What is certain, however, is that the Dahlem director wished to use it to sell "the science card" to the influential Reich Ministry of the Interior, Gütt's bureaucratic home. The crux of Fischer's argument was that the effectiveness of future racial policy measures depended on their scientific grounding—the kind that an internationally renowned Institute like his own could deliver. "The various governmental offices need once and for all to have researchers and institutions from which they can acquire objective scientific underpinning," Fischer maintained. State committees would naturally be responsible for all future racial policy laws. But "scientific research must always be available to the government to clarify certain preliminary questions." Although this scientific expertise would not come cheaply, it was imperative that all "population policy measures be grounded in science in such a way that they remain irreproachable and can work for the good into the distant future." Only by proceeding in such a manner could the dreams of the "ingenious Führer" be realized, Fischer concluded in his report.[56] The 220 journal articles and numerous books that were on display at the *Kuratorium* meeting must have impressed upon Gütt the enormous research productivity of the Fischer team. The director's annual "Activities Report" would have demonstrated that the racial genetic research undertaken at the Fischer Institute was just the kind required by governmental offices in the planning stages of crafting racial legislation.[57]

13 The KWIA with the Nazi flag. Its director, Eugen Fischer, had already made the "Faustian bargain" and placed his Institute at the service of the National Socialist government. Photo courtesy of the Archiv der Max-Planck-Gesellschaft, Berlin Dahlem.

What did Fischer learn from his unpleasant denunciation campaign? It would appear that it and his experience with Gütt at the first *Kuratorium* meeting after the "national revolution" helped Fischer better understand the true nature of the National Socialist state. They constituted his "wake-up call," so to speak. He realized by then that he was not dealing with an ordinary coalition government of the variety that had existed throughout the Weimar Republic. The denunciation campaign and his talks with Gütt almost certainly helped him recognize the parameters within which he would now need to operate if he wished to remain a Kaiser Wilhelm Institute director with any semblance of influence as well as a key spokesman and science manager for human genetics.

Yet despite's Fischer's ability to learn from the experience, we should not underemphasize the toll that this campaign took on him. By the time of the "seizure of power," Fischer certainly felt himself politically astute enough to operate within the new state as he had in the Weimar Republic. With the scientific prestige of the Kaiser Wilhelm Society behind him, Fischer had won the battle in the 1920s against those opposed to the establishment of the Institute. The advent of the Third Reich, with its German Nationalist–Nazi coalition, patriotic ideology, and emphasis on racial improvement, should have provided an even more congenial political environment for the director to operate. But what Fischer realized through the denunciation campaign and his conversations with Gütt was that his national conservative convictions—ironically respected throughout the democratic Weimar Republic—were not sufficiently conducive to winning over the new regime to his side. This must have hit the renowned racial anthropologist especially hard.

Why did Fischer seal a deal with the Nazi state? At fifty-nine, the Dahlem director was no longer a young man in 1933. He had already secured his scientific reputation and no longer needed to climb the career ladder. Moreover, the Dahlem director certainly realized that he would pay a price for selling his Institute to the new Nazi masters—although it is doubtful that he recognized just how high the price would eventually be at the time. Nonetheless, Fischer did offer his Institute's services to his new political taskmasters, and he might have done so believing that he could still secure the upper hand in shaping racial policy if he appeared to comply with their needs. After all, he still had an important card in his pocket, or so he thought: his international scientific reputation. And he was anything but shy about using it as intellectual bait for the new regime. As Fischer stated in an address at the University of Berlin, "please do not consider me immodest when I say here that, at present, there is hardly another man in Germany who is as useful for the state government in the field of genetic and racial science . . . , especially with regard to judgment abroad, as I." Moreover, in an almost threatening tone he told representatives of the Nazi state that they should think twice about muzzling him on the question of race. Fischer more than hinted that this would result in negative political fall-out for the state. "I believe I know," he remarked with a healthy measure of political self-confidence, "what [the international scientific community] would think if I were *no* longer allowed to speak on the subject [of race] in public."[58] The Dahlem director might have still hoped that he and his fellow scientific experts in the field of human heredity at the Institute as well as at the various other Kaiser Wilhelm Institutes and universities in Germany would ultimately win the day and guide Nazi racial policy. After all, they, not the uninformed Nazi officials, had the expertise to make important decisions on issues concerning racial hygiene and racial cleansing. We know that professionals in many fields during the Third Reich accepted some of the unpleasant facets of the Nazi state in the belief that it was better for a "non-Nazi" to stay at his post than have a hard-core, card-carrying National Socialist gain control, especially one without the requisite professional experience. He also might have stayed on to protect and nurture his protégé, von Verschuer, still quite politically vulnerable at the time.

But perhaps Fischer was so happy that a government had finally come to power that would be better able to fund his Institute and his research in racial genetics than the Weimar Republic that he could not foresee any potential danger of such an alliance. In early 1933, it might have appeared to the Dahlem director that a new day had finally dawned. During the late Weimar Republic, Fischer constantly complained about not receiving enough money for his Institute. In 1931, for example, he threatened that much of the research that was currently undertaken would have to cease unless he received additional funds, although in reality no major project was ever terminated on financial grounds. In one letter, Fischer angrily stated that the Institute's monetary difficulties placed his "scientific reputation . . . on the line" He tried to secure additional financial resources from the RF by painting the fiscal picture of the Dahlem Institute in the bleakest terms possible and selling it as a major scientific resource. "I am aware," Fischer told the American philanthropic organization, "that my request [for funds] is not a modest one—but I consider the tasks of this international and almost unique Institute to be so extremely important, and on the other hand the danger of having to close the Institute for lack of means to be so great, that I herewith present my petition, trusting it will be granted." Fischer made it perfectly clear that he refused to accept the responsibility for a diminished status of his Institute owing to a lack of adequate financial resources. He told the Kaiser Wilhelm Society in no uncertain terms that he would indeed close the Institute if more money "from new sources" was not forthcoming.[59]

Although we will never know exactly why Fischer chose to sell his Institute, two things are certain. First, the parameters within which he would now have to work left him sufficient room for professional self-fashioning. He would accept the realities of doing scientific business under the swastika and effectively exploited the "science card." Indeed, not long after the *Kuratorium* session, the director composed a memorandum outlining the biomedical research at the Dahlem Institute that was allegedly imperative for the new regime along with its price tag.[60] As we will see, it was typical of the way in which he would obtain much needed financial support throughout his career. Second, Fischer's decision to place the Dahlem research center in the service of Gütt and the Nazi state was merely the beginning of a mutually beneficial symbiosis that would continue even after he stepped down as director in 1942.

14 Fischer with then Kaiser Wilhelm Society president and world-renowned physicist Max Planck. Photo courtesy of the Archiv der Max-Planck-Gesellschaft, Berlin Dahlem.

"The Means and Size of the Institute Do Not Reflect Its Current Importance"

The enormous expansion of the Dahlem Institute in terms of money, materials, and personnel in the first years after the Nazi seizure of power demonstrates just how adept Fischer was at selling his science, human genetics, as a national "resource." In the pursuit of funding, Fischer left no stone unturned. For example, in one proposal sent to the Reich Ministry of the Interior, Fischer went so far as to request additional money in a manner that did not quite meet with the Kaiser Wilhelm Society administration's approval. He pointed out that the budget of the Kaiser Wilhelm Institute for Brain Research, another biomedical research center at the outskirts of Berlin, was higher than his own and rhetorically questioned whether this other institute was really more important for the national community than his. But one could not argue with results. In 1933 the Fischer Institute received 75,711.95 RM (reichsmarks) from state sources. Just one year later, however, state contributions amounted to 127,235 RM—nearly a 60 percent increase in governmental support. The prewar high appears to have been reached in 1937 when the Kaiser Wilhelm Institute for Anthropology received 168,100 RM—a 75 percent increase over the state's 1934 level of financial commitment. In addition to this generous annual funding, Fischer also deftly negotiated what turned out to be a onetime 170,000 RM contribution from the Reich Ministry of the Interior for a major addition to his Institute. He did so by making reference to Gütt's own feeling that "the means and size of the Institute do not reflect its current importance."[61]

There were numerous reasons behind the positive financial development of the Dahlem Institute during the Third Reich. To begin, the Kaiser Wilhelm Society itself offered the Fischer Institute more money partly because of governmental pressure. Fischer also encouraged the Society to ask the government for supplemental funds. For example, when Fischer showed Gütt around the Institute at the opening of a series of teaching seminars for a handpicked group of medical students in 1934, he made it clear that the research center's quarters were much too small to assemble "a workplace for twenty new men." Fischer mentioned this incident in a memo written to the Society's General Secretary Glum, who immediately communicated Fischer's request to Gütt after which the director received his money.[62] In addition, Fischer was still able to obtain outside funding for Institute research, just as he did during the Weimar Republic. Moreover, as we will see, the Dahlem Institute enriched itself through the racial testimonials that it wrote on behalf of individuals needed to prove their "Aryan" lineage. And finally, between 1933 and 1935, the money available to the Institute could literally not be completely spent. The explanation for this rather strange state of affairs was that until Fischer gave up his post as Rector of the University of Berlin in 1935, he did next to no research and thus refrained from extensively dipping his own hand into the Institute's budget. In addition, other Institute members were engaged in their own "service to the Reich" and likewise placed little financial strain upon its purse. The money saved during the first few years of the Third Reich owing to a lack of time to pursue scientific work was set aside and accumulated. It would be put to good use by Institute personnel in the second half of the 1930s. As will be evident in the next chapter, the contrast to the Munich Institute under Rüdin could not be more glaring.[63]

Fischer's ability to secure generous Kaiser Wilhelm Society and governmental financial backing from the right quarters also enabled him to enlarge the Institute's scientific personnel. For example, between 1933 and 1934, the number of "working scholars" present at the Dahlem Institute swelled from thirty-seven to fifty-six.[64] More importantly, additional money allowed the Dahlem director to strengthen already existing research foci and to expand the Institute's research orientation after 1935. Following the forced removal of Muckermann in 1933, Lenz, a long-time national conservative racial hygienist, Nordic enthusiast, and coauthor (with Baur and Fischer) of the *Grundriss*, was appointed Eugenics Department head.[65] Considering Lenz's earlier and recent open support for Hitler, Fischer could be fairly certain he had selected a person who possessed impeccable scientific credentials for the post and who would also pass the political litmus test. Given his position as a member of the Reich Ministry of the Interior's Sachverständigenbeirat für Bevölkerungs- und Rassenpolitik (Expert Advisory Council for Population and Racial Policy), Lenz was also a significant political resource for Fischer and his Institute. Owing to their earlier ambivalent views toward certain facets of Nazi racial policy, neither Fischer nor von Verschuer had been asked to serve on this important Council. Lenz's importance for the reputation of the Institute notwithstanding, he was a difficult personality who tended to pick fights over scientific issues and lacked the tact to know when to keep silent. Although he would remain at the Institute until its demise, Lenz and Fischer would never develop a warm relationship.[66]

In 1935, Fischer's favorite, von Verschuer, was called to head up a new "daughter institute" for *Erbbiologie und Rassenhygiene* (Hereditary Biology and Racial Hygiene) in Frankfurt in 1935.[67] Although he was still not "brown" enough for some bureaucrats to receive a post at the University of Berlin—Germany's signature university—important Nazi officials like Gütt and Walter Gross (1904–1945) of the Rassenpolitisches Amt (Racial Policy Office) had no problem with an appointment in Frankfurt. From his comments on eugenics made during the Weimar Republic, they could take heart that von Verschuer probably found National Socialism far more congenial than democracy: "Racial hygiene can never be effectively put into practice under the spiritual flag of untamed Individualism. It demands a way and view of life that is ready to sacrifice . . . the individual good for that of the commonweal [and accepts] that the *Volk* is more valuable than the individual."[68] In fact, a general sympathy toward the new regime was all that was really required for most purposes. His nonparty status might indeed be a plus. Von Verschuer's "objective and essentially scientific, apolitical demeanor could have an especially convincing effect on skeptics" of the regime's racial policies. As such, his "appointment could also have a valuable effect from the standpoint of propaganda," Gross maintained.[69] Clearly, von Verschuer's behavior was an intellectual resource for Nazi medical bureaucrats like Gütt and Gross, even if his commitment to all facets of National Socialist ideology was too ambivalent for him to teach at Germany's leading institution of higher learning or to speak at party rallies.

With von Verschuer's departure, the Department of Human Heredity at the Dahlem Institute was abolished and its research projects were taken over by Lenz and Fischer. Von Verschuer continued his twin research in Frankfurt with great success.[70] He remained a scientific member of the Kaiser Wilhelm Society, and stayed in very close contact with Fischer. Both felt the loss resulting from the geographical distance that separated them. Fischer used the vacancy, however, to create a Department of Hereditary Psychology, a field of research that had become increasingly central to Fischer owing to the importance of an "objective scientific study" of the inheritance of mental traits for "any racial population policy"—in other words, for Nazi racial policy. The psychologist Kurt Gottschaldt (1902–91) was hired to head this new research discipline.[71] It was not until just prior to the war, however, that further changes and research diversification occurred.

By and large, scientists hired by Fischer to advance his research agenda in the broad fields of racial science and medical genetics employed similar research methodologies: pedigree studies (examining ancestry and family trees), twin studies, embryological studies, and animal models—whereby the last two methods were considered state-of-the-art techniques to assess the role of heredity in human populations at the end of the 1930s and early 1940s and played an important role in the field of developmental genetics.[72] As we have seen, many of these methodologies undertaken at the Dahlem Institute were used successfully in other countries.

According to the international standards of the day, much, if certainly not all, of the research undertaken at the Fischer Institute must be considered respectable science—however much we might find this science politically and morally distasteful today. Its scientific respectability for the time notwithstanding, its research was political in at least three distinct ways.[73] First, the ideological contexts and institutional structures of Nazi Germany encouraged the hundreds of racial and medical genetics studies in Dahlem. In some cases, it made the research possible in the first place, especially through its racial policy. Fischer certainly could not have secured the kind of "racial portraits" he included in his highly anti-Semitic coauthored book *World Jewry in Antiquity* in the absence of

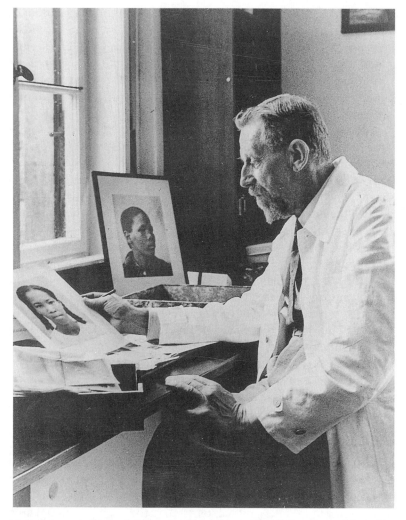

15 Fischer engaging in genetically based racial research, ca. 1938. Photo courtesy of the Archiv
der Max-Planck-Gesellschaft, Berlin Dahlem.

Jewish ghettos, in this case the Lodz ghetto in occupied Poland.[74] Second,
although no science is politically neutral, much of the research at the
Dahlem Institute after 1933 was specifically undertaken or adapted to
serve the needs of National Socialist racial policy, a point expressed quite
bluntly by von Verschuer in a letter to C. B. S. Hodson, honorary secretary of the British Bureau of Human Heredity, regarding his own area of
expertise, medical genetics. "German research," von Verschuer explained
to his British colleague, "is especially well represented in this field, as we

were faced with the urgent necessity of supplying the scientific underpinning for practice-oriented racial hygiene legislation." Indeed, even seemingly "harmless" scientific investigations, like those demonstrating the inheritance of racial differences of the ear or fingerprints, were pursued to support the racial policies of the state.[75] And finally, all Institute research that attested to the power of nature over nurture, especially if it was high quality, functioned to legitimize the Nazi racial state both nationally and internationally—even if the origins of this research predated the Third Reich. When Fischer sold his Institute and its research potential as a "resource" to Gütt, and later to other National Socialist Party and state officials, he dangerously strengthened the research-propaganda-instructional complex that first began to emerge during the final years of the Weimar Republic.

"In the Service of the Reich"

In 1935, two years after the fateful *Kuratorium* meeting that marked the beginning of the Faustian bargain, Fischer proudly reported that his Institute had placed itself in the service of the new state, sometimes even at expense of its "purely scientific tasks."[76] In this and future activity reports to the Society, the Dahlem director outlined what this "service" entailed. It was nothing short of an extensive network of state advisory, legal, instructional, and propaganda activities in Nazi racial policy that, when taken together, consumed much of the Institute members' time and efforts.

Among the Institute scientists' most important obligations during the Third Reich was the preparation of state-sanctioned expert testimonials. The Law for the Prevention of Genetically Diseased Offspring (1933), which legalized mandatory sterilization for those deemed hereditarily "unfit," created a need for expert testimony that went beyond anything existing during the Weimar years. Fischer could claim indirect credit for this law since it was based on the 1932 draft of a voluntary sterilization edict for which he, Muckermann, and other Institute members were in no small measure responsible.[77] Although none of the Dahlem scientists were as preoccupied with the sterilization question as their Munich colleague Rüdin, Fischer, Lenz, and von Verschuer provided the expert eugenics testimonials that frequently resulted in rendering the individuals examined infertile. Approximately four hundred thousand people fell under the knife in the borders of the old Reich alone. Between five and six thousand women as well as six hundred men died from the operation. For all but a small fraction of the individuals sterilized, however, any definitive proof that they suffered from a genetic disease was lacking. Many deprived of their right to procreate were sterilized simply because they were deemed socially inadequate from the Nazi point of view.

These problems, however, were not appreciated by most human geneticists in the Third Reich, especially those working at the Kaiser Wilhelm Institutes in Dahlem and Munich. In a 1934 progress report to RF Officer Alan Gregg, Fischer stressed that the material support given by the American philanthropic organization to the Kaiser Wilhelm Institute for Anthropology enabled him and his colleagues to provide the "scientific underpinnings" for the Sterilization Law. He communicated this evidence proudly at a German Gynecological Congress.[78] Fischer's protégé, von Verschuer, had nothing but praise for the new law. Indeed, he thought that Germany was courageously following the excellent example of the United States. "The good experience made in California [regarding mandatory sterilization]," von Verschuer reported, "allows us to hope that the Law will also be a blessing in Germany." Although "liberal or egotistically-inclined people had their reservations" about sterilization in the past, these concerns "have fortunately been totally eliminated with the victory of the National Socialist world view," he proclaimed. Von Verschuer was a leading exponent of the Sterilization Law in Germany; he used his influence to spread the gospel to physicians through his textbook, *Erbpathologie* (*Medical Genetics*). Indeed, he outlined new norms of professional ethics for would-be doctors: "The medical role of the physician in today's *völkisch national socialist state* is completely different, and much broader [than before]. Today we view the *Volk* as a spiritual and biological entity. We owe it to our *Führer, Adolf Hitler,* that we are once again *a spiritual, united Volk.* . . . The *new task of medicine* today is: *care of the national body* through the preservation and advancement of healthy genes, the elimination of unhealthy genes and through the conservation of the specific racial character of our people through *hereditary and racial care.*" Von Verschuer also spread the message in the pages of his journal, *Der Erbarzt* (*The Genetic Doctor*). Moreover, the devout Protestant communicated his views on racial hygiene to religious sympathizers through institutions such as the Innere Mission, a Protestant German welfare organization.[79] That having been said, von Verschuer's enthusiasm for mandatory sterilization was moderate compared to the fanaticism that his Munich colleague Rüdin demonstrated for the procedure, as we will see in the next chapter.

Most of the time, Dahlem scientists operated within the law when providing expert testimony—as morally problematic as the law was. In the case of the sterilization of the "Rhineland bastards," however, Fischer and his Institute took part in a project that fell outside of the legal parameters of the draconian Nazi Sterilization Law.[80] The phrase "Rhineland bastards" refers to the children born of German women and French colonial soldiers, mainly indigenous Africans, who were stationed in the German Rhineland following Germany's defeat in World War I. The occupation of the Rhineland was seen by many German nationalists as a humiliation, and to have it occupied by African colonial soldiers was an even greater slap in the face for those who saw the occupiers as an "inferior race." Needless to say, the sterilization of these "Rhineland bastards" was based on their racial composition rather than on hereditary defects, the latter being the stipulation for mandatory sterilization under the 1933 law. It is estimated that 385 such individuals were sterilized with the help of testimonials from the Fischer team. Researchers outside of the Dahlem Institute, however, also played a role in this stealth undertaking.[81]

In addition, both von Verschuer and his mentor served on the newly established Erbgesundheitsgerichte (Hereditary Health Courts) in Berlin, one of 205 such courts established throughout Germany to hear cases regarding the Sterilization Law. Fischer even served on one of the Reich's eighteen prestigious Obererbgesundheitsgerichte (Appellate Hereditary Health Courts)—the final court of appeal for a person whose reproductive capacities were threatened.[82] Needless to say, service on such courts violated a physician's duty to keep a patient's medical condition confidential—and both von Verschuer and Fischer were trained in medicine. Here we have a conflict between two commandments: that of the state demanding that physicians report individuals who fell under the provisions of the sterilization mandate and the doctors' responsibility to uphold medical confidentiality. There is yet another ethical dilemma stemming from the execution of the Nazi Sterilization Law. Gütt made sure that pertinent information collected on all cases coming before the Hereditary Health Courts was archived so that it might serve to further human genetics in Germany. Fischer and von Verschuer were two of the three biomedical scientists, the third being Rüdin, who were allowed access to this valuable research "material."[83]

The Sterilization Law and the availability of data on cases appearing before the Hereditary Health Courts also had a large impact on the type of human heredity studied at the Institute, again demonstrating how the symbiotic relationship between human genetics and Nazi racial policy could alter the nature of scientific research. Prior to 1933, the Dahlem Institute spent virtually no time investigating those diseases that eventually fell under the Law for the Prevention of Genetically Diseased Offspring—for example, neurodegenerative illnesses, mental retardation, psychiatric illnesses, epilepsy, and deafness. After 1933, however, members of the Institute were busy at work, determining the genetic component of these disorders and communicating their results in scientific publications. As Fischer himself expressed it in his Institute's yearly activities report, the need to examine clinical diseases following the Sterilization Law encouraged "the Institute [to take] a completely new step." "A promising new task group has been formed" on this topic, he added. Frequently, twin studies were used by those in this "task group" as a methodological vehicle to demonstrate the importance of heredity in these illnesses.[84] As should be obvious, these new investigations served as an intellectual resource for Nazi racial policy; such politically relevant genetic science also brought professional advantages to those engaged in it.

National Socialist anti-Semitic policies, and more specifically the infamous 1935 Nuremberg Laws, which defined an individual's race according to his or her ancestry, also increased the demand for expert testimony–this time in the form of racial testimonials. Normally, parish records or birth certificates sufficed to prove a person's "Aryan" or "non-Aryan" status by confirming the race of an individual's parents and grandparents. However, in ambiguous or contested cases the Reichsstelle für Sippenforschung (Reich Office for Genealogical Research) would get involved and frequently send the individuals in question to the Fischer Institute for an examination.[85] Sometimes people wishing to clarify their racial status sought out the Institute on their own initiative, but this was not a practice Fischer wished to encourage. Indeed, Fischer constantly complained about the large number of time-consuming expert testimonies his Institute had to write compared to other institutes. According to an estimate by KWIA scientist and SS member Wolfgang Abel (1905–97), a "higher scientific expert" at the Reich Office for Genealogical Research, Fischer and his colleagues completed about eight hundred such racial testimonies during the Third Reich. Despite the Dahlem Institute's competitive prices for these and other expert testimonials, they provided a handsome supplement to the research center's state-financed operating budget.[86] They also occasionally served as raw data for research undertaken by Fischer and his team. Not many of the Third Reich's "biological enemies," however, would have agreed with von Verschuer when he wrote in 1941 that "today every person in Germany has the greatest interest in an objective identification of his or her blood ancestry."[87] By that time, the outcome of such racial or paternity expert testimonies could spell death for those deemed by science to be "racially alien."

Fischer and von Verschuer also served the state by holding talks on the "Jewish question." They did so under the auspices of the "Reich Institute

16 Wolfgang Abel carrying out racial research on French colonial soldiers after 1940. Abel joined the SS in 1935 and worked for the SS Racial and Settlement Main Office as well as the Reich Genealogical Office. He was involved, as was his boss Fischer, in the notorious "General Plan East"—a Nazi program designed to depopulate parts of the Soviet Union in order to make room for additional individuals of "German blood." Photo courtesy of the Archiv der Max-Planck-Gesellschaft, Berlin Dahlem.

for the History of the New Germany," which established a special "research division" on the "Jewish question" and had von Verschuer serve as an "expert advisor." As part of the lecture series, von Verschuer held a talk in 1937 entitled "What Can the Historian, the Genealogist and the Statistician Contribute to the Research on the Biological Aspects of the 'Jewish Problem'?" His thesis: the crossing of any "alien race," be it the "Negroes, Mongols . . . or Jews" leads to a change in the "special biological composition of the *Volk*" and to its "unique culture." One year later, Fischer delivered a lecture on the "Racial Biology of the Jews." Although most of his talk concentrated on the so-called racial origins of the Jews in antiquity, when his address turned to their "spiritual-mental" characteristics, he fell back on crude anti-Semitic stereotypes. "One observes in the early history of the Jews the hate and atrocities that frequently spilled into to a blood vendetta." The Dahlem director attributed these unfavorable traits to a racial characteristic of the "oriental-racial livestock herders." At this same meeting, von Verschuer held a talk in which he attempted to dismiss the idea that one can tell a Jew by his nose. Instead, he separated Jews out "scientifically" on the basis of medical genetics. Using the latest findings of population genetics, he argued that Jews were more susceptible to a number of diseases, including diabetes, flat feet, heredity deafness, and Tay-Sachs syndrome.[88] Von Verschuer's talk was reported in the party paper *Völkischer Beobachter* (*The People's Observer*). The lesson of the human geneticist's lecture, the paper asserted, was the "necessity of the separation between Germans and Jews."[89]

The interesting question to ask of Fischer and von Verschuer's participation in this lecture series is what they themselves hoped to achieve. From the private correspondence between von Verschuer and Fischer, it appears that the two geneticists aimed to place the regime's anti-Semitism on a more "rational" scientific footing. "It is important," von Verschuer wrote to his mentor, "that our racial policy—also the Jewish question— receives an objective scientific foundation that is recognized in wider circles." In a later letter to the director requesting that Fischer give a talk in the lecture series, von Verschuer appraised his own speech as successful and something that could lead to a "calmer view of the topic." This, he believed, made his and Fischer's participation in the lecture series a duty. Von Verschuer even went as far as to mention that one of the talks he held, while generally well received, did not satisfy the "extreme anti-Semites." Such statements suggest that the two scientists were trying to establish a racial hygienic viewpoint at odds with the extreme strain of Nazi anti-Semitism; this is probably in some sense true. But we must also keep in mind that, in addition, their goals were probably self-serving. Indeed, they may have also been attempting to protect their reputations abroad by appearing "objective" while simultaneously fulfilling their "service to the Reich" at home. After all, in the same letter that von Verschuer wrote to his mentor stating that he hoped Fischer's participation would lead to "a calmer view of the topic," he stated boldly that "international Jewry knew which side we are on; participation at such a meeting doesn't make a difference." Clearly, their talks were an important intellectual resource for the regime's anti-Semitic racial policies. The two human geneticists were certainly not worse off professionally as a result of their participation,[90] although von Verschuer's active involvement at these conferences would come to haunt him in the immediate postwar years.

Along with providing expert testimony on Nazi racial policy-related issues and holding talks on the "Jewish question," Fischer and other Dahlem scientists were saddled with a wide variety of in-house and external teaching duties. Beginning as early as the autumn of 1933, the Institute members offered a large number of educational courses in "Hereditary and Racial Care" for select groups of biomedical personnel staffing hospitals, administrative offices, hereditary health courts, and other institutions with a connection to Nazi racial practices. Given the intensity and frequency of these courses, Institute scientists would have needed a calendar to remind them of their instructional obligations. According to Fischer's own reports, at least 1,100 medical personnel attended one or several of such training courses by mid-1935. Dahlem personnel also staffed special half-weekly educational courses for biologists, judges, and clergy. Von Verschuer himself remembered that by the time the Sterilization Law went into effect on January 1, 1934, virtually all public health officers had been schooled at the Fischer Institute.[91] The in-house courses taught by Fischer and his coworkers were usually held there, although it would not have been uncommon for a young physician, usually wearing party insignia, to receive instruction in human genetics in a building almost adjacent to the Dahlem Institute, the Harnack House. Many of these educational duties, however, also took place outside the Institute for such organizations as the National Socialist Teachers' Association, the National Socialist Doctors' Association, and the State Medical Academy. On June 20, 1934, for example, Fischer held a lecture at a special Nazi "community camp"—an obligatory retreat—for junior members of the legal profession sponsored by the Ministry of Justice.[92]

In mid-1934, twenty-one SS physicians were sent to Fischer's Institute to attend what would become a series of year-long courses in "Genetics, Racial Science and Racial Hygiene" paid for the Reich Ministry of the Interior. The men attended daily lectures and seminars led by the Fischer team and were also involved in all "scientific investigations" undertaken at the Institute. Allegedly, their presence in Dahlem contributed to a "negative atmosphere" at the Institute.[93] It is unclear why Fischer would agree to provide such time-consuming courses, unless he had been compelled to do so. What is certain, however, is that Fischer was not above using this new service obligation as an excuse to demand more money for his Institute. There is also no doubt that the "scientific training" that these SS men acquired in Dahlem was meant to compliment the "political training" that they would receive from the Racial Policy Office of the NSDAP, Gross's organization. Gross expected that their elite education (both scientific and political) would render these doctors especially good candidates for important positions in the party.[94]

Several men attending the course later joined the Fischer team as assistants. In some cases, they proved to be a valuable political asset. For example, Herbert Grohmann, a graduate of the year-long course, was Fischer's assistant in 1938–39. Beginning in 1939, Grohmann served as director of the Division of Hereditary and Racial Care in an important health agency in the occupied Polish city of Lodz. This SS health official almost certainly made it easier for Fischer to obtain the Jewish "racial portraits" from the city's ghetto in early 1940, which he used for his above-mentioned coauthored book. In one case, an SS medical man who received scientific training at the Kaiser Wilhelm Institute for Anthropology was later involved (to an unknown extent) in human experiments on concentration camp victims; another served as one of numerous medical experts who determined whether particular mentally handicapped asylum patients would be killed in Nazi Germany's "euthanasia" project.[95] This nefarious Dahlem-SS liaison notwithstanding, Fischer never made the fatal mistake of forging a direct alliance with this dreaded Nazi institution, as his rival Rüdin did to his disadvantage.

Perhaps the Dahlem scientists' most valuable service to the Third Reich was to confer legitimacy to the entirety of the Nazi racial project—to publicly bestow their professional blessing on the ideal of "the racial state."[96] At home, the Fischer team lent an air of respectability to the regime's dystopian biomedical vision in several ways. In addition to delivering papers at normal scientific meetings, offering talks at overtly Nazi-organized conferences or holding official speeches at various German universities, members of the Institute frequently gave public lectures in their role as members of the world-renowned Kaiser Wilhelm Society, both in the Harnack House and in large lecture venues in other cities with ties to the Society.[97] In the case of these Society-sponsored lectures, Dahlem scientists not only bestowed respectability on National Socialist racial policies by popularizing their research to an elite audience; they also simultaneously helped legitimize the Society as an organization for those skeptical Nazi officials who were still not inclined to recognize the value of science and scientific institutions for the "national revolution."

But the most important battle for the credibility of National Socialist racial policy would be fought in the international arena, and one of the key ways Fischer and the Dahlem researchers ensured its favorable outcome was by hosting and participating in international conferences. International professional conferences, at least in racial-policy-related biomedical fields, were viewed as political and treated as such by the Nazi state. Only politically reliable scientists were allowed to attend such highly coveted conferences. In addition to the role that Institute scientists played at such meetings, they also bestowed international legitimacy on the National Socialist state by representing Germany at official foreign functions and by serving as the regime's scientific and cultural ambassadors. Kaiser Wilhelm Society–sponsored national lectures as well as numerous professional talks and conferences in the international arena are such a critical dimension of this symbiotic relationship that they will be discussed in greater detail in chapter 4.

Nonetheless, such conferences and talks were only one of many ways, in addition to teaching and service functions, in which the Dahlem Institute placed itself directly in the service of the Reich. And given how beneficial this service proved to be for the Institute, affording Fischer the opportunity to ask for and receive more governmental funding and recognition, we must contemplate the ethical dilemma that the Institute's scientists may have faced. On the one hand, these individuals were doing simply what they had been doing under the Weimar Republic: cooperating with the ruling government to sustain the Institute's prestige (as well as their own) and to increase its budget. After all, these biomedical professionals in Germany were long-standing members of the civil service and as such they had a special duty to serve the government. Loyalty to the state was something that most German civil servants took extremely seriously, even in the Third Reich. But on the other hand, does any individual, including a scientist, have an obligation to serve an immoral regime? And did Fischer and his colleagues view the activities of the Third Reich, particularly in the racial policy arena, as unethical? Whatever these individuals may have thought at the time, one cannot argue that the ultimate result of their willingness to serve the Reich was one that garnered the moral opprobrium of posterity.

Paradigms and Politics

The outbreak of Nazi Germany's "racial war" in September 1939 witnessed numerous changes in state and society. As one might expect, it did not leave the Kaiser Wilhelm Society or the Dahlem Institute unaffected. It was during this time that Fischer laid the groundwork for a new scientific paradigm at his Institute—one that would have far-reaching ethical consequences for the Devil's second director in Dahlem, especially after 1942.

Fischer's decision to alter the Institute's research focus was in line with an international recognition of the limitations of classical Mendelian genetics. The idea that traits were inherited as single units—either as recessive or dominant characteristics—made way for what was known as "higher Mendelism," which presupposed that the transmission of traits was far more complicated than earlier imagined. Genes could not be viewed in isolation. They must be understood in their relationship to other genes—indeed in the context of the organism's entire genome—as well as to environmental influences. By the end of the 1930s the international community of human geneticists, including German researchers in this field, accepted "higher Mendelism." In particular, von Verschuer's Frankfurt Institute for Hereditary Biology and Racial Hygiene was at the vanguard of this new perspective in human genetics. One of von Verschuer's coworkers, Bruno Rath, offered the first undisputed example of crossing-over in humans.[98] Von Verschuer himself laid bare the constellation of new questions facing human geneticists in 1939 at a talk he held in Breslau: "What influence does a gene have for development? At what time and at what place does it manifest itself? What changes in the tempo or in the stages of developmental processes does it affect? Is it possible to prevent deleterious [developmental] processes? How do individual genes work together? Do specific [genetic] traits demonstrate external differences according to race or constitution?"[99] "Human beings are thoroughly researched objects of the human sciences," von Verschuer told his audience. Yet much work still needed to be done.

Fischer incorporated the newest developments in human heredity at Dahlem by embarking on a new scientific paradigm: *Phänogenetik* or developmental genetics. The task of developmental genetics was to separate out genetic from nongenetic factors in the development of the organism and to analyze the impact of genes and environment (in Fischer's broader understanding of the term) in their relationship during embryological growth.[100] Impressed by the potential for his own research interests in this new field, the director had apparently decided to pursue this new approach as early as 1938—although by 1935 he already recognized that the study of gene expression in hereditary diseases was the "next great task of twin studies."[101] In 1938 Fischer dedicated an entire section of the meeting of the German Society for Genetics to the topic of "Developmental Genetics in Humans."[102]

In a confidential letter written to von Verschuer in 1940, Fischer first proposed the introduction of developmental genetics to the study of both "normal" and pathological traits as a new scientific frontier for his Institute. It would first be necessary, however, to set up a new division in Dahlem with a researcher already skilled in the new paradigm. Fischer mentioned his intention to hire the animal geneticist Hans Nachtsheim for the post. Using his experimental work on epileptic rabbits, Nachtsheim, Fischer argued, would help the Institute move beyond the now outdated genetics of fruit flies (*Drosophila*) to better understand how hereditary pathological traits became expressed in mammals, especially at the embryonic stage of development. Ties to clinical research on humans would continuously be made, Fischer assured von Verschuer. This new experimental approach using animal models and embryos would also continue earlier investigations done at the Kaiser Wilhelm Institute for Anthropology into the inheritance of normal (i.e., racial) traits. His institutional reorganization was a "bold plan," as Fischer himself confessed.[103] The director's decision to create a Department of Experimental Hereditary Pathology with Nachtsheim at the helm would turn out to be supremely important.

For Fischer, the new institutional paradigm had numerous advantages, perhaps the most important of which was that it accommodated studies on an aspect of human heredity compatible with the plant and animal genetics research undertaken at the nearby Kaiser Wilhelm Institute for Biology and was also amenable to experimental work.[104] But there were also practical considerations for a more complete understanding of race and hereditary illnesses. As the director later pointed out in a January 1941 Institute *Kuratorium* meeting, they did not yet know when a "Negro embryo begins to differentiate from a European embryo." Moreover, human embryological material with verifiably diseased genes was very difficult to come by.[105] It would be extremely useful to employ Nachtsheim's artificially induced diseased rabbits as an appropriate animal model to understand human genetic disorders. Fischer's decision to offer biologist Karin Magnussen (1908–97)—a specialist in the field of the genetics of eye pigmentation and iris structure—a stipend to work at the Dahlem Institute in the context of this new research program would have far-reaching ethical consequences for his Institute in subsequent years.[106]

As it turns out, the Dahlem director had, by this time, acquired his new party patron, Conti. The SS-Brigade Leader was not only a medical man; he was appointed Reich Health Leader, the most important medical post in the regime. Never enjoying a close personal relationship with Gütt, Fischer was probably not all too unhappy when his star began to fade and Conti became the most important spokesman on medical issues.

Recognizing the usefulness of having Conti as a patron, the Dahlem director worked behind the scenes to ensure that he would serve on the Institute's *Kuratorium*. In fact, Conti was named *Kuratorium* chairman by the Society in October 1940. In November 1940, Fischer and Conti had a personal meeting. In all likelihood, the Dahlem director used the opportunity to acquaint Conti with his plans for the Institute. At the January 1941 *Kuratorium* meeting, Conti used his new position to help secure the financial resources needed by Fischer at a time when the Society was cutting financial resources elsewhere.[107] The Dahlem Institute—the "first and most important" research center in the field of heredity and racial care—"must serve as a model and inspiration for other institutes," Conti told the other *Kuratorium* members.[108] This would require additional funding. The motives for the Reich Health Leader's actions, however, still need clarification. The SS-Brigade Leader apparently had "a plan for a new order of the entire health system complex."[109]

As Germany's most important health czar, Conti could not have been indifferent to the ever-rising number of cases of tuberculosis in the population. Indeed, this disease was the Reich's greatest health epidemic. Lowering the number of cases of TB played a role in his interest in aiding Fischer's institutional plans. Karl Diehl, von Verschuer's long-term research partner, was now formally a member of the Institute. His work on tubercular rabbits appeared promising as a partial solution to this serious national health problem. We know that Conti was impressed with Diehl's research from a letter written by von Verschuer to his Dahlem associate. He told Diehl not to fret about money for his work; he would get as much as he wanted. "Seeing as your work was recognized and declared essential by important individuals, especially . . . the Reich Health Leader [Conti] at the *Kuratorium* meeting of the Dahlem Institute," von Verschuer reassured Diehl, "you do not need to worry about the future."[110]

But there might have been an additional motive for Conti's help. At this time he was also involved with the Nazi regime's ethnic Germanization policy in the occupied East. Fischer might well have convinced Conti of the possibilities of employing developmental genetics to "improve" and "modernize" the construction of racial testimonials useful for the regime's racial "sorting out policy." Developmental genetics could thus help determine who could be "Germanized" in the East and be accepted as a "national comrade" and who would live out his or her life as a helot for the master race. In other words, there were likely important racial and health policy implications for this new paradigm, and it was probably not funded simply in the interest of some "abstract" or "neutral" biomedical science.[111]

The Faustian Bargain in Berlin: Act II

"A Leap Back as Head of the Institute"

By the start of 1942, the nearly sixty-eight-year-old Fischer could look back on a long and successful career. He was not only the recipient of numerous scientific honors—one of the most important of which was his election to the Prussian Academy of Science in 1937—but was also esteemed in elite intellectual circles in Berlin and enjoyed the good fortune of being loved and respected by his students and coworkers. Fischer weathered the turbulent first two years of the Third Reich to become what his ideological critics and jealous colleagues had tried to prevent: the undisputed academic spokesman for racial science under the swastika. However, once it became clear to Fischer and officials in the Reich Ministry of the Interior that arrangements between the Dahlem Institute and the Nazi state were mutually beneficial, tensions began to die down. Although the "non-Nazi" director officially became a party member in January 1940, his membership seems to have altered little. More important to the party than Fischer's own affiliation was probably the favorable growth in the number of National Socialists among his junior Institute colleagues. Von Verschuer's Frankfurt Institute, it should be noted, was even more of a haven for SS and other party members interested in human genetics and its political implications than the Dahlem Institute under Fischer. Von Verschuer's own acquisition of party membership in July 1940, however, appears to have had greater personal consequences than that of his mentor. It was probably a necessary step to alleviate any potential obstacles to his appointment as Fischer's successor, although von Verschuer argued in the postwar period that his party membership came to him automatically. Given that von Verschuer and Fischer always discussed matters of scientific and political importance—Fischer congratulated his colleague on his official party status in a letter—it is unlikely that von Verschuer was merely a passive recipient of his party membership.[112]

Von Verschuer's appointment as future head of the Institute was a long time in the making. In a letter soon after von Verschuer's call to Frankfurt in late summer 1934, Fischer promised that this "great leap" from Dahlem to Frankfurt would be followed "in only seven or eight years" by a leap "back as head of the Institute." Indeed, the institutional groundwork had been prepared for the von Verschuer transition well before the time Fischer finally stepped down on September 30, 1942.[113] Despite Fischer's influence, had von Verschuer's scientific reputation not been exceedingly high,[114] it is unlikely that he would have received the post. During his years as director of the Frankfurt Institute for Hereditary Biology and Racial Hygiene, he became known as one of the world's experts on twin research as a methodological tool in the relatively new subspecialty that is known today as medical genetics—a fact attested to by his invitation to lecture on the subject at the Royal Society of London just prior to the war.[115] From his private correspondence with Fischer, one gets the sense that the fifteen years of intellectual and professional grooming by his mentor were well worth the effort; by 1942 he was ready to make this long-planned, negotiated "leap." Although von Verschuer had serious misgivings about handing over the directorship of his own

Frankfurt Institute to Heinrich Wilhelm Kranz (1897–1945)—a fanatical National Socialist whose good party connections, rather than scientific abilities, launched his career—he was far too ambitious to pass up an opportunity to follow in his mentor's footsteps. As Fischer attempted to reassure him in a letter, Dahlem, not Frankfurt, would be the place where von Verschuer would establish his scientific legacy.[116]

Biomedical Research Transcending All Ethical Boundaries

There is no question that von Verschuer left a scientific legacy in Dahlem—a most uncomfortable one—and one with which the world is still grappling today. Although there was certainly plenty of morally problematic research and "service to the state" undertaken under Fischer's watch—more than enough to condemn the Institute in the eyes of future generations even had it closed its doors in 1942—scholars have naturally tended to focus on the unethical research and heinous medical crimes that occurred during the von Verschuer years. There were, sadly, numerous cases of research involving the exploitation of victims of concentration and extermination camps as well as "euthanasia" hospitals during the Third Reich; three, however, definitively involved official members of the Dahlem Institute.

The first such research project was initiated by von Verschuer himself in 1943. It was done in association with his former assistant from Frankfurt and "guest researcher" in Dahlem, the notorious "Angel of Death" at Auschwitz, Josef Mengele.[117] The project was designed to investigate "specific serum proteins." As his obsession with his dissertation topic demonstrated, von Verschuer had a keen interest in biochemical problems. The aim of this particular joint research project with Mengele appears to have been to serve as part of a plan to come up with a new test to classify the race of individuals based on the biochemistry of blood serum. This new test would augment or replace more laborious forms or racial diagnoses then used in expert racial and paternity testimonies.

Interestingly, this was not the first time that at least a former member of the Kaiser Wilhelm Institute for Anthropology was implicated in the construction of such a test. Lothar Loeffler, former assistant at the Dahlem Institute during the early years of Fischer's directorship, was involved, during the war years, with an Austrian-born physician, Karl Horneck, in trying to come up with a serological racial diagnosis. At any rate, by the time von Verschuer applied for money at the Deutsche Forschungsgemein-schaftungs (German Research Council), the funding agency that replaced the Emergency Association of German Science, for work on his "specific

17   The young Josef Mengele (*right*) with other medical students at von Verschuer's Frankfurt
Institute for Hereditary Biology and Racial Hygiene, ca. 1938. By this point, Mengele already
possessed a doctorate in anthropology. He was working under von Verschuer to earn a
second doctoral degree in medicine, which he completed in March 1939. Photo courtesy of
the Archiv der Max-Planck-Gesellschaft, Berlin Dahlem.

protein" project with Mengele, there was clearly a race over who would
come up with the first viable serological racial diagnostic text. There can
be no doubt that von Verschuer knew of the scientific developments
undertaken in this direction. Long desiring such a test and recognizing
the institutional competition for a discovery with such political import,
von Verschuer had little qualms in undertaking research on this subject
with his former doctoral student.[118] The two men would have scientific
advantages using the enormous number of "racially diverse" inmates at
Auschwitz that no other potential competitor could even dream of. The
blood serum taken from over two hundred "racially diverse individuals"
by Mengele was sent back to Dahlem for analysis by protein serum expert

Günther Hillmann of the nearby Kaiser Wilhelm Institute for Biochemistry. This institute was headed by the future Nobel Prize winner and Max Planck Society president, the biochemist Adolf Butenandt.[119]

Another project grossly transcending the ethical boundaries of normal science during the second Dahlem Devil's directorship was one involving investigations into the genetics of eye pigmentation and iris structure initiated by the fanatical Nazi, Magnussen. A party member since 1931, Magnussen worked in direct collaboration with Mengele while the latter was at Auschwitz. Their purpose was to demonstrate the racially determined hereditary differences in iris structure—information that would also serve as the basis of a new "iris table." This new table would replace the outdated "eye color charts," which helped to classify an individual's race by identifying his or her eye color. Mengele sent at least four pairs of eyes taken from one Sinti family (an ethnic group that, along with the Roma, is commonly lumped together as "Gypsies") in Auschwitz to Dahlem. It is certain that Magnussen had previously examined some members of this family before they were deported. Mengele himself killed four pairs of twins from this family so that the eyes could be delivered to his female colleague at Dahlem for her research. It is also almost certain that Magnussen, given both her party membership and her good contacts in high places, knew the nature of Auschwitz—her postwar denial of this fact notwithstanding.

Magnussen's request to Mengele to have heterochromatic eyes from a Sinti family sent to her in the event some of them died would then implicate her—albeit indirectly—in their death. Mengele mentioned to her that he knew a Sinti family at Auschwitz "thoroughly infected with tuberculosis." Under the general conditions of Auschwitz—of which Magnussen was in all likelihood aware, if perhaps not in every detail—this communication can be viewed as Mengele's "discrete offer to aid in these individuals' death" for his female colleague's (and his own) scientific benefit.[120] While at Auschwitz, Mengele also conducted experiments designed to change eye color using the stress hormone adrenaline. These investigations were likely undertaken to test the effects of hormonal changes on eye pigmentation.[121]

And finally, a third example of highly immoral research by members of the Kaiser Wilhelm Institute for Anthropology under the second director involves Nachtsheim. He, along with Gerhard Ruhenstroth-Bauer (1913–2004), a colleague from the Kaiser Wilhelm Institute for Biochemistry, initiated experiments designed to perfect a way to differentiate between individuals with hereditary and nonhereditary forms of epilepsy—knowledge that would be useful in the context of the Nazi Sterilization Law. The two scientists tested six epileptic children from a "euthanasia" hospital in Brandenburg-Görden at low air pressure in Air Force chambers to see if they would experience seizures. Earlier experimental studies on rabbits suggested that those with hereditary epilepsy would have them sooner than those suffering from the nongenetic form of the disease. Now this hypothesis could be tested on humans. Although the children left the experiment unharmed, it was performed without consent. Only one of these children in Brandenburg-Görden is known to have definitively survived the war.[122] It should also be mentioned that Lenz was involved in drafting a "euthanasia law" for the Nazi government. Fearing popular resistance to such a measure during wartime, Hitler, however, did not wish to make this project legal. As such, Lenz's efforts to eliminate the "night and fog" nature of this operation (not the operation itself), came to naught. Historically, he has thus been spared from being as actively implicated in medical crimes as his other three colleagues at Dahlem. We also know of two additional Dahlem researchers who worked at Auschwitz, Siegfried Liebau and Erwin von Helmersen. Their precise scientific pursuits while engaged at this most notorious death and slave labor camp, however, have not yet been sufficiently clarified.[123]

Dahlem Research and Germany's Racial War

The egregious examples of unbridled scientific research undertaken during von Verschuer's directorship still beg the question: why did human heredity at Dahlem take the criminal path it did? To address this query we must place the Institute's scientific investigations in the context of Nazi Germany's racial war. Indeed, at no other time did this Faustian bargain so transform the research practice of human genetics at the Dahlem Institute than during the particularly brutal and brutalizing war unleashed and perpetrated mercilessly by the National Socialist state.

This transformation could not have been achieved had Fischer not decided to institute the new paradigm of *Phänogenetik* by hiring Nachtsheim in 1941, already two years after the outbreak of World War II. As we mentioned earlier, Nachtsheim's developmental genetic approach used in mammals (in his case, rabbits) provided far greater insight into gene expression in humans than previous investigations in fruit fly genetics. In addition to experimenting using animal models, *Phänogenetik*, at least as envisioned at the KWIA, necessitated the establishment of "biobanks"[124] of "human material," ostensibly to compare with the pathological animal organisms bred at the Institute. In discussing his plans for a future Embryology Division with a scientific colleague a few months before taking over the directorship from Fischer, von Verschuer expressed his views concerning a suitable researcher to head it. Von Verschuer did not wish to hire a zoologist for the position, as "he wanted to build a bridge to humans" and use "human material" in his research. As such, only a human embryologist would do. Several months later, as Dahlem director, he wrote a revealing letter on the same subject. In order to acquire the necessary "human material" for his institute's investigations, von Verschuer told a Cologne colleague, he would "set an organization in motion . . . whereby, in all of Germany, embryos would be collected in gynecological clinics from hereditarily diseased women who [were forced to abort] and be given to us." He considered the new comparative genetic approach made possible by acquiring human embryos in this fashion to be "so great and promising that it would be worth the investment of a researcher's life work."[125] Von Verschuer's plan to establish a human embryo bank in Dahlem would have only been possible in conjunction with one of the organizations involved in the children's "euthanasia" project that began with Germany's unleashing of WWII.[126] Although the contingencies of war did not permit von Verschuer to establish a new Embryological Division, other forms of "human material," as we have seen, found their way to Dahlem.

To accompany the sophisticated theoretical laboratory work in the field of *Phänogenetik* of researchers like Nachtsheim, von Verschuer and his colleagues performed clinical studies done on humans. This necessitated more human research subjects than ever needed before at the Institute. But as the years of fighting in Europe progressed, and as more citizens became involved in the war effort, the Institute would be met with a shortage of available subjects to undergo such studies. Lenz bemoaned the situation in a report written in 1944: "Especially since the summer of 1943, research has been greatly hampered owing to the war. It is very difficult, in part impossible, to acquire enough observational material for certain scientific and practical questions. Genealogies and twin studies are hardly possible as a result of the evacuation of women and children. Questionnaires can also no longer be undertaken."[127] The director himself lamented the same point that year. "The continuing twin and genealogical research," von Verschuer complained, "is very reduced."[128] What would solve this problem, and eventually alter the practice of biomedical science during the later war years such that medical crimes became a distinct possibility was the sudden availability of a great reservoir of "racially diverse" potential human subjects for research purposes. These were people who, owing to uniquely National Socialist racial policies, were incarcerated in concentration and extermination camps, as well as "euthanasia" hospitals, and stripped of all rights. Such individuals—either dead or alive, either as a source of "genetically interesting" organs or as guinea pigs for direct experimental purposes—could serve to replace the war-related shortages in the supply of research subjects available through "normal" channels.

Specifically in the case of twin studies, research under ethically normal conditions was always a time-consuming and expensive project. It was frequently difficult to get parents to agree to have their children serve as subjects, even prior to the war. Moreover, there were some research problems that could only be solved by dissecting internal organs. This would require having access to twins who died at the same time, a very infrequent occurrence. Although biomedical research institutes could count on a large number of corpses delivered from the Gestapo and SS—again the product of the interface of human heredity and National Socialist politics—twins were rarely among them.

Given that Auschwitz was one of the few places in Europe capable of providing valuable organs or blood serum from "racially diverse" twins in the large quantities necessary for developing new racial diagnoses, it is hardly surprising that there was an interest in exploiting this notorious death and slave labor camp for Nazi racial policy aims.[129] By so doing, Dahlem Institute scientists could provide the necessary human clinical supplement to the ongoing experimental work in the developmental genetics of normal and pathological traits in animals—work that itself was originally undertaken to serve Nazi Germany's racial policy. Again, this ethically precarious step to employ the bodies or body parts of victims without rights was viewed as necessary for the new paradigm of *Phänogenetik* to serve Nazi racial policy. To do so required a combination of experimental research on animals *and humans*. And after all, even prior to the war, von Verschuer noted that German human geneticists were in fact under pressure to advance biomedical science relevant to National Socialist biological goals.[130] There was certainly not less pressure placed on the Dahlem scientists during the war. That having been said, there is no reason to assume that von Verschuer and his colleagues were opposed to undertaking such tasks at any time during the Third Reich. Indeed, the opposite is far more likely the case. Regarding his own research with Mengele—work designed to come up with a more modern racial diagnosis—von Verschuer could report in 1944 that the "gathering of material has already begun."[131]

As mentioned earlier, we know that at least three Institute scientists—von Verschuer, Magnussen, and Nachtsheim—benefited from the opportunity provided by National Socialist politics. In the case of von

Verschuer and Magnussen, their ability to exploit this new "research material" was made possible by their personal and professional connections to SS physician and "guest researcher" Mengele, who was in a position to grant them access to these subjects who were stripped of all rights. It is not known exactly how Nachtsheim gained access to his "subjects." What is clear, however, is that without the help of a biomedical-military network established during the war, it is unlikely that Nachtsheim could have carried out his experiments.[132] Although one will never be able to fully fathom why these scientists were willing to take advantage of this opportunity, their actions were certainly partially motivated by research fanaticism. This is the most frequently offered reason given by scholars for the immoral activities of these individuals, if one is given at all. It is understandable why people are fixated on this one explanation. After all, Mengele himself bluntly remarked that it would be a "sin," a "crime," not to use the "chance" that Auschwitz made possible for biomedical research.[133] We will also recall how fanatical von Verschuer was while undertaking research for his dissertation.

We should ask whether unbridled research enthusiasm is the only explanation for the German biomedical scientists' descent to the moral depths indicated above. If this is the case, we need to address the query of whether von Verschuer and his colleagues would have subjected healthy, politically upstanding "Aryan" Germans to deadly human experiments. We have no evidence that they ever did so. It is one thing for von Verschuer to have used "Aryan" twins for his research into determining the role of heredity in disease in the early 1930s: it is another to employ the blood serum of Auschwitz victims for *Phänogenetik* investigations designed to win the race to construct a viable serological racial diagnosis needed for Nazi racial policy. This would be exceptional even if von Verschuer really did not know the true (or full) meaning of Auschwitz, as reported in the postwar period.[134] Turning to the lowest ebb of the moral spectrum, it seems difficult to imagine that any human geneticist—even one in Nazi Germany—might have been implicated, albeit indirectly, in deliberately killing healthy "Aryan" Germans in order to utilize their eyes for research purposes, as Magnussen in her use of "racially inferior" Sinti victims appears to have done. One wonders whether even Mengele would have killed those he considered his genetic and racial equals, had he possessed such power, merely to advance his science. The critical question regarding the constellation of motivations behind these and other German biomedical scientists' actions will be more fully addressed in the conclusion.

In a 1943 article entitled "Heredity as Destiny, Tasks of Human Heredi-tary Research," the recently retired Dahlem director Fischer expressed his pleasure at the seemingly unencumbered unity of scientific theory and political practice under the swastika. "It is a rare and special good fortune for a theoretical science to flourish," Fischer maintained, "at a time when the prevailing ideology welcomes it, and its findings can immediately serve the policy of the state." This statement has been quoted by scholars as one of the best expressions of the Faustian bargain forged between Dahlem scientists and the National Socialist state. What has not been noted, however, is that it was published in the same newspaper, albeit "co-ordinated," where former Kaiser Wilhelm Society President von Harnack was quoted as having attempted to assure skeptics that "true racial sci-ence will bring segments of a nation closer together, not divide them [in a hostile manner]" some fifteen years earlier.[135]

Looking back at his statement today, we have every reason to believe that von Harnack was sincere in his sentiments. Although trained in the-ology, he, like many of his generation, trusted in the liberating power of science. He probably could not have imagined that a research program based on racial genetics initiated by a respected scholar such as Fischer could become far more reprehensible than even skeptics and opponents during the Weimar Republic initially believed. History has certainly not borne out the first Kaiser Wilhelm Society president's hopes articulated in that newspaper in 1927. Racial science and the Kaiser Wilhelm Institute most directly associated with it not only became part of a state-sponsored "sorting out policy" that separated out "desirable" from "nondesirable" members of the German nation; they also prepared the way for the lat-ter's "social death," and, in many cases, physical annihilation.

In examining the relationship between the science of human genetics and politics at the KWIA, we cannot help but note the mutually beneficial symbiosis between the Institute and the Nazi regime. As we have seen, human heredity was viewed as a scientific resource for the Nazi state. The regime was quick to provide financial inducements for its advance-ment in the hope that biomedical research would serve the interests of racial policy goals. The more funding Fischer, and later, von Verschuer, were given, the greater the expectations of its patrons, National Socialist Party and state officials. Wishing to lose neither prestige nor funding, and applauding most, if not all, of the goals of the Nazi regime, Fischer, von Verschuer, and their colleagues produced more and more scientific studies relevant to racial policies. They did this in an attempt to advance their own research agendas as well as "deliver the goods" to the regime.

National Socialist politics served as a resource for them in the prewar years, enabling them to acquire their research material more easily, thereby making it easier to serve their own professional interests as well as those of the state. More money flowing into the Dahlem Institute also meant that more "racial experts" could be trained for the Nazi bureaucracies that oversaw the implementation of racial policies. Hence, the regime could push forward with its racial project more quickly and efficiently.

This interplay between human heredity and politics initiated by Fischer's bargain with National Socialist officials resulted, during the war years, in a radicalization of the racial politics of both the first and second directors of the Dahlem Institute. One only has to compare the statements of Fischer on Jews during the first months of the Third Reich and his pronouncements and actions on this subject during the later years of the war. Whereas he was viewed as being soft on the "Jewish question" in 1933, by 1944 few Nazi diehards would have had reason to complain. In that year he wrote to Alfred Rosenberg, the Nazi Party ideologue on racial matters, "that it was high time [that action on the part of scientists was taken against the Jews] since the Jews have been leading not only a political, but also a spiritual, campaign against us for decades."[136] Fischer's protégé was equally radical in his anti-Semitism during the war. As we have demonstrated, this radicalization process led von Verschuer and other members of the Institute to embrace concentration and extermination camps as well as "euthanasia" hospitals as legitimate sources of "scientific material" to advance their research and career agendas, as well as Nazi racial policies.

Indeed, the case of the Dahlem Institute shows us how complicated the relationship between human heredity and National Socialism really was. With each step that the Institute's directors, Fischer and von Verschuer, took in securing governmental funding and prestige, they became, wittingly or unwittingly, more complicit in the regime's inhumane and, ultimately, genocidal policies. And as the next chapter will show, the Dahlem Institute was not, in essence, an exceptional case. Rüdin's German Research Institute for Psychiatry also made a Faustian bargain with National Socialism; indeed, it made not one, but two, "pacts" with the devil, so to say. In so doing, Rüdin's Institute also engaged in morally compromised research.

THREE

The Munich Pact

On March 20, 1933, Therese Rüdin wrote a letter to her close friend Anita Ploetz. After discussing the usual matters one normally finds in personal correspondence among women well acquainted with each other, the letter turned to political issues. "As you can well imagine," Therese informed Anita, "Erni [Ernst] and I are elated about the national government! It is a wonderful feeling to see the old [Imperial] flag again. . . . The swastika is waving proudly at the Research Institute; it was mandated by the Kaiser Wilhelm Society." "To tell you the truth," she added, "it was high time that order was created." Those thoroughly disgusting films and plays are now prohibited; "the hospitals in Berlin that were completely staffed by Jewish physicians have been cleaned up. . . . I believe that there was a tremendous fear of Bolshevism among the people. . . . We fervently hope that the new government will be convinced that racial hygiene is imperative and sound," Therese reported. At the end of her letter, Therese's husband took the opportunity to write a few lines. Continuing the theme of what, if any, impact the regime might have on the practical effects of eugenics in Germany, Rüdin remarked that although it would take time for significant financial resources to be released for racial hygiene research, one could hardly overlook "the entirely new spirit that is already evident."[1]

This joint letter is remarkably revealing. It positions Rüdin squarely in the conservative nationalist camp—the political group that welcomed an authoritarian Nationalist-Nazi coalition government as a means to destroy the culturally progressive Weimar Republic, curtail the alleged excessive influence of Jews in German life, and eliminate the political left. In this respect Rüdin's political outlook was similar to Fischer's. Unlike his Berlin colleague and scientific competitor, it was not politically necessary for the Munich director to hide his right-wing sympathies prior to 1933. During the Weimar years, Munich, unlike Berlin, was a center for nationalists and *völkisch*-oriented individuals. Yet as much as this letter demonstrates Rüdin's political receptiveness to the new order, it hints at something even more significant for understanding the checkered history of the German Research Institute for Psychiatry (GRIP): the director's need to acquire sufficient money to support its research initiatives, especially those of its Genealogy Department.

As this chapter will demonstrate, Rüdin continually found it impossible to function within the Research Institute's operating budget during the Third Reich. Indeed, much of his time was spent trying to secure additional funding for what he never tired of insisting was state-imperative research. The pursuit of money for the Munich Institute runs like a red thread throughout Rüdin's fourteen-year directorship. Fischer also bemoaned the lack of money for his research center during the final years of the Weimar Republic. However, his fiscal appetite—at least on the whole—appears to have been satisfied after 1933. And when Fischer needed additional financial resources to "modernize" his Institute in 1941, he was savvy enough to select a reliable and relatively lenient political patron to help him, Conti. Rüdin's need for funding, however, knew no limit. Perhaps the complex legal status of his Institute was in part to blame for his financial difficulties. The Munich Institute was not a state-supported Kaiser Wilhelm Institute like the one in Dahlem; legally, it remained an independent institution. As we have seen, it was originally funded by the German Jewish Loeb family with the assistance of several wealthy German donors like Gustav Krupp von Bohlen und Halbach; the Bavarian state government later supplied the additional necessary revenue.

Even after it came under the Kaiser Wilhelm Society's umbrella in 1924 and things stabilized for the German Research Institute for Psychiatry, money and space for scientific investigations remained an issue. Despite these uncertainties, the GRIP became an international locus for scientists interested in the newest trends in neuropathology and psychiatry even before Rüdin assumed its directorship in 1931. Foreign researchers came to acquire training in Rüdin's methodology for investigating the genetic origins of psychiatric disorders: the empirical hereditary prognosis. Yet prior to the Third Reich, the Research Institute was not identified solely with Rüdin and his Genealogy Department. How did the German Research Institute for Psychiatry go from a multidepartmental institute where the division heads were equals to a research center where Rüdin became its *Führer* in every sense of the word? The answer, however complex in the details, was anchored in the new political realities after 1933. Rüdin was inclined to serve the "national government" out of conviction and was rewarded monetarily for his Institute's important service to the state. Ironically, however, the racial policies Rüdin legitimized, if not created, ultimately resulted in the removal of important sources of his revenue: the cautious Rockefeller Foundation (RF) and the Loeb estate.

During the prewar years Rüdin successfully compensated for this financial shortfall by applying for money from various state and party institutions. However, when Rüdin was later unable to acquire sufficient funds through these channels, he decided to forge an alliance with the dreaded SS. It soon became apparent that SS Chief Himmler (and his minions with whom Rüdin would have more direct contact) required a higher price for their patronage than the Munich director originally anticipated. This final "Munich Pact" proved to be far less profitable for Rüdin than he imagined. It only exacerbated the conflicts he continually had with his Institute colleagues and added to the disharmonious atmosphere that existed there even prior to his directorship—problems that neither Fischer nor von Verschuer ever confronted. Indeed, by the time he extended his hand to the SS, Rüdin was in an unenviable situation. As will become evident, he quickly became the most prominent spokesmen on a key Nazi policy near and dear to his heart and his science: mandatory sterilization. During most of the Third Reich, his stature as "court" expert on racial policy programs and his role as "leader" in professional organizations were at least equal to Fischer's, if not greater. Given Rüdin's penchant to exhaust his Institute's funds, it is hardly surprising that he would serve the Nazi state to such an extent. Even his daughter admitted in an interview decades after his death that Rüdin "would have sold himself to the devil in order to obtain money for his institute and his research."[2] He certainly seems to have done just that, if not simply through his involvement (and the precise nature of this "involvement" needs to be discussed) with Nazi Germany's "euthanasia" project. And as we will see, this was not the only manner in which Rüdin implicated himself and his Institute in the crimes associated with Nazism. Before we address these issues, it is first necessary to return to a time prior to Rüdin's arrival in Munich—before he officially resigned from his position in Basel.

"May . . . this Institute Be the Starting Point of a New Epoch in Healing"

It was customary that the president of the Kaiser Wilhelm Society deliver a speech at the opening of a new Institute. Although the German Research Institute for Psychiatry was less traditional than most Kaiser Wilhelm Institutes in terms of its legal status, President Adolf von Harnack was proud to offer a few words during the dedication ceremony for the Institute's new building on June 13, 1928, just as he did at the opening of the Fischer Institute in Dahlem less than a year earlier. The festivities notwithstanding, it was a day of mixed emotions for the honored guests attending the ceremony. The Herculean campaign to win the support from the Rockefeller Foundation—the American philanthropic organization that provided the lion's share of the funds needed to make the research center a reality—had finally borne fruit. Kaiser Wilhelm Society General Secretary Friedrich Glum's early fears that the RF would not finance an institute in Germany proved to be ill founded, and foundation officer Alan Gregg's assessment that the Munich Institute was "one of the best single organizations worth study that I have seen" went a long way toward convincing the remaining skeptics. As we have seen, the German Research Institute for Psychiatry officially became a member of the Kaiser Wilhelm Society in 1924. By late 1926, the RF had pledged the enormous sum of $325,000 for the Institute's new building; the remaining $300,000 was provided by a relative of Loeb, the Bavarian government, and the Kaiser Wilhelm Society. James Loeb, a trustworthy friend and patron of the Munich Institute from the outset, offered another $150,000 as a "special fund" to cover unusual expenses that could not be taken from the Institute's normal operating budget. But the man whose tireless efforts, international scientific reputation, and close connections to Alan Gregg made the building possible—the eminent Emil Kraepelin—did not live to see the new structure completed.[3] He died in 1926, precisely when the RF agreed to fulfill the dreams of the "logical leader" of German psychiatry.[4] The question, however, remained: would the Research Institute now housed in the *Kraepelinstrasse* remain true to the vision of its founder and first director?

According to the RF report of the inaugural ceremony, President von Harnack had little doubt that the Institute would be "the starting point of a new epoch in healing." Krupp von Bohlen and Halbach, a major benefactor of the Institute, reminded the audience that more than two hundred thousand insane people were housed in institutions throughout Germany, and while other medical sciences made great strides, "psychiatry has remained in the rear." It was his sincere hope that those working at the Research Institute remain conscious of their responsibility "for the future of humanity" and the "august and sacred task" imposed upon them.[5]

But it was probably Dr. Felix Plaut (1877–1940), a close research associate of Kraepelin who traveled with him to America to seek RF funding, who offered the most moving address. In his speech he stressed the vision of the late director. "This Institute was to be a foster-mother to the sciences related to clinical psychiatry," Plaut told his audience. Only those subjects linked to clinical psychiatry but not adequately represented at the university should be housed in the Institute. This will ensure, Plaut continued, that "the different sides of the common enemy"—psychiatric disorders—are adequately tackled. But Kraepelin realized that this lofty aim presupposed a unique institutional structure where department heads were equal, subordination to a director was absent, and where the members worked toward a common end. "Scientific tolerance and versatility," Plaut emphasized, were the intellectual foundation stones of Kraepelin's Research Institute. Only by remaining true to the founder's vision will the Institute achieve its ambitious goals.[6]

Given the heavy-handed leadership role that Rüdin assumed at the German Research Institute for Psychiatry during the Third Reich, Plaut's comments are worth our attention. Was the Research Institute in fact ever really harmonious? Would the Institute remain a bastion of "scientific tolerance," free from any compulsion on the part of any future director? Before analyzing these queries in the course of this chapter, some of the major divisions and some of the key personnel that comprised the Munich Institute by 1928 must be examined.

Perhaps the most influential department at the Munich Institute was that of Walther Spielmeyer (1879–1935), which was also the largest division prior to 1933. He and his colleagues in the Department of Neuropathology investigated general questions regarding the histopathology of the central nervous system, especially functional disorders of cerebral circulation caused by toxic substances; numerous scholars from around the world came to what came to be known as an "international mecca of neuropathology."

An extension of this cerebral neuromicroscopic work was undertaken by the prospector (a person responsible for dissections) Karl Neubürger at Eglfing-Haar, a mental institution not far from Munich. Felix Plaut, the distinguished head of the Department of Serology, undertook investigations laying the groundwork for the modern study of neuroimmunology. Closely connected to Plaut's division was the Department

MPIP-HA

18 Professor Felix Plaut, director of the Department of Serology. In 1935 he was forced to vacate his position, as he was a "racial outsider." Plaut worked as a researcher in London until his death in 1940. Photo courtesy of the Historisches Archiv des Max-Planck-Instituts für Psychiatrie.

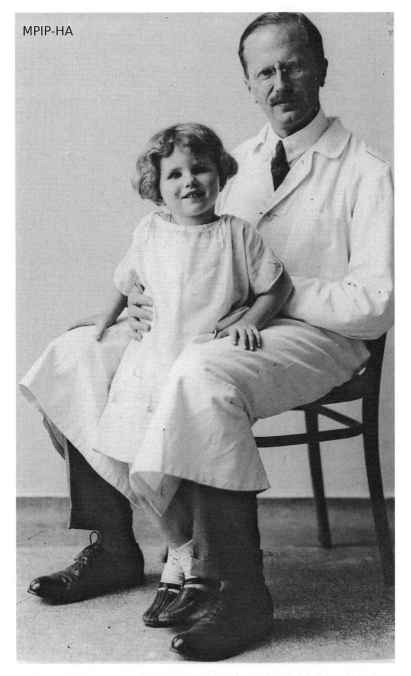

19 Professor Walther Spielmeyer, director of the Department of Neuropathology, with his
    daughter, Ruth, ca. 1930. His anti-Nazi sentiments and his Jewish spouse brought him
    constant trouble during the early years of the Third Reich. When he died in 1935 from an
    illness, some thought he had committed suicide. Photo courtesy of the Historisches Archiv
    des Max-Planck-Instituts für Psychiatrie.

of Spirochaetal Research headed by Franz Jahnel (1885–1951). Formerly a laboratory attached to Plaut's department, this division investigated the etiology and treatment of infectious bacterial diseases of the central nervous system, especially spirochaetal (parasitic or freely living bacteria harmful to humans and animals) infections such as syphilis. Johannes Lange headed the Clinical Department with its base of operation at the Psychiatric Unit of the Schwabing City Hospital in Munich until 1931. The Genealogy Department was run by Rüdin.[7] According to an annual report from 1929, the Institute boasted thirty-seven academic members, twenty-four of whom were German and thirteen of whom came from foreign countries.[8]

To be sure, these departments were independent. However, whether they (and their departmental heads) ever worked in tandem for a common goal either before or after Kraepelin's death is questionable. As one RF officer noted in his diary after touring the Research Institute, "I have the feeling that I have visited an excellent medical research institute in which there simply happens to be a rather large interest in neurological and psychiatric phenomena." The "Kraepelin Institute," he confessed, stands in stark contrast to the "singleness of direction and unified plan of work, centered around one idea," that was the hallmark of Oskar Vogt's (1870–1959) Kaiser Wilhelm Institute for Brain Research in Berlin-Buch, another RF-financed German research center.[9]

Even before Rüdin negotiated his return to Munich, when his exorbitant salary demands placed pressure on the Kaiser Wilhelm Society vis-à-vis other department heads at the Institute, there were serious conflicts among its members.[10] The construction of a new home for the Munich Institute appears to have eased the tensions; it was also the major precondition for Rüdin's return from Basel. His decision to leave Munich in 1925 notwithstanding, he always had faith that Kraepelin's name would serve as an intellectual resource to secure money in the United States for the much-needed new building.[11] The long negotiations that preceded Rüdin's return to Munich with two of his coworkers, Hans Luxenburger and Adele Juda, on November 1, 1928, demonstrated that the head of the Genealogy Department, like his Berlin colleague Fischer, was cognizant of his own worth. He was certainly not shy about driving a hard bargain with the Kaiser Wilhelm Society. As he stressed in an unofficial letter to Spielmeyer and Plaut—then acting codirectors of the GRIP—a prerequisite for his resignation in Basel was his appointment as head of a department whose research expectations went far beyond what "any good psychiatric clinic could muster." It would have to be outfitted, Rüdin added, so that he would play the "leading and creative role" in a department bent on tackling the broadest possible range of problems in psychiatric genetics. Not only was such a materially privileged department important in order to "confront the causes of mental disease in the German Reich"; it was also necessary "to remain competitive with existing and future institutes as well as maintain and advance the reputation of the [Genealogy] Department and the [German Research Institute]."[12] Despite Glum's warning not "to push his luck too far," Rüdin ultimately received the lavish sum of 46,734 RM for his department, a threefold increase over its former operating budget. His salary alone amounted to more than half of the money earmarked for the department. The 27,156 RM Rüdin secured for himself was equivalent to more than one and a half times the normal salary of a full professor at a Bavarian university. Moreover, his Department could also count on extra support from the Emergency Association of German Science as well as financial backing from the RF. Among Rüdin's requests was money for two new "traveling physicians" to carry out genealogical-demographical studies beyond Munich. Even his demand for a departmental automobile—a request made to facilitate travel necessitated by Rüdin's and his coworkers' research—was granted.[13] Perhaps surprisingly in light of later developments, Spielmeyer, who had become sole director of the German Research Institute for Psychiatry by 1928, recognized that extravagant material concessions would have to be made to Rüdin. Spielmeyer believed that Rüdin was indispensable for its prestige, and he communicated these sentiments to Glum in no uncertain terms.[14] Rüdin was now the head of the vanguard department for psychiatric genetics that he long desired.

Recall that Rüdin's investigations aimed at assessing the hereditary patterns of psychiatric disorders, ascertaining the number of Germans affected by mental diseases, and, most importantly, preventing future psychiatric disorders from sapping the efficiency of the state through the implementation of a thorough-going eugenic policy. Chapter 1 made clear that Rüdin's research and his racial hygiene goals were intricately linked to a degree absent in the work of any other leading German human geneticist; Rüdin's science and his eugenic-inspired politics were two sides of the same coin.[15] Although the new building with the bronze bust of Kraepelin in the stairwell was heralded as "the starting point of a new epoch in healing," Rüdin's Department was the only one whose existence was not directly tied to therapeutics. Interestingly, Kraepelin himself never expected the Institute and its departments to remain static; he anticipated—indeed hoped—that the Munich Institute would change with the times.[16] With the aggrandizement of the Genealogy Department and the reinstatement of Rüdin at its helm, it is tempting to posit that the Institute was already moving in a eugenic direction as early as 1928. After all, Kraepelin was himself a strong advocate of racial hygiene.

"A Matter of Survival": In Hot Pursuit of Data

Rüdin and his coworkers' genealogical investigations were predicated upon the availability of a vast network of information, not only on the subjects of their research but on their extended families. Rüdin's empirical hereditary prognosis calculated the statistical probability that hereditary diseases (or nonpathological traits) would appear in the descendants and siblings of patients. The results would then be compared to an entire population. This presupposed that researchers had sufficient data on a wide variety of character traits at their disposal—data that could not always be procured by merely examining the subject and his or her relatives.

Even prior to his official return to Munich, Rüdin attempted to do all in his power to acquire increased access to personal data. He enlisted the support of von Harnack to secure court, criminal, welfare, disability, and health records of subjects and their relatives housed in various state and Reich ministries when his own efforts to collect this data came up short. In his memorandum to von Harnack, Rüdin explained that personal data were "a matter of survival" for his Department. A detailed examination into the "very carefully filed records" of individuals is imperative to gain knowledge of their "character traits," Rüdin argued. In his justification to von Harnack for placing the interests of science ahead of the protection of individuals' rights, Rüdin sold his research as something that serves the "common weal"—an enterprise whose worthy goal is the "investigation of diseases and their prevention as well as racial hygiene." The Kaiser Wilhelm Society president reassured skeptical ministerial bureaucracies that the use of personal information would have no negative impact on the individuals in question; "the character of the Research Institute is the guarantee."[17]

Von Harnack lent his support to Rüdin and his Department. From a letter Rüdin wrote to Glum in 1931, we know that the Kaiser Wilhelm Society made a serious attempt to secure records from numerous ministries at the state and Reich level that were not all in vain. Unfortunately, Rüdin was not able to report "a complete success."[18] The resistance by some ministries notwithstanding, the willingness of the Reich Chancellery, the Foreign Office, the Reich Economic Ministry, and the Reich Ministry of the Interior to hand over their records to Rüdin speaks volumes about the erosion of private data protection already occurring during the last years of the Republic. By this time, governmental rule on the basis of Article 48 of the Weimar Constitution—a provision giving the Reich president extensive powers—had seriously undermined democracy in Germany. Even institutions for the mentally ill apparently had few of the qualms that some state and Reich ministries had toward the release of their records; they readily agreed to provide the Genealogy Department with the data it requested.[19] Needless to say, the Munich Institute (and to a much lesser extent the Dahlem Institute) capitalized on the Weimar state's readiness to put its public social welfare concerns before the rights of its citizens. If the Dahlem Institute was already operating in an ethical gray zone prior to 1933 by using school and hospital records for von Verschuer's twin research, how much more problematic was their widespread use in Rüdin's Department?

As was mentioned, Rüdin was not the only member of his Department in need of data. Rüdin's longtime coworker, Luxenburger, required information for his pioneering research in psychiatric genetics. By the time he returned to Munich with Rüdin in 1928, he was already known for his pathbreaking work on the relationship between tuberculosis and schizophrenia. In addition, Luxenburger had recently completed a massive study on the distribution of psychiatric disorders in paralytic patients, which required the examination of no less than 580 siblings of one hundred paralytic couples. He also announced the results of his large-scale twin study. Based on the high coefficient rate of identical twins, Luxenburger confirmed the genetic basis of schizophrenia; however, given the great divergence in the trajectory of their illnesses, Luxenburger showed that the specific effects of the disorder and the specific prognosis for the afflicted individuals depended upon environmental factors. Like von Verschuer, Luxenburger championed the importance of twin studies for the science of medicine.[20] In the following years, Luxenburger remained reliant upon a large database, and he contacted numerous German psychiatric institutions for records. He viewed "exact statistics" and probability studies as his method of choice. This was not a rejection of Mendelism for the study of the inheritance of psychiatric illnesses. But Luxenburger, like Rüdin, was convinced that an orientation toward inductive empirical statistical studies was the most promising avenue for "progress" in psychiatric genetics.[21]

Juda, too, who began her research in the tradition of Rüdin, required a large database for her work. While still in Basel, she initiated an investigation known as the "Most-Talented Study," a statistical project analyzing

20 Hans Luxenburger worked in Rüdin's own Department of Genealogy and Demography. He was known for his work on the heredity of schizophrenia. Although he supported the Nazi sterilization law, he felt that Rüdin was far too aggressive in its application. Photo courtesy of the Historisches Archiv des Max-Planck-Instituts für Psychiatrie.

the relationship between psychiatric disorders and exceptional talent, which she continued after her return to Munich. Schulz, the other major researcher in the Genealogy Department and its "acting head" while Rüdin was in Basel, used the empirical hereditary prognosis to investigate the frequency of psychiatric illnesses in normal populations as well as in distant relatives of psychiatric patients. He and other coworkers, such as the Viennese Friedrich Stumpfl (1902–94), hired in 1930 to establish a criminal-biological research emphasis in the Department, also needed access to sufficient data.[22] It goes without saying that the generously funded RF-sponsored five-year anthropological survey of the German population—for which Rüdin's Department received approximately one-third of the $125,000 allotted to seventeen university and research centers—garnered useful information for the Department's researchers. Although Fischer was instrumental in initiating the project, in 1930–31

Rüdin's Department received two and a half times the amount allotted to the Dahlem Institute.[23]

Sold to the RF as an essential eugenic study, this anthropological survey underscores the importance of racial hygiene at the Genealogy Department. All of its members were strong supporters of eugenics during the Weimar years. In 1929 Rüdin addressed the Kaiser Wilhelm Society and bluntly informed it that the goal of psychiatric research was less to heal sick individuals than to serve the commonweal through the promotion of prophylactic measures—to prevent the defective from being born.[24] As we have seen, this was hardly a new position for Rüdin. Luxenburger, Rüdin's closest associate, wrote numerous articles on the importance of eugenics for both medical and popular audiences. As he asserted in one article, "the only really ideal prophylaxis for hereditary diseases is eugenic prophylaxis." "Therapeutic procedures and eugenic procedures must work together," Luxenburger continued. In the future, he added, "there should no longer be physicians who are not simultaneously eugenicists."[25]

The Munich Pact: Prologue

As we have seen, it was precisely during the Weimar Republic's final years that eugenic sterilization as a means of lessening the financial burden of the "unfit" became a serious topic for medical bureaucrats and legislators alike. Although other countries, including one canton in Rüdin's native Switzerland, already enacted mandatory sterilization laws by 1929, Rüdin generally did not openly support the use of force to prevent dysgenic births at that time. Like Lenz, a nationalist-*völkisch* member of the Munich chapter of the German Society for Racial Hygiene, Rüdin believed that the political conditions in Germany were not yet ripe for such measures. But Rüdin, like Lenz, was extremely active in calling for voluntary sterilization. This was the logical extension of decades of eugenic-inspired research. In fact, both men drafted a petition for a parliamentary committee of the Reichstag in 1929 that would effectively legalize sterilization with the patient's consent.[26] Although nothing came of this petition, as Germany's financial situation worsened owing to the Great Depression, more and more medical organizations and state bureaucracies used Rüdin's and his departmental colleagues' research to lobby for this eugenic measure. Although Rüdin did not take part in the Prussian State Health Council meeting that drafted the 1932 voluntary sterilization law, his views and research left a clear mark on the session. Muckermann, who played a leading role at the fateful Council meeting on July 2, 1932, called Rüdin, Luxenburger, and Lange "benefactors of humanity" for their work on the "hereditary prognosis."[27] The failed 1932 eugenics legislation served as the basis for the Nazi Sterilization Law Rüdin would soon champion in word and deed.

The last years of the Weimar Republic also witnessed changes within the Munich Institute. In 1931 Spielmeyer resigned his post as director, anticipating that the RF would fund a large-scale interdepartmental project that presupposed his active participation freed from administrative duties. Following the suggestion of Loeb, the Foundation Council nominated Rüdin to serve as director for a three-year period. Loeb, the German Jewish philanthropist, could not have anticipated what a Rüdin directorship would mean only two years later.

Unfortunately for Rüdin, troubles awaited him from the start of his appointment; Germany's economic crisis posed serious problems for the Munich Institute. Although Spielmeyer and Plaut enjoyed funding from the RF tied specifically to them and their research,[28] the new building with its additional rooms and laboratories could only be maintained by money outside the normal operating budget—resources that the public purse could not afford during these lean times. The German Research Institute's operating budget continuously fell from a high of 350,000 RM in 1929–30 to 100,000 RM in 1933. The Kaiser Wilhelm Society's contribution to the Munich Institute decreased by a third in this same period; the Bavarian state government also greatly curtailed its subsidies.[29] To make matters worse, only weeks before the "national government" was set in place, the RF voiced its concern over the financial demands made by Rüdin for the Institute. In a letter to Foundation Officer Robert Lambert, Gregg argued that the support given to the Munich Institute was "quite adequate for the purposes for which it was requested." "Despite our interest in psychiatry," Gregg concluded, "I do not feel additional aid for a place already so much supported . . . is easy to justify." It appeared that the RF wanted Rüdin to find out what his financial "base line really [was]."[30] Even before Rüdin realized that additional RF funding was not likely, the director attempted to communicate his Institute's bleak fiscal reality in a letter to Loeb dated November 26, 1932. After listing all the financial reductions from the Kaiser Wilhelm Society, various state governments, and the Emergency Association of German Science, the director made it clear what this shortfall meant for Loeb's beloved Institute: the Departments of Neuropathology, Serology, and Spirochaetal Research experienced more than a 50 percent reduction in their operating budget;

the Chemical Department, barely up and running, had to be closed completely. His own Department was in dire straits. Many important investigations would have to be terminated, Rüdin threatened, if new sources of funding were not forthcoming. "The money that remains," he informed his benefactor, "doesn't begin to adequately cover the temporary and long-term genetic and genealogical-demographic topics, especially the empirical hereditary prognosis investigations." However one looked at the fiscal situation, Rüdin informed Loeb, "the total level of productivity of the Research Institute will sink sharply" if help did not come soon.[31] Given the Institute's precarious financial situation as well as the community of interests that its director shared with the National Socialists on a range of issues, including racial hygiene policy, the time seemed ripe to cast the first Faustian bargain.

The Munich Pact: Act I

Broadening His Base and Expanding His Space

As was already noted, Rüdin wholeheartedly embraced the new political order. Although Fischer shared Rüdin's optimism that the "national revolution" would be a political boon for Germany and provide a financial boost for research, Nazi Party members interested in racial hygiene and National Socialist medical bureaucrats, especially Arthur Gütt, placed far more trust in the Munich psychiatric geneticist than in his Berlin colleague. It was probably not merely Rüdin's close proximity to the Nazi Party's birthplace in the Bavarian capital that accounts for this support. Rüdin's research was inseparable from his desire for state-sponsored racial hygiene, and it was known to be so.[32] Although Fischer and Rüdin were probably equally renowned internationally, the former's decision to find a professional modus operandi with the Centrist–Social Democratic stronghold of Prussia during the hated Weimar Republic made Fischer suspect in many Nazi quarters. Moreover, his first speech under the Nazi banner in which he appeared to question the negative effects of racial mixtures hardly elicited enthusiasm in National Socialist circles.

It is almost certainly the case that Rüdin's membership in the *völkisch* and pro-Nazi Munich chapter of the German Society for Racial Hygiene as well as his close ties to the aging champion of German eugenics, Ploetz, were important factors in his favor. During the early years of the Nazi regime, Rüdin never faced the humiliation and professional threat of a denunciation campaign as did the Dahlem director. If Gütt played the honest broker between Fischer and Nazi Party officials who wanted to destroy the Berlin anthropologist completely during the first eighteen months of the Third Reich, it was largely because he believed that the new regime could not implement its racial policy without the aid of the KWIA. Gütt and Rüdin had a more personal and cordial relationship.[33] Whatever the constellation of factors that resulted in Rüdin's initial (if not long-lasting) triumph at Fischer's expense, the Munich director put it to good personal use; the general sentiment among National Socialist officials that Rüdin was probably the most pro-Nazi human geneticist in Germany proved to be an intellectual resource that he exploited for professional and financial gain.

Perhaps the first event demonstrating the trust that Nazi Party racial experts had in Rüdin was his nomination to replace Fischer as president of the German Society for Racial Hygiene. Karl Astel (1898–1945) and Bruno K. Schultz (1901–97), both appointees of the Rasse-und Siedlungshauptamt-SS (SS Race and Settlement Main Office), appeared to have strong-armed this change at a meeting of the Munich chapter of the German Society for Racial Hygiene prior to the end of May 1933. The current members of the Executive Committee in Berlin—comprised of individuals committed to the old order—were to resign in favor of racial hygienists whose "allegiance to the State of Adolf Hitler is not in doubt." According to a confidential memo sent by Astel and Schultz to a select group of party comrades, the two authors of the report as well as "the great Ploetz" believed Rüdin to be well qualified to head the Society. Rüdin's scientific qualifications, his character, and his "*völkisch* outlook" were impeccable, the three men asserted. Perhaps Ploetz was taking revenge for Fischer's pragmatic decision to add the term "eugenics" to the name of his society during the last years of the Weimar Republic. Rüdin, we should note, was hardly a passive pawn in this vote of confidence that bestowed additional professional power and prestige upon him; it was Rüdin himself who suggested sending his brother-in-law Ploetz, along with Astel and Schultz to Berlin to break the news to Fischer at the June 10, 1933, meeting of the Society.

Most of the members of the Munich regional group present at the meeting calling for the removal of Fischer were affiliated with the German Research Institute for Psychiatry. Interestingly, of the two who opposed nominating Rüdin as the new *Führer* of the long-standing Society, one was Luxenburger.[34] Gütt, as we have seen, did nothing to prevent this change of leadership that had the appearance of grassroots support. Although the German Society for Racial Hygiene was never intended to offer mass public lectures on racial policy, it was hoped that the Society

21 Professor Ernst Rüdin, director of the German Research Institute/Kaiser Wilhelm Institute of Psychiatry and head of its Department of Genealogy and Demography, standing in front of his file cabinet, ca. 1936. Photo courtesy of the Historisches Archiv des Max-Planck-Instituts für Psychiatrie.

would serve as a "fighting force" for eugenically minded Germans and expand into as many new regions as possible; Rüdin himself was now in charge of approving the Executive Committees of these new groups.[35] In a 1934 memo, Rüdin specifically mentioned that the German Society for Racial Hygiene sought a "harmonious working relationship with all State and Party offices"; he was anxious to know how the Society could continue to serve as a resource for the regime.[36] Both Rüdin and Nazi officials could hardly have been disappointed with the new arrangement: between 1933 and 1939 membership in the Society increased from 950 to 4,500; during that same time the number of regional chapters increased from twelve to sixty-three.[37]

From the standpoint of procuring additional funding for the Genealogy Department and affording him increased professional visibility, no post was as important to Rüdin as his position as chairman of Task Force II of the newly created Expert Advisory Council for Population and Racial Policy. Sometime after mid-June 1933, Gütt asked Rüdin to join the Council he recently set up at Wilhelm Frick's request within the Ministry of the Interior. As its name suggests, the Council's task was to tackle the government's important population and racial policy issues by bringing together state, party, and academic experts.[38] On June 28, 1933, Frick gave a programmatic speech to the entire Council in which tried-and-true eugenic arguments were given a new sense of urgency. "Feebleminded and defective" strains of the population were reproducing two to three times faster than the "talented, valuable, class," Frick warned the Council members; moreover, the Reich Minister continued, the cost of maintaining the "asocial, defective, and hopelessly genetically ill" now far exceeds what could reasonably be expected by a state hoping to maintain itself in the struggle for survival. In addition to Rüdin, Lenz, and Ploetz, the self-styled racial anthropologist H. F. K. Günther and population policy expert Friedrich Burgdörfer were honored with an invitation to this important meeting and an offer to serve on the Council.[39] Fischer and von Verschuer were conspicuous by their absence both at the June speech and on the Council.

Task Force II dealt specifically with racial hygiene and racial policy matters; it was designed to serve as a link between the government, the universities, and the Kaiser Wilhelm Society. Its advisory agenda was quite broad. Among other things, Task Force II examined laws from the standpoint of genetics and eugenics; it also offered suggestions for acceptable university professors and gave advice on the execution of racial policy. In addition, the task force drafted plans for courses in racial biology for teachers, physicians, welfare workers, police, and members of various party organizations. It also discussed medical solutions to the problem of "hereditary criminals" and "sexual criminals."[40] Of immediate importance, Task Force II was charged with hammering out the final details of a draft mandatory sterilization law already under discussion among party dignitaries for several months. Apparently, Frick gave Task Force II only one day to discuss the measure—less than two weeks before the Nazi sterilization decree, the Law for the Prevention of Genetically Diseased Offspring, was made public on July 14, 1933.

Rüdin, the chairman of Task Force II, came to Gütt's attention through his public lectures.[41] The Munich director's political sympathies and professional expertise made him the perfect internationally renowned academic to lead this working group of the Council. Although Rüdin, together with Gütt and the SS lawyer Falk Ruttke, wrote the commentary to the Nazi Law when it went into effect on January 1, 1934, Rüdin's specific role in its formulation remains unclear, his own later testimony that he was "one of its creators" notwithstanding. As already mentioned, the Law was based on the failed 1932 Prussian draft; its mandatory character and general formulation was probably put in place largely by Gütt.[42]

Although Rüdin may not have actually drafted the Law, we do know that he attended an important March 1933 meeting of the Federation of the Bavarian District Governing Assemblies in which the topic of mandatory sterilization was raised. While there, he expressed his "delight" that eugenics had made such progress that it was now possible to put its knowledge into practice. It wasn't enough, Rüdin maintained, "merely to advance teaching and research; something positive had to happen." Sterilization was clearly the answer; in cases where the individual was seriously "defective" or obstinately refused to recognize the need to refrain from having children, force was acceptable to achieve the desired end, Rüdin believed. In other words, although the Munich director did not promote mandatory sterilization for cases at the Federation meeting, he had made it clear that he was open to the idea.[43]

To understand the extent of Rüdin's culpability for the Law's creation and its consequences, it might be necessary to consider factors other than authorship. For example, his stature as its medical commentator and the decade-long influence of the Genealogy Department's empirical hereditary prognosis research in psychiatry left a mark on the all-important interpretation of the Law. In a speech designed for psychiatrists—those physicians for whom the Law would have the greatest professional impact—Rüdin laid bare his broad constructionist commentary of the measure. For the Munich director, it represented only "a beginning in the direction of hereditary health." After listing the nine criteria for which an individual could be sterilized, he went into detail explaining what the Law meant by "genetically ill." Someone who is "hereditarily diseased," Rüdin explained, had at one time or another suffered from one of the nine categories of illnesses mentioned in Section 1 of the Law (hereditary feeblemindedness; schizophrenia; manic-depressive insanity; hereditary epilepsy; Huntington's chorea; hereditary blindness; hereditary deafness; serious hereditary bodily deformities; and serious alcoholism). In addition, the person's pathological traits have been demonstrated to be genetic because they were either definitively passed on as either dominant or recessive Mendelian traits, deemed to be so inherited on the basis of hereditary prognostic investigations on a large number of diseased families, or because they manifested themselves abnormally in the relatives of an individual family—for example, in the extended family of the person to be sterilized. Rüdin emphasized that only one of the three "proofs" sufficed for an individual to fall under the surgeon's knife.[44] It turns out that about 95 percent of the roughly four hundred thousand sterilizations performed in Germany were done on the first four grounds.[45]

Rüdin and his Department's research methodology—the empirical hereditary prognosis—became a part of National Socialist racial policy. Rüdin's colleague Luxenburger delivered a talk on the role of the empirical hereditary prognosis in psychiatry for physicians working in this field.[46] That evidence of heritability using this method was based on statistics and probability was not a problem for Rüdin. Not only was it unnecessary to prove a trait to be genetic with complete certainty; in Rüdin's estimation, categories of illnesses listed in Section 1 of the Law were so threatening to the health of the *Volk* that any chance of their heritability justified sterilization. Moreover, since the new government initially could only deal with the most serious cases, other milder or more dubious forms of degeneration would have to be prevented—at least for the time being—through marriage counseling, celibacy, and birth control.[47]

Another way in which we might assess Rüdin's responsibility for the Law is to examine whether he derived any professional profit from it for his research. Here we find that the director for whom access to data was "a matter of survival" reaped rich rewards from his position on the Appellate Hereditary Health Court in Munich. As we have mentioned, Gütt was determined that the valuable information on human heredity gathered because of the Law for the Prevention of Genetically Diseased Offspring be used for research purposes. To that end the Emergency Association of German Science held a meeting on June 20, 1934, attended by Rüdin, Fischer, and von Verschuer to discuss how to evaluate this material and who would pay for the delivered expertise. As a result of the talks, Rüdin

22 Diagram used by Rüdin for his lectures on recessive-heterosomatic patterns of heredity. Photo courtesy of the Historisches Archiv des Max-Planck-Instituts für Psychiatrie.

was able to secure an unspecified number of assistants for his Institute to examine this body of data. He tried to get as high a stipend for them as possible since "married researchers" with "established specialized training" were needed for this critical task.[48] Undoubtedly, the Hereditary Health Courts provided researchers like Rüdin with a source of material for their research. Nonetheless, because of the Nazi government's interest and sponsorship of racial hygiene research, the Munich director gained access to relevant private records from other numerous sources, as well.[49]

Rüdin's role on the Council likely did more to enable him to secure badly needed funds for his Institute than any of his other professional activities. Even before he was formally asked to serve on the Council, the Munich director sold his Institute to Reich Minister of the Interior Frick as an intellectual resource to obtain additional funding. "I am the Director of the German Research Institute for Psychiatry," Rüdin informed Frick, "which, as long as it has been in operation, has worked on the scientific genetic foundations that today's state under Adolf Hitler's leadership desperately needs to put the racial hygienic part of his program of national renewal into practice." Although the German Research Institute enjoyed the support of the Emergency Association of German Science, "considering the scope and importance of my Institute for the urgent needs of today's state," Rüdin reported, "its normal budget is relatively small." A significant portion of this money goes to pay the postage for the Institute's voluminous correspondence. Rüdin asked the Minister whether he could recover the approximately 5,000 RM in mailing costs. On July 11, 1933, less than a month later, Rüdin received a note from Kaiser Wilhelm Society President Max Planck promising to grant the requested sum. The money was secured from the Reich Postmaster.[50]

By 1935, after accepting the chairmanship for Task Force II and writing the medical commentary to the Sterilization Law, Rüdin felt secure enough to contact the Kaiser Wilhelm Society to request additional funds. In March 1935, the Munich director asked General Secretary Glum for 3,000 RM for an additional vehicle for his anthropologist as well as for additional typewriters and other equipment, "as the broadening and deepening of my areas of research" requires supplemental funding. His request was met. Five months later, Rüdin wrote an even bolder letter to the Society petitioning it to write the Reich Ministry of Education on his Department's behalf to procure additional financial support. Again, he used the science card—but in a more direct and self-assured manner than in earlier memos:

The work of the genetic division of the German Research Institute for Psychiatry in Munich was essential for the creation of the Law for the Prevention of Genetically Diseased Offspring and other public and private racial hygienic measures in the Third Reich. The patterns of inheritance of many serious human diseases are still not solved; they can, however, be so through the research methods . . . of my Institute in the near future, if it receives the necessary operational funds. *Volk*, Party, and government have the right to expect that this form of research—work designed to provide the scientific foundation of racial hygienic action—is not bogged down.

It was costly to keep the "genetic research factory" going at the Genealogy Department, Rüdin added.[51] After all, supporting the army of female file clerks alone was not cheap. Rüdin's 1935 request for material support

MPIP-HA

23 Women recording genealogical information in the Department of Genealogy and Demography, ca. 1935. Photo courtesy of the Historisches Archiv des Max-Planck-Instituts für Psychiatrie.

was answered; the tireless fund-raiser was granted an additional 50,000 RM from the German Research Council.[52]

This supplementary funding enabled the Munich director to enlarge the Genealogy Department at the expense of the other Departments—a change noticed by many. In a 1934 letter to RF Officer Daniel O'Brien, former Institute director Spielmeyer complained that Rüdin's Department "has already occupied the 11 or 12 new rooms that have be reconstructed especially for this type of [genetic] work." This additional attention to research on heredity, Spielmeyer added, "has naturally necessitated an increase in the scientific and technical personnel; there are now eight doctors and thirty typists . . . at our genealogical division."[53] By 1936, 60 percent of the 124 members of the German Research Institute worked in the Genealogy Department.[54] Rüdin's importance to the new order as a key scientific resource, along with the professional perks he received from it, obviously also enabled him to expand his space at the Munich Institute. This additional space was political in more ways than one.

Like Fischer, Rüdin became an ever more important research manager during the Third Reich. He was soon head of the "coordinated" and enlarged Gesellschaft Deutscher Neurologen und Psychiater (Society for

German Neurologists and Psychiatrists)[55]—a position that placed him in close professional contact with physicians directly involved in Nazi Germany's "euthanasia" project. His role as chairman of Task Force II, his excellent relationship with provincial and Reich ministries, and his control over his field's professional organization enabled Rüdin to influence the appointment of like-minded psychiatric geneticists to university posts and research institutes throughout Germany. For example, Rüdin made it possible for his friend and colleague, the "euthanasia" psychiatrist Kurt Pohlisch (1893–1955), to establish what amounted to a smaller version of the Munich Institute in Bonn.[56]

The Munich director also organized the courses for physicians in his field. Perhaps the most important of these was "Genetics-Racial Hygiene Training Course for Psychiatrists" held at his Institute in January 1934. In his opening address, Rüdin mentioned that psychiatrists who do not embrace the Sterilization Law with "inner conviction" should resign from their posts. Over 130 psychiatrists attended the nine-day seminar; they heard talks not only from members of the Genealogy Department like Luxenburger, Schulz, Stumpfl, and the director himself, but from Rüdin's network of colleagues throughout Germany. Even Fritz von Wettstein (1895–1945), an eminent plant geneticist at the Kaiser Wilhelm Institute for Biology in Dahlem, instructed seminar participants in "the genetic foundations of racial hygiene."

At the end of the training course, there was an interesting two-hour question and answer session headed by Luxenburger. Asked by one colleague which groups constituted the "feebleminded," he replied that in addition to children attending special schools for the learning disabled, even those having great difficulty in normal elementary schools and "who fail in life" should be viewed as such. In response, another psychiatrist asked whether the state didn't need a certain number of feebleminded to execute menial tasks. Luxenburger's rejoinder: "even after sterilization there will be enough hereditary feebleminded individuals to serve as coolies." Parts of the seminar were open to the general public. One of the most important anthologies on human heredity during the Third Reich, *Erblehre und Rassenhygiene im völkischen Staat* (*Genetics and Racial Hygiene in the Völkisch State*), published the popular and technical speeches delivered there.[57]

Rüdin also held talks at courses organized by other institutions, like one he delivered at the Military Academy of Berlin in October 1937. This speech is worth examining since he prepared it for medical professionals who would come to play an increasingly important role during the Third Reich, especially during wartime: military sanitation officers. It also gives us insight into the mindset of the Munich director at the height of his professional and political influence and reveals his talent for employing the most effective linguistic resources for his purposes.

Since a "defensive war" was unavoidable, Rüdin began, it was necessary to adequately prepare for it during peacetime. Racial hygiene considerations were central to such preparations. As Hitler himself proclaimed that racial hygiene was "the greatest revolution experienced by Germany" at a recent Nazi Party rally, it was not necessary, Rüdin remarked, "to stress [its] importance further." Although population quantity was significant for the defense of the nation, so, too, was its genetic quality; to achieve this end, negative and positive eugenic measures were necessary. Rüdin did not doubt that the present German military represented a "good manly selection." In order that Germany's military men return from their tour of duty healthy "in body and soul," it was important to avoid sexually transmitted diseases. Soldiers needed to ensure that they had sexual relations only with women who were their social equals; it was not only dangerous but "dishonorable" for soldiers stationed near brothels to make use of them.

In some respects, however, a proper racial hygiene perspective required altering long-standing military traditions. At least as far back as the reign of "old Fritz" (King Frederick the Great of Prussia [1712–86], an honored German military leader), married officers were viewed as only "halfway combat ready." The Prussian king allegedly derided officers who sought wives as "not worth any gunpowder." Married officers might have divided loyalties; winning a battle needed to be the soldier's only concern. After all, "the hussars [members of light cavalry units dating back to the fifteenth century] have to find their happiness through the saber, not through the vagina." But since the time of "old Fritz" much has changed, Rüdin conceded. Indeed, it was largely married soldiers and officers in the early 1870s who secured Germany's ascension to nation-state status; they knew what they were fighting for, the Munich director assured his audience. Shouldn't the same be true, he asked, now that Germany was "threatened by a war of occupation and oppression from Bolshevism"?

Rüdin was obviously preoccupied with the possibility of war, and he confronted the issue of the next outbreak of hostilities head on. Some say, he remarked, that "the next war will be a so-called total war where every man and woman, normal or abnormal . . . has to be used where [he or she] can best contribute to victory." If the best men sacrifice their genetically valuable lives and semen to defend home and hearth, what about the "unfit"? We must have the courage of conviction to demand that the "abnormal" also contribute to the war "as intensely as possible according to their abilities," Rüdin asserted. As we now know, "it is a misfortune for many abnormal individuals as well as many who are sick to protect them as was earlier the fashion," to leave them with nothing to do and support them through the public purse. "Work therapy," Rüdin concluded, "has had glowing success." The Munich director closed his speech to Germany's military medics by remarking that even among them much propaganda was still needed.[58]

Rüdin's rhetorical talents are also exemplified in his indefatigable efforts to secure additional money for his Department. By this time, Rüdin had equated the Institute's fiscal needs with that of the Genealogy Department. The importance of his research notwithstanding, the Munich director was told as early as late 1935 that he could no longer expect the exorbitant sums he received from the German Research Council—one of his main sources of revenue. Glum tried to help Rüdin by encouraging him to emphasize the expensive nature of his research that "was of special importance for the National Socialist government."[59] Glum's encouragement notwithstanding, Rudin feared that he would be in financially dire straits for the 1936 fiscal year. He had already decided to take a daring step: on November 22, 1935, he wrote directly to the *Führer* requesting 20,000 RM. His funding from other sources was not enough to cover the essential racial hygienic projects of his Department, Rüdin complained. The money he sought would not merely be used to lay a more solid foundation for the Sterilization Law; it would also provide the "scientific, genetic tools" so that the "state marriage counseling centers as well as the State and Party health offices could carry out their noble work on a scientifically surer footing." Rüdin was amazingly brazen in what could almost be described as an open threat: he informed Hitler that if additional money was not forthcoming, he would have to dismiss numerous workers in the Genealogy Department. The Munich director softened his tone by inviting the *Führer* to see the important work at his Department firsthand. Such a visit, he added, would not only be a "great honor"; it would convince Hitler that such funding would be well spent. In signing the letter, Rüdin was sure to include all his titles: chairman of Task Force II of the Council, head of the German Society for Racial Hygiene, and President of the Society for Neurologists and Psychiatrists.[60]

Although the *Führer* appears to never have accepted Rüdin's offer to visit his Institute, the latter received his money. The reference written by his friend and patron Gütt undoubtedly did not hurt matters. As he mentioned in his letter of support, "there are probably few fields of research that are in a position to legitimize and further the goals of the National Socialist government more than those pursued by Professor Dr. Rüdin."[61]

MPIP-HA

24 Ernst Rüdin freeing a "research automobile" from the snow, ca. 1936. Photo courtesy of the Historisches Archiv des Max-Planck-Instituts für Psychiatrie.

One year later, Rüdin again asked Hitler for an additional 38,000 RM. He did so in "the name of science and his Institute," both of which knew no higher goal than to achieve the "broadest and most secure scientific legitimation" of Hitler's plan for the "racial health of the German people." Again, Rüdin secured the resources he requested.[62] The last financial aid that the Munich director received from the *Führer's* Chancellery was in January 1939. He was granted the sum of 40,000 RM after his Institute allegedly suffered from an 80,000 RM shortfall.[63] In the future, he would turn to another party organization for financial support: the SS.

Race, Politics, and Philanthropy

Ironically, Rüdin's financial difficulties during the Third Reich were exacerbated by the regime's notorious anti-Semitic racial policies. Although Fischer was forced to "coordinate" his Institute and dismiss the Jesuit eugenicist Muckermann, the Berlin anthropologist never had "racial enemies" in important positions in his Institute; apparently his anti-Semitic sentiments prevented Jewish students from coming to him in the first place. Fischer deftly used this fact as proof of his good political qualifications during his denunciation campaign.[64] Rüdin could not claim that his Institute was free of Jews. Not only was the Munich Prospector Karl

Neubürger Jewish; the world-renowned head of the Department of Serology, Felix Plaut, was a "non-Aryan." Plaut had been a close colleague of the late Emil Kraepelin and accompanied him to America to secure Rockefeller Foundation support for the Institute's new building. To make matters worse, Plaut was appointed deputy director of the Institute in 1931 when Spielmeyer retired. In other words, after Hitler's "national revolution" Rüdin faced a situation where an eminent Jewish scientist with close ties to, and substantial financial support from, the Rockefeller Foundation was second in command at the Munich Institute. Moreover, Rüdin's internationally respected head of the Department of Neuropathology, Spielmeyer, although "Aryan," was an anti-Nazi and married to a woman who had "Jews in the family." Spielmeyer also had good relations to RF officers and, like Plaut, enjoyed a healthy amount of funding from America's leading philanthropic organization.[65]

When the National Socialist government passed the notorious Gesetz zur Wiederherstellung des Berufsbeamtentums (Law for the Reestablishment of the Professional Civil Service) in April 1933—a measure designed to eliminate Jews and politically suspect individuals from civil service positions in Germany—Neubürger and Plaut were initially exempt. Apparently, Kaiser Wilhelm Society President Planck intervened to avoid causing Plaut any "unnecessary embarrassment," especially since he enjoyed the financial support of the Americans.[66] Such actions were typical of the Society in the early years of the Third Reich; Germany's most prestigious research society attempted to hold on to world-renowned German Jewish scientists as long as possible. In the case of the German Research Institute for Psychiatry, the Kaiser Wilhelm Society could rightly claim that the Institute was not entirely state supported, and therefore Jewish scientists employed there could legally be exempted from immediate expulsion. However, the Society offered no large-scale resistance to the Nazis' escalating anti-Semitic policies; it always accepted the "inevitable." After the passing of the Nuremberg Laws in September 1935, Rüdin terminated the positions of Neubürger and Plaut.[67]

Even prior to the official "retirement" of the Institute's Jewish scientists, Spielmeyer and Plaut kept the RF abreast of the unhappy turn of events in Munich. The correspondence between them and RF officials as well as the private reactions of the latter toward German racial policies early in the Third Reich provide an interesting perspective on the scientific politics of America's premier philanthropic organization. It raises thorny ethical issues as to whether political considerations should have played a role in the Foundation's funding of biomedical science under the swastika.

As early as September 1933, RF Officer Robert Lambert noted the political changes at the Rüdin Institute in his diary. Spielmeyer had apparently informed him that the "present atmosphere is not conducive to the peace of mind necessary for objective research" there. "Developments in the past six weeks," he told Lambert, "have made it clear that the [government] intends to carry out an educational program under which universities and separate research institutes must serve the immediate interests of the National-socialist regime." Lambert reflected upon Spielmeyer's situation: here was a fifty-four-year-old scientist with a "partly Jewish" wife and an eight-year-old daughter. What should he do?[68] Alan Gregg's diary entries testify to the growing uneasiness at the Munich Institute. During his meeting with Spielmeyer in June 1934, the neuropathologist informed Gregg that he was "ashamed of the . . . silence [among academics] regarding the developments in Germany." Gregg also reflected upon his lunch with Plaut. While pouring coffee for his guest, the serologist remarked that he might not be able to do so by next November. As it turned out, it took a bit longer than Plaut expected for him to be officially retired from his post. Plaut commented in a communiqué to RF Officer O'Brien that when Rüdin extended invitations to him, it was merely to "create a favorable impression with the RF." However, he was not deceived by these gestures; indeed Plaut told Rüdin that "he would never risk the possible embarrassment of going to a German hotel."[69]

We will recall Spielmeyer's comments on the expansion of Rüdin's Department within the Institute. He informed Gregg that the RF-supported Chemical Division would have to be sacrificed because "Professor Rüdin needs the space." Spielmeyer's Department budget was cut by 30,000 RM. Gregg was also apprised of the "coordination" of the Society of German Neurologists and Psychiatrists; in the future, Spielmeyer remarked, everything having to do with psychiatric research will fall under the "Governmental Commissar for Racial Matters, that is, under Ernst Rüdin."[70] By January 1935, the neuropathologist had informed Gregg that he could not "stand the climate" at the Institute any longer.[71] Spielmeyer had become so depressed about his situation that when he suddenly died in the wake of a severe bout with flu on February 6, 1935, the RF speculated that he had taken his own life.[72] Spielmeyer's successor, Willibald Scholz (1889–1971), developed friendly relations with Alan Gregg but was never able to challenge Rüdin's position at the Institute. With Spielmeyer's death and the forced removal of Plaut, Rüdin became the unchallenged Führer of the Munich Institute. The interests of the Genealogy Department and that of the German Research Institute for Psychiatry virtually became one and the same.

From the above comments it is clear that the RF knew of the politically motivated organizational changes at the Institute almost immediately after the Nazi takeover, and the officers were certainly not in the dark regarding the actions of its director. How did the RF react? We mentioned in chapter 1 that at least Lambert recognized the degree of anti-Semitism at German universities even prior to the "national revolution." On the other hand, there was a naively optimistic view of German science just months before Hitler assumed the Chancellorship. This naïveté continued in some quarters even after January 1933. The lack of evidence for the existence of a "Nordic race" in the publications of the RF-sponsored anthropological survey of Germans seemed to signal to a couple of officers in the Social Science Division that biology under the swastika was not tainted by politics.[73] This upbeat position, however, was in stark contrast to Gregg's view that Germany would not provide "much of a milieu for science in the near future." RF Officer O'Brien, also from the Medical Division, shared a similar assessment. As early as March 1933, he recognized the precarious nature of the Jews in Germany and the fear of the population toward expressing any "political views that were against the Party in power." Although skepticism in this domain was well founded, Gregg's belief in the instability of the Nazi regime as late as June 1934 and his musings about the possible positive economic effects of a shift toward the communists reveal that the Foundation as a whole had no clear picture of the political landscape of the country into which it continued to pour its research dollars.[74]

Interestingly, the question of whether the American philanthropic organization was funding politically tainted research came to a head in late 1933 or early 1934 when its New York Office received "several caustic inquiries" into its dealings with the Munich Institute. In December 1933, Bruce Bliven from the *New Republic* wrote that "a would-be contributor" to the magazine informed him that money from the philanthropy was funneled to support the German Research Institute for Psychiatry. This Institute, he claimed, has "now largely lost its scientific character" and became a center for the National Socialist worldview. On February 2, 1934, an article published in *The American Hebrew* reported that a Kaiser Wilhelm Institute in Munich was "spreading Nazi propaganda under the cloak of science" with the financial support of the RF and the Loeb estate. Apparently, a day before the article appeared, an American rabbi also accused the RF of supporting racism.[75]

At this point, RF Officer Thomas Appleget wrote to the philanthropy's major attorney explaining his side of the story. He mentioned that the Kaiser Wilhelm Institute in question was the "Kraepelin Institute" and that the RF had indeed paid for its new building. In addition, $89,000 was given to support the work of Spielmeyer and Plaut. The two men, Appleget continued, were still "on the job" and were "conducting the research as originally planned in a thoroughly objective and scientific spirit." Rüdin, Appleget conceded, continued to be a member of the Institute, but he never received any RF grants for his work. "Rüdin's present political affiliations were not under the control of the Institute or the Kaiser Wilhelm [Society]," he added. Perhaps Rüdin and others at the Munich Institute wrote scientifically dubious articles in the building funded by the RF. Nothing could be done about that. The grant to Spielmeyer and Plaut was itself subject to revision if either left the Institute, Appleget reassured his lawyer. However, there was no indication of that happening. Under these circumstances, Appleget continued, "I think it is quite untrue to say that the Foundation funds are being used to subsidize race prejudice." Later in the month Appleget wrote a memo to a fellow RF officer in the French Office admitting to receiving "a number of inquiries from various liberal groups" regarding the RF's connection to the Munich Institute.[76]

What are we to make of Appleget's explanation? The only fact that must be contested is his assertion that the RF never supported Rüdin. Although the Medical Division had not done so, the Social Science Division supported the anthropological survey of the German people. This error aside, the Foundation's continued support of research within Rüdin's Institute—even the "good science" practiced by Plaut and Spielmeyer— leads to questions concerning its culpability toward funding an institution so complicit with Nazi racial ideology. Indeed, it raises the thorny issue of whether any philanthropic organization should fund "good science" in otherwise unacceptable circumstances, or whether it should take a political and moral stance by withholding its money. The RF's official position was that politics was irrelevant as long as it was funding "good science."

Regardless of the Foundation's responsibility toward its funding of the German Research Institute, what remains beyond question is that RF officers in the Medical Division were profoundly saddened by Spielmeyer's death. In addition, money tied to his research was terminated despite Rüdin's pleas for continued funding. When Plaut was officially "relieved of his responsibilities" on October 10, 1935, the RF funds he received were not given to his successor, Franz Jahnel.[77] Whatever the official RF policy might have been, it was clear by late 1935 that the forced removal of Plaut and the proliferation of Nazi racial policy measures had definitely soured the Medical Division officers toward Rüdin and his Institute. As Lambert wrote O'Brien, "one might think from reading Rüdin's letter that nothing had happened except Spielmeyer's death. But he certainly must know that from our viewpoint Germany's Nazification of science has created a situation totally different from that which existed in 1930 when the grants for Spielmeyer and Plaut's work were made. And he certainly must realize the effect on Foundation Officers and trustees of Plaut's dismissal."[78] It would be interesting to know whether the RF officers' reluctance to further support the Rüdin Institute had more to do with the Munich research center's politics, a more reflective understanding of the relationship between science and governmental policy, or with the belief that the science under the swastika was no longer "good." By early 1936, the Foundation Council of the Munich Institute recognized "the difficult situation" it faced owing to the government's anti-Semitic policies.[79] At any rate, RF funding to the Institute was finally and completely curtailed at the end of 1937. By that time, the RF could no longer hide its contempt for Rüdin and the racial policies he stood for; it saw his attempt to secure money for the allegedly "non-political scientific needs or activities of one of the constituent departments" of the German Research Institute as irritating.[80] RF officers from the Medical Section helped underwrite Plaut's salary in a London hospital in 1939 after he left Germany. The despondent émigré committed suicide in 1940.[81]

"Liberal quarters" in the United States not only pointed a finger at the RF; they also accused the trustees of the Loeb estate of supporting research that could have been used against the great German Jewish benefactor had he not died in May 1933. Although Loeb left a significant amount of money to the German Research Institute, practical difficulties in executing the will as well as the possible intervention of Loeb's "Aryan" stepson, Joseph Hambuechen, prevented Rüdin from profiting from it.[82] In late 1935, Kaiser Wilhelm Society General Secretary Glum promised the Munich director that he would do everything possible to keep Hambuechen in the Foundation Council—although he did not think it would be easy "considering the catastrophic impact that the Nuremberg Laws had, especially in the Anglo-Saxon countries."[83] The only Loeb revenue that Rüdin had access to, and was anxious to protect, was the sum earmarked as "special funds." In fact, in 1940, fearing that America would enter the war against Germany, Rüdin did all in his power to make sure the money was "not completely lost"; he suggested a transfer of the funds to a Swiss bank as quickly as possible.[84]

Considering the flack over the support of racist research at the Munich Institute, it is worth discussing just what role "race" played in the scientific work at the Genealogy Department—the only department where "race" in the anthropological sense of the term was an issue. In contrast to the Dahlem Institute where Fischer from the outset instituted a research program based on the genetics of normal and pathological racial differences, Rüdin's Department, both before and after 1933, was dedicated to psychiatric genetics. Race, anthropologically speaking, was not a part of its research agenda. We will recall that the word "race" had multiple meanings at this time; its use during the first half of the twentieth century was not considered unscientific or politically incorrect. Race could either be used to denote any interbreeding population, including the *Vitalrasse* (the entire human race), or it could be understood as a *Systemrasse* (an individual racial group) within a population. In almost all cases where the term "race" was used by Genealogy Department researchers, it was employed in the first sense. Scientists there were interested in mental illnesses of the German *Volksgemeinschaft*, whereby the latter term did not implicitly have any anti-Semitic connotation as far as these researchers were concerned. Interest in the distribution of mental illnesses among individual racial groups was not a significant part of their research project. However, as early as 1911, Rüdin accepted that "racial mixtures" could lead to degeneration; he was in favor of "racial purity" as a prophylactic measure.[85]

What this all meant for Rüdin and his Genealogy Department coworkers in the context of the Third Reich—when "race" was imbued with a clear political meaning—was that the term was employed in different ways at different times. For professional consumption, *"Rasse"* was usually understood in terms of *Volksgemeinschaft;* scientists intentionally employed the term vaguely.[86] When used for popular consumption or applying for research money, biomedical practitioners almost always included "racial purity" as part of their discussion. For example, when Rüdin gave a formal talk at the Military Academy in Berlin in 1937, he stressed the importance of achieving racial purity in his definition of the task of racial hygiene. In his letters to the *Führer,* Rüdin always mentioned that he and his Institute were working in Hitler's interest—implying that he dedicated his research not only to improving the genetic quality of Germans, but ridding them of foreign (i.e., Jewish) blood.[87] Given Rüdin's interest in promoting himself as an authority on Nazi racial policy as well as securing research funding for his Institute, it should come as little surprise that his use of the word *Rasse* was flexible; it was defined to fit the moment and circumstance.

However, if one asks whether Rüdin was anti-Semitic and approved of Nazi anti-Semitic policies, one finds oneself in a gray zone. Certainly, Rüdin did not applaud racial mixtures. He fired Plaut and Neubürger when requested to do so—apparently with no questions asked. He did, however, help Jews unofficially, especially if they were in his field of research.[88] Apparently, he intervened at his own Institute. In 1936, Rüdin asked Glum whether the "25% non-Aryan" daughter of a former psychiatrist in Heidelberg could be employed in the Department of Neuropathology. The General Secretary replied that if the chairman of the Foundation Council of the Institute had no objections, neither would he, "as it is only a question of a technical assistant"—a comment that says as much about the official attitude of the Kaiser Wilhelm Society on the "Jewish question" as it does about Rüdin.[89] It should also be pointed out that no racial genealogies of the kind carried out at the Dahlem Institute were undertaken in Munich;[90] although this was not the Institute's area of expertise. Rüdin, as we noted from the letter described at the opening of the chapter, shared the anti-Semitism prevalent among conservative nationalists at the time. The Munich director certainly opposed the crudest anti-Jewish rhetoric and practice common during the Third Reich in such publications as *Der Stürmer*. When asked by a colleague whether a person submitting a manuscript for a professional journal should have to specify his or her racial background, Rüdin insisted that such information should be gathered informally.[91] That Rüdin worked with Jewish colleagues prior to 1933, apparently without incident, would suggest that there was nothing unusual about his form of anti-Semitism. In sum, Rüdin profited both professionally and financially from the multifaceted meaning of "race" during the Third Reich; the word served him as an invaluable linguistic resource. Although it is difficult to assess the nature of Rüdin's anti-Semitism, it appears similar to what was typical among conservative nationalists at the time. The important difference, of course, was that the Munich director was not just any conservative nationalist during the Third Reich. Moreover—whether out of commitment or convenience—Rüdin joined the Nazi Party in 1937.[92]

Conflicts and Controversies

By the time the Munich director became a member of the NSDAP, he was a serious player in the field of Nazi racial policy. As has been mentioned, he was the head of the leading professional organization in his field, the *Führer* of the German Society for Racial Hygiene, and "court" racial policy expert for the government. He also held a chair at the University of Munich and was the president of the International Federation of Eugenics Organizations. The German Research Institute for Psychiatry housed scientific guests from numerous countries, including Denmark, Great Britain, Italy, the Netherlands, Poland, Spain, and the United States. The

25 Rüdin in his circle of female assistants in the Department of Genealogy and Demography, ca. 1938. Photo courtesy of the Historisches Archiv des Max-Planck-Instituts für Psychiatrie.

eminent Swedish genealogist and paternity expert Erik Essen-Möller and the renowned English psychiatrist Eliot Slater worked in Rüdin's Genealogy Department during the prewar years. Although many of these foreign guests, like Slater, were critical of Rüdin's role under the swastika, their willingness to put political differences aside and journey to Munich insured that the Institute was not professionally isolated—at least prior to the war.[93] Rüdin was certainly happy to house foreign researchers at the Institute, as long as they did not politically embarrass him. As Rüdin stated in a letter to a famous Dutch geneticist, he could ill afford to have someone at his Institute who "had an anti-German outlook." He chastised Slater for his article in the British *Eugenics Review*, since it painted the Sterilization Law in a negative light. Slater, Rüdin bemoaned, did not alter the wording of the piece as he suggested. The director expected his longtime British guest to understand his official position and that of his Institute; he obviously viewed Slater's actions as a professional slap in the face.[94]

Leaving aside the occasional problems he faced from his foreign guests, during the prewar years Rüdin entangled himself in a number of professional and personal conflicts and controversies. One, involving a close colleague in his own Institute, was never publicly aired; other conflicts pitted the Munich director against important human geneticists at other research institutes. Rüdin also became part of a larger Nazi state-Party controversy—the kind that frequently marred the functioning of laws and policies in the Third Reich.

As we mentioned, Luxenburger was one of Rüdin's oldest and closest colleagues. Known internationally for his work on the genetics of schizophrenia using twins as subjects, the Catholic eugenicist—perhaps owing to religious conviction, perhaps owing to scientific caution—was initially loath to endorse mandatory sterilization. While this was not an issue prior to the proclamation of the Law for the Prevention of Genetically Diseased Offspring, it certainly was afterward. An article written by Luxenburger scheduled for publication after August 1933 (i.e., after the Law was made public) contained statements seemingly opposed to the new measure. As it turns out, the heretical article was nipped in the bud, done so paternally, if sternly, by Rüdin's patron, Gütt. The highest-ranking Nazi health minister told Rüdin in no uncertain terms that Luxenburger's statements were now unacceptable. He even hinted that perhaps Luxenburger was dispensable. Upon receiving Gütt's letter, the Munich director immediately responded by thanking his friend for sparing him any embarrassment by apprising him of the matter. Luxenburger clearly wrote the article in which he criticized forced sterilization as a measure "opposed to the professional ethics of a physician" prior to the passing of the Law, the Genealogy Department head emphatically assured his patron. Luxenburger saw his error and now stood "completely on the side of the government," Rüdin replied. No additional "measures" were necessary.[95]

Although Rüdin shielded his colleague from denunciation, if not a worse fate, his intervention could not hide the differences between them. In what could only be viewed as a courageous act, Luxenburger publicly stated in 1934 that the regime's eugenic goals could only be accomplished in an atmosphere of scientific freedom. "Science," the twin expert asserted, "had to . . . remain free; it should not be the servant, but rather the master, of its own house." Only in this way, he continued, could it truly be free to serve society.[96] Although there is no evidence that his boss reprimanded him for these sentiments, they could not have been entirely welcome. To say the least, Rüdin did not give the impression that his research was shackled by Nazi ideology.

In a very revealing correspondence between Luxenburger and his fellow twin specialist von Verschuer, the former vented his frustrations over the tensions at the Institute; he also commented on the professional tightrope he was constantly asked to walk. The Luxenburger–von Verschuer exchange suggests an ambiance at the Munich Institute that hardly matched Kraepelin's desire for "scientific tolerance" and "versatility."

In a letter referring to a manuscript that Luxenburger submitted to von Verschuer's journal *Der Erbarzt,* the latter complained that his Munich colleague was too radical in his views on negative eugenic measures. According to the Frankfurt director, Luxenburger's alleged willingness to counsel many people who were not suffering from a genetic disease (but who might have a deleterious recessive trait) from having a family was wrong. "Our *Volk* cannot afford this," von Verschuer wrote. He believed that Luxenburger undervalued the significance of policies designed to encourage the "fit" to have larger families.

In his written response, the Munich psychiatric geneticist let loose his pent-up anger. First, he thought it odd that one had the right to forcibly sterilize individuals but encourage their children to marry. In Munich, Luxenburger sarcastically remarked, the inherent eugenic logic was clearer than it apparently was in Frankfurt. He also found it more than a bit ironic that von Verschuer criticized his strictness, while Rüdin blamed him for being a softy. "For months," he bemoaned to von Verschuer, "I have had the most difficult time merely because I have declined to forbid all siblings of schizophrenics the right to have children—difficulties which I do not even wish to mention." "Even today," the Munich geneticist lamented, he got into a heated argument because he refused to forbid a "healthy warm-hearted girl to have children, only because she had a schizophrenic aunt." And now, Luxenburger complained to his colleague, he had the audacity to declare his position to be too radical "when I constantly risk losing my [financial] existence (and perhaps more) because I am considered too lax and yielding."

The Munich geneticist continued his litany of complaints. He constantly tries to moderate things here, but only succeeded in hurting himself "in ways you could not imagine," Luxenburger caustically informed von Verschuer. If he dared to state that, as a rule, children of schizophrenics should be allowed to have families, "I would be hanged." That might not be so bad, Luxenburger continued, but then "I could no longer be of help. At least now, I am sometimes able to prevent a sibling or a child of a schizophrenic from being denied the right to procreate—people, who, if they went to any other counselor here, would be strictly forbidden from marrying." The beleaguered Luxenburger ended his letter by informing von Verschuer that the latter had the luxury of passing judgment within the relatively free atmosphere of Berlin and Frankfurt. The Fischer protégé could hardly understand the difficulties Luxenburger faced in Munich; nobody, he argued, who had not been forced to "breathe the Munich air" had the right to utter a negative comment about him. Two days later, von Verschuer responded to his colleague's long letter. Although he did indeed realize that the "Munich air" was different than that further north, the Frankfurt twin specialist was shocked to hear how radical the racial hygienic views were in the Bavarian capital. Perhaps, von Verschuer mused, "the time was not yet ripe for a moderate standpoint."[97]

Luxenburger's boss also had conflicts with researchers outside his own Institute. Rüdin himself butted heads with von Verschuer and with Fischer over the use of personal data and research subjects. However, it suffices to limit ourselves to the Rüdin-Fischer conflict.[98]

In June 1937 Fischer complained to Rüdin that a member of his Munich Institute was collecting data on twins in Berlin for a study on the role of heredity in diabetes—a topic that the Dahlem director viewed as falling within the research locus of his Institute. Fischer was "unpleasantly surprised" that the Munich director "intruded" on a research topic supposedly far removed from Rüdin's sphere of scientific interest; in essence, Fischer remarked, Rüdin's assistant—with the Munich director's blessing—was attempting to steal "scientific material" Fischer worked so hard and so long to collect. According to the Dahlem director, "it is not practical for us to compete over these [human subjects]." Although Fischer would not, and could not, prevent anyone from working on diabetes, he did not "consider it correct, not to mention, collegial," to steal data. He suggested to Rüdin that they respect each other's spheres of scientific influence. He promised to allow the Munich director and his coworkers access to his material dealing with diseases that, "owing to their rarity," could only be investigated using data on twins. Fischer ended his arrogantly worded letter hoping Rüdin would understand his position. It is possible that Fischer's caustic remarks stemmed from an earlier promise by Rüdin to respect the scientific interests of other human geneticists whose work entailed twin studies, which he gave in a letter to von Verschuer.[99]

Fischer's communiqué was followed by an acerbic response by the Munich director. After denying that he was in any way a scientific "intruder" in Fischer's research field, Rüdin told the Dahlem director in no uncertain terms that nobody had a "monopoly on material or areas of research." The Genealogy Department head first mentioned that he didn't complain when one of Fischer's assistants worked on a topic in criminal biology—a field allegedly in his bailiwick. As a result, Rüdin assumed that their earlier gentlemen's agreement to avoid overlapping research was null and void. In any event, earlier divisions between the two Institutes' research fields had to do with politics no longer relevant in the Third Reich. "Today," Rüdin emphasized, "I am happy about the larger degree of [scientific] freedom in this regard." He then declared that studies of the inheritance of disorders connected to the central nervous system lay entirely in his scientific domain; obviously diabetes fell into this category. Twin studies, the empirical hereditary prognosis, and family studies were critical for his work. Perhaps the distancing between the Dahlem and the Munich Institutes following the "Muckermann affair," Rüdin continued, explained why Fischer did not realize that his Institute was also working on diabetes. Indeed, this research was supported by the Reich Ministry of the Interior. And anyway, Rüdin added, "competition for the best and quickest results can only be in the interest of the state and the German people." Rüdin ended his less than friendly letter in the hope that Fischer would soon realize that he was not the "dangerous" scientific "'intruder'" depicted by the Dahlem director.[100]

The Fischer-Rüdin controversy—although revealing for what it tells us about the mind-set of the two directors—also raises interesting questions about the role of social convention and acceptable competition in scientific work. Whether or not researchers should have a monopoly on certain areas of investigation once they have made their mark in a field; whether in cases where human subjects and financial resources are limited, it would be both practical and in the best interest of scientific advancement for institutes to articulate clear demarcations of research domains; or, as Rüdin suggested, whether it is not more advantageous to have stiff competition among institutes and their researchers—all these are important queries that transcend this particular debate between two biomedical practitioners in Nazi Germany.

As a case in point, we can reflect on a somewhat similar, if far more famous, case of scientific propriety in the controversial work *The Double Helix*. In his "personal account" of the "race" to discover DNA, James Watson describes how he and Francis Crick were interested in nucleic acid, but unfortunately found themselves at an institute in Cambridge, Great Britain, that had agreed to restrict its attention to the structure of proteins. DNA was "the personal property" of Maurice Wilkins at King's College in London, although it was his "assistant," Rosalind Franklin, whose painstaking empirical research made the discovery of its structure possible. During the financially lean postwar years in Great Britain, it was common for scientific institutes to have clearly demarcated research domains. It was not considered good form to work on an area of research assigned to another institute; to do so as part of a stealth operation—as did

Watson and Crick—was an affront against scientific convention at the time. One can certainly view them as having stolen Rosalind Franklin's data—the very information they needed to construct the correct Tinkertoy-like model of DNA.[101] The Fischer-Rüdin conflict as well as that between the scientific players in Great Britain in the early 1950s reminds us that such questions about acceptable scientific practice remain both contentious and unresolved.

In addition to intra- and interinstitutional conflicts, Rüdin became indirectly embroiled in a far more significant state-party clash over the execution of the Sterilization Law. We will recall that the Nazi state was plagued by rivalries between ministries and the individuals who headed them. Gerhard Wagner, leader of the Nationalsozialistischer Ärztebund (National Socialist Physicians League), often complained that the authors of the Law had not taken party considerations into account. Party members were not immune from the Law's inhibitions on procreation; some individuals, Wagner insisted, had made useful political contributions but were placed in the nebulous category of "feebleminded" by state medical officials. For Wagner, many of the medical referees on the Hereditary Health Courts were "shortsighted genetics fanatics" pursuing a "science alienated from the life of the people." In 1937, he demanded that the Hereditary Health Courts be responsible to the party, something that Gütt and Rüdin were naturally opposed to. The situation between party and state became so tense, that the *Führer* was asked to intervene. It is probably not accidental that Rüdin became a party member precisely when the conflict over who would be in charge of this key element of Nazi racial policy came to a head.[102] Gütt's response to Hitler's effort at reconciliation between the state medical bureaucracy and the party apparatus is instructive for what he thought about his scientific right-hand man, Rüdin.

After informing the *Führer* that he and the Munich director were "very angry" over Wagner's attempts to discredit them, Gütt explained that by selecting Rüdin as his trusted assistant to write the commentary to the Law, he chose "a worthy researcher and pioneer . . . who also enjoyed the greatest reputation abroad." Gütt did admit that the Munich director's original view of the Law's wording was stricter than his own. As a "practical man," he "sought the middle way." Apparently, Rüdin and Gütt reached agreement only after difficult negotiations. But now, Gütt asserted, the Law and commentary were viewed both at home and abroad as a "standard work." Indeed, Gütt informed Hitler, he specifically saw to it that Rüdin attended international congresses to defend National Socialist eugenic policy. That the Genealogy Department head was successful in this endeavor was recognized by Reinhard Heydrich, second in command to Himmler in the SS.[103]

Although many of the complaints Wagner brought forth in his attack on the Gütt-Rüdin-Ruttke interpretation of the Law were purely partisan and designed to transfer decisions to sterilize from state medical bureaucrats and racial hygiene academics to party hacks, the broad constructionist interpretation of the Law had denied many the right to be parents on dubious grounds, to say the least. There is evidence that by 1937 pressure was placed on the courts to rethink the way decisions regarding sterilization were made; Wagner's complaints obviously did not go unheeded. For example, in the district of Jena, the Appellate Hereditary Health Court required physicians and judges sitting on the lower courts to become familiar with the demands of agricultural and manual labor in the region; they were asked to gain some grassroots experience by speaking with the foreman of small factories about the work requirements of particular jobs. This information should be considered in cases where individuals coming before the courts were diagnosed as "borderline-feebleminded." The decision to require sterilization would be made on the basis of whether individuals were "fit for life"—that is, if they could work efficiently.[104]

There were other serious problems with how the Law was executed that raised concerns. Although the Law had strict provisions for protecting the sterilized from discrimination, this was certainly not always followed. R.R., a man sterilized in 1937 at age twenty because of his alleged heredity blindness, wrote a letter in 1942 to the Hereditary Health Court in Munich demanding that the judgment be reversed. The reason: R.R. had suffered untold professional and social hardships owing to his forced sterilization. At the time he was assured that the "procedure was viewed as a sacrifice for the good of the society and would be viewed as such." Since R.R was a minor when the decision was made, his father, who had no knowledge of genetics, accepted the operation with the proviso that it would not harm his son later in life. The decision to prevent his son from having a family was based on one piece of evidence from the University of Munich Eye Clinic. There was little else the father could do but acquiesce, since he could not afford a lawyer knowledgeable in the Law. After the operation, several eye specialists informed R.R. that his blindness was not genetic; in other words, his illness did not fall under the provisions of the Law. That having been said, R.R. listed the disadvantages he faced as a marked man: his stipend to attend a school for the blind was rescinded; when he wanted to marry he was told that such a union was not in Germany's interest; he was denied party membership when his sterilization became known; the term "genetically ill" was in all his personal files; and finally, he doubted that he would gain employment as an elementary school teacher's assistant. R.R. was applying to have the decision reversed, not, of course, so that he could once again be fertile; but he saw himself as a "completely worthy, productive member of society" and wanted the stigma of "genetically ill" removed, once and for all, from his records.[105]

Although we cannot state whether R.R. was successful in his effort to destigmatize himself, we do know that a compromise was reached between the state and party; in 1938 a third and final version of the Law was drafted that allowed the party to intervene in cases coming before the Hereditary Health Court. Owing to Wagner's waning influence, this change never came into effect. More significant for the future, however, was the establishment of an Ausschuß für Erbgesundheitsfragen (Committee for Questions of Genetics) within the Reich Ministry of the Interior. It would soon serve as the organizational basis for an operation far more radical than sterilization: "euthanasia."[106]

The Munich Pact: Act II

The Bargain with the Nazi Black Suits, the SS

By 1939 Rüdin was at the height of his power. His sixty-fifth birthday was marked by numerous commemorative volumes; his colleagues paid tribute to the importance of the empirical hereditary prognosis for psychiatric genetics and Rüdin's contributions to eugenic legislation during the Third Reich. He received the prestigious Goethe Medal for Arts and Sciences from the Reich Ministry of the Interior as well as numerous other awards for his accomplishments in racial hygiene.[107] The Munich director had also won the full confidence of the party. When asked whether he should be allowed to give talks on the Sterilization Law, the party judgment was clear: "His political trustworthiness can not be doubted."[108] For his part, Rüdin continued to demonstrate his loyalty to the regime. In a letter thanking Frick for his warm birthday greetings, the Munich director emphasized that he would continue to serve both the Reich minister and the *Führer* "as long as his energy allowed."[109]

Rüdin's role as "court" racial policy advisor notwithstanding, he still struggled to find sufficient financial backing to support the ever-expanding research projects undertaken in his Department; according to one of his assistants, there were no less than twenty medically trained workers and fifty to sixty secretarial and office staff at the Genealogy Department by 1939.[110] Unable to live within his budget since the start of his directorship, he managed to cover his shortfall during the prewar years of the Third Reich by appealing directly to the *Führer*. Well before this unusual source of revenue no longer became available to him, Rüdin was questioned about his inability to curb his fiscal appetite. For its part, the Kaiser Wilhelm Society warned him as early as 1936 that responsible parties were wondering why the price tag for his Department's research was so exorbitant. The general feeling at the German Research Council— at the time a significant source of revenue for the Institute—was that Rüdin should accomplish more for less money. After all, half of all money earmarked to support genetics at the Research Council for the fiscal year of 1937 was already funneled to the Munich Institute. Even the official Minutes of the German Research Institute for Psychiatry's Foundation Council's meeting mentioned that Rüdin's Department "has been attacked in some quarters" for its extravagant monetary demands.[111]

By 1939, but before the outbreak of war, Rüdin thought himself to be in an untenable financial situation. He expressed his dismay in a letter to his trusted patron Gütt. The German Research Council did not have sufficient funds to support the Institute; the Reich Chancellery insisted that Rüdin secure money through normal channels. The Munich director believed that the Kaiser Wilhelm Society had promised to secure the funding he requested, if necessary from special sources. It turned out, however, that he had misunderstood the new general secretary, Ernst Telschow. The Society would be cutting 160,000 RM from his budget. After listing the research projects and personnel that would have to be axed if immediate help were not forthcoming, Rüdin vented his exasperation at the fiscal crisis confronting him. "I cannot believe that the state agencies whose *Führer* never tires of emphasizing that racial efficiency and genetic health are the foundation of the German state," the Munich director wrote Gütt, "would allow the Institute that dedicates itself full-time to compiling the scientific groundwork for this goal to [simply] languish."[112]

If the state ministries and funding agencies were unwilling or unable to come through for Rüdin, perhaps a more influential source might serve as his financial savior. Just days before he wrote his letter to Gütt, the Munich director made the fateful decision to appeal directly to the Black Suits. By this time the SS was the most powerful and, for the "enemies of the state," the most dreaded Nazi organization. Perhaps recognizing that Gütt no longer had the clout he once commanded, Rüdin met with the Reich manager of the SS-Ahnenerbe (SS-Ancestral Research Society),

Wolfram Sievers, in Berlin on June 19. Founded by SS-Chief Himmler in 1935 to promote interdisciplinary "scholarship" to foster Germanic culture throughout Europe, the Ahnenerbe was certainly well endowed.[113] It was the logical association within the growing SS empire for Rüdin to seek out. He certainly assumed that Sievers would find the work undertaken at the Institute relevant to the racial concerns of his boss.

As it turns out, Rüdin's primary contact person in the Ahnenerbe would be a dean and the Rector of the University of Munich, Walther Wüst (1901–91). A high-ranking SS officer with excellent scholarly credentials and close personal ties to Himmler, Wüst was the *Kurator* (academic director) of the organization. Days after his initial meeting with Sievers, Rüdin wrote Wüst indicating his scientific and financial motives for approaching the Ahnenerbe. This important letter constitutes the second, and for Rüdin ill-fated, "Munich Pact." It was initiated solely by him.

After praising Wüst's new Institute and the scholarship undertaken by the Ahnenerbe, the Genealogy Department head suggested that work on "protoplasmic" (genetic) ancestral history would provide a "welcome addition" to the organization's emphasis on cultural ancestral research. What Rüdin specifically proposed was research currently undertaken in his Department of Genetic Talent and Hereditary Health employing the empirical hereditary prognosis; such work naturally fit into the research portfolio of the Ahnenerbe, the Munich director assured Wüst. Rüdin informed the *Kurator* of the several Black Suit researchers already working in his Department, including a woman, Dr. Käthe Hell, who was very close to the SS "in her outlook." In discussing the colleagues who had been with him from the beginning, Rüdin went out of his way to mention that Dr. Adele Juda, despite her last name, was 100 percent Aryan and "looked Nordic." More importantly, she was undertaking significant work on the inheritance of talent. Although his other long-standing coworker, Bruno Schulz, was not a "flaming National Socialist," he was "a national-mined man through and through" whose usefulness for the Ahnenerbe he could assure without reservation. Interestingly, Rüdin said nothing about Luxenburger. Perhaps the Munich director felt that it would be disadvantageous to mention him given the twin specialist's earlier conflict with the crude anti-Semite Streicher, and because of the party view that, as a Catholic, Luxenburger was an outsider who could not be trusted.[114] In any event, Rüdin clearly felt that his Institute had something to offer Wüst.

At this point, Rüdin related his dire financial circumstances to Wüst: 160,000 RM had been cut from his budget. He told the *Kurator* that he would need money immediately to keep ongoing research alive. The

Munich director apologized for combining a declaration of desire for "long-term contact" with the Ahnenerbe with his "plea for [financial] help" in his letter, but the "news regarding the catastrophic budgetary cut-back" of his Institute hit him "like a bolt of lightning out of the blue." Whether out of sincerity or calculation, Rüdin concluded by reminding Wüst that "my concern is also your concern; and both are the concern of the Third Reich, its *Führer* and also one of his main adjutants, the *Reichsführer-SS* [Himmler]."[115] This "concern" was, of course, the racial and genetic improvement of Germany.

That Rüdin was driven to the SS out of financial desperation is clear; what is more ambiguous is whether he truly supported the growth and aims of this "state within a state." On the one hand, the Munich director seemed no stranger to the SS. Heinz Riedel, an assistant in the Genealogy Department, had been working under the Munich director since 1935; he held a rank equivalent to first lieutenant in this elite unit and was a member of the influential SD, the Security Office, within the SS. Although Riedel would soon become embroiled in a vicious controversy with Rüdin, in early 1939 he sang his praises in an article honoring the Director's sixty-fifth birthday. Rüdin also attended at least one meeting at the SS doctor's *Führer* School in Alt-Rehse, although when and why he was there remains unknown.[116] In early 1940, after the bargain between Rüdin and Wüst was sealed, the Munich director offered a detailed research plan, complete with expense budget, on ethnic Germans in the newly annexed "east," more specifically, in occupied Poland. Such work would ensure the scientifically correct Germanization of the area. The scheme, however, never materialized.[117] On the other hand, there is no evidence that Rüdin approached the SS prior to this time of financial desperation. He was, of course, on excellent terms with SS members like Gütt and Ruttke. But this new connection was on a more personal level. As long as he could adequately fund his Institute through more traditional channels, he did not beg for money on the doorsteps of the SS. Of course, it was only at this time that the SS became the powerful organization it would remain throughout the war. Perhaps Rüdin did not think the Ahnenerbe was in the financial position to aid him during the earlier years of the Third Reich. What is clear, however, is that the Munich director had no intellectual reservations about extending his hand to the SS; and he certainly expected to profit from it.

As we know, it takes two parties to seal a "Faustian bargain." The interesting query is not so much why Rüdin turned to the SS, but why the Ahnenerbe was keen to finance the Munich Institute. This is especially puzzling in light of the devastating critique Rüdin received from Rudolf

Mentzel (1900–87), an SS man and head of the German Research Council. In a discussion that took place on the day Rüdin first approached Sievers, Mentzel warned against supporting Rüdin's lavishly designed research factory. The Munich director's fiscal appetites knew no limit, Mentzel asserted. More importantly, the "money already used is disproportionate to the desired results." Fischer and von Verschuer, Mentzel added, were working on similar topics, and Rüdin was constantly embroiled in controversies with them, since he could not limit himself to his own scientific and geographic territory. This scathing commentary notwithstanding, Wüst, with the blessing of his boss Himmler, favored supporting the Munich Institute. Naturally, Wüst received a seat on the Institute's Foundation Council.[118] But why would the two men be so anxious to support the German Research Institute?

In this context, Wüst made an interesting comment: given SS-Brigade Leader Dr. Leonardo Conti's "plan for a new order of the entire health system complex," it was "especially important to gain influence at the Rüdin Institute."[119] Conti, we will recall, was the Reich Health Leader at this time; he was also an important patron for Fischer. It was Conti who managed to get the Dahlem director money to expand and modernize his Institute in 1941. Although he occupied an even higher position in the SS than Wüst, Conti allowed Fischer (and later von Verschuer) to control the KWIA as he saw fit, trusting that the money secured for the Institute would serve Conti in his new role.

Unlike Conti, Wüst and his boss Himmler did not offer money with no strings attached. Unbeknownst to Rüdin, the *Kurator* of the Ahnenerbe was determined to set up what amounted to an SS research institute within the Genealogy Department (not unlike the state within a state that characterized the way the SS operated in Nazi Germany) with trusted SS scientists and assistants. Those who formed a part of this research "site" would be supported by the Ahnenerbe, although not openly, and they would make life uncomfortable for the old guard, including Rüdin. Heydrich, second in command to Himmler and the man who praised Rüdin's accomplishments in the international arena only a few years earlier, now wanted the director replaced—something that was never accomplished. Heydrich was killed in a partisan raid in occupied Czechoslovakia in 1942. Whether Himmler ever insisted on actually ousting Rüdin is unlikely, since he was never removed from the helm. Perhaps the SS-head found it more useful to keep him officially in charge of the German Research Institute; discarding him could have caused trouble with the Kaiser Wilhelm Society. What is certain, however, is that the

Reichsführer-SS was not content that the "soul-destroying," overly academic work that characterized the research at the Genealogy Department be allowed to continue unabated. Scientists who met the Black Suits' political and moral qualifications and worked on appropriate topics would be setting the tone from now on. As we will see, one SS researcher deliberately wrote a caustic article attacking Luxenburger, Rüdin, and the work of the Department in the journal *Der Biologe* (*The Biologist*)—a publication with a wide readership that had recently come under the control of the Ahnenerbe.[120]

Exactly why Himmler was interested in Rüdin's Institute when Conti's plans for a "new health order" were underway remains unclear. Perhaps Himmler saw his organization's involvement in the German Research Institute for Psychiatry as a way of extending the grasp of its tentacles deep within the Reich's health sector, heavily implicated as it was with Nazi racial policies. Whatever the reason, one fact cannot be disputed: Rüdin was far less an equal partner to the Faustian bargain made with Wüst than Fischer had been with the Nazi state medical bureaucrats. Perhaps Rüdin's lack of bad experiences with the party rendered him too naive to predict what a bargain with the Black Suits would entail. Fischer's early hand slapping in the denunciation campaign may have left the Dahlem director far more politically savvy than his Bavarian colleague and competitor. After the second "Munich Pact" was sealed, times would be tough for the non-SS members the Genealogy Department. Rüdin would soon learn that acquiring financial support did not guarantee that he and his senior coworkers would be able to pursue their research in peace.

The Devil's Price

The SS authorized a first payment of 30,000 RM to be transferred to the Munich Institute as soon as the deal was finalized. On February 4, 1940, Rüdin attended a meeting with Wüst and Walter Greite, editor of *Der Biologe* and an active member of the Ahnenerbe who was now involved with the Munich Institute. The SS conditions for financing the GRIP were clearly laid out. The memo reporting this meeting also mentions that Luxenburger, Juda, and Schulz—the old guard at the Genealogy Department—were "unacceptable" to Himmler's organization. There is no record of Rüdin attempting to defend them.[121] The then Kaiser Wilhelm Society general secretary, Ernst Telschow, was extremely happy about this second Munich Pact. In a memo to Wüst, he expressed his pleasure that Himmler "was interested in the work of the German Research Institute for Psychiatry, since that was equivalent to showing an interest in the work of the Kaiser Wilhelm Society."[122] The venerable Society also appeared to be moving with the times.

As detailed internal memos regarding the SS-German Research Institute connection indicate, an SS cell was immediately established at the Munich Institute—a fifth column of sorts—under the nose of Rüdin. Unofficially, it was known as the Forschungsstätte für Erbpflege (Research Site for Hereditary Care).[123] The group was comprised of three SS-stipend recipients, Erwin Schröter, Käthe Hell, and Heinz Riedel. This elite research group was charged with working on "studies of abnormal personalities and difficult life situations." More specifically, they investigated the genetic quality of children born out of wedlock (one of Himmler's favorite topics), the role of nature and nurture in feebleminded twins, and the inheritance of mental illness in children of psychopathological families. Other studies dealing with the genetics of homosexuality and criminal biological studies—both of which were of interest to Reinhard Heydrich, Himmler's second in command—never bore fruit. The war frequently interrupted the work of the SS scientists.[124]

Far more significant than their research was their ability to wreak intellectual and political havoc at the Institute—just as they were supposed to do. The main instigator was Riedel, a man who, prior to the SS decision to use the Rüdin Institute for its own purposes, appears to have gotten along with everyone. In early 1940 he attacked Luxenburger, Rüdin, and the Institute in an article published in *Der Biologe,* a journal designed more for high school biology teachers and generalists than psychiatrists. This attack was quickly followed by another in the journal of the National Socialist Physicians' League, *Ziel und Weg (Goal and Path).* Although the details need not concern us, suffice it to say that Luxenburger was made to appear as an "enemy of the state" and Rüdin was accused of tolerating such a person; moreover, the work of the Munich director and his Institute was berated and the Genealogy Department head was charged with having a "positive attitude toward Jews."[125]

This attack naturally did not go unanswered. Rüdin furiously initiated a hearing within the Nationalsozialistischer Dozentenbund (National Socialist Professors' Association) in which virtually all members of the Department were asked to offer testimony on the matter. Nearly all the Institute members—both the longtime colleagues of Rüdin as well as the more recent so-called Austrian group surrounding the racial biologist and party member Karl Thums—condemned the actions of Riedel.[126] According to one testimonial, Riedel was alleged to have stated that Rüdin was neither "personally or scientifically fit to carry out the duties of the

Directorship." The Munich Institute's representative in the Association, Hermann Grobig, reported that Riedel confided to a former guest in the Serological Department that Rüdin was little more than "an old man . . . who is completely incapable and who only wrote one [real] scientific study."[127] Interestingly, it was one of Rüdin's coworkers intricately connected to the "euthanasia" action who offered the clearest assessment of this smear campaign against Luxenburger, his boss, and the Institute. Riedel's actions in late 1940, asserted psychiatrist Julius Deussen (1906–70), must be considered within the context of his charge "to initiate a fundamental personnel and intellectual change at the German Research Institute."[128] Finally, Rüdin attempted to sue Riedel in court for libel. The court, however, did not feel that this was the most appropriate time for solving a "professors' dispute" in which both sides probably shared blame.[129] Germany was in the midst of war.

Rüdin's situation was aggravated still further by an article written in the *Archiv* in 1942, apparently authored by someone with close ties to a clandestine political Catholic organization. He claimed to have proof that children born out of wedlock were morally inferior to those whose mothers were married. This assessment flew in the face of Himmler's well-known views on the subject. As such, it caused a political scandal. It is not clear whether the Munich director approved of this article (its publication could have been an oversight by Rüdin, who had plenty of other things to worry about at that time); it sufficed, however, that he was the prestigious journal's editor and failed to prevent it from appearing in press. Rüdin was severely reprimanded by both party and state authorities. Indeed, the journal itself was in danger of being shut down.[130]

The upshot of these attacks—especially from within his own house—was that the Institute became polarized to a degree that made earlier internal conflicts appear trivial. If the Munich Institute never quite worked in harmony as envisaged by its founder, Kraepelin, it was now a travesty of that vision. Normal research, as it had been energetically pursued prior to 1940, became far more difficult. With the onset of war, the distinguished foreign scholars who worked in the *Kraepelinstrasse* did so no more. Rüdin's oldest and most trusted colleague, Luxenburger, would have to be sacrificed to the SS's more radical form of "coordination." To be sure, the twin specialist had long since been an uncomfortable colleague. If his deeply held Catholic beliefs made him suspicious to the party, he appeared nothing short of a state enemy to the SS. A sincere eugenicist, Luxenburger had come to accept mandatory sterilization once the die was cast. He certainly did not view himself as an opponent of the regime; indeed, he appears to have shared the anti-Semitism typical of many in his confession—a fact initially used by Rüdin in his defense![131] As we will recall, Luxenburger did, however, make a pitch for freedom of research—something that he, unlike his boss, saw as waning rather than growing at the Institute. His complaint to von Verschuer about the stifling "Munich air" he was forced to breathe was again testimony to his difficulties with the Munich director.

By mid-1940, it was clear to Luxenburger that he would have to leave the Genealogy Department, although Rüdin and the new Foundation Council (with Wüst as a member) assured him that he would be given time to find another position before he would leave. From a revealing letter in which Luxenburger addressed the alleged problems he posed to the Institute, it is clear that Rüdin was grasping at straws to bring charges against him. The attempt to paint Luxenburger as a "political Catholic" or someone disloyal to the regime could be dismissed out of hand; why would he be allowed to give so many talks on psychiatric genetics and racial hygiene, and why would his own boss ask him to fill in for him if the twin specialist was politically unreliable? No one except the members of the SS clique, he insisted, could blame him for the polarized atmosphere at the Institute. Luxenburger eventually found a position with the *Luftwaffe* (Air Force) which he held throughout the remainder of the Third Reich.[132] It is not clear why Rüdin's other longtime nonparty colleagues at the Department, Juda and Schulz, were spared expulsion. They had been likewise viewed as "unacceptable" by the SS. Perhaps Juda and Schulz were loath to openly express their opinions, unlike Luxenburger; maybe they were simply not important enough for the SS and the Foundation Council to take action against them.

There is every reason to believe that by 1942 Rüdin had come to realize the price he would have to pay for his second Munich pact. Although there is no evidence that he was seriously in danger of being ousted as director after Heydrich's death, his constant attempt to place the German Research Institute completely under the Kaiser Wilhelm Society was undoubtedly designed to secure his position. Although all sides appeared to have desired "clarity" regarding the financial position of the Munich Institute, this was not achieved on paper until 1944. In July of that year, the Institute suffered serious damage from an air raid; all work effectively ceased. The German Research Institute remained in legal limbo until it was transformed into a Max Planck Institute after the war—but without Rüdin.[133]

In January 1943, the Munich director wrote an article for the *Archiv* entitled "Zehn Jahre nationalsozialistischer Staat" ("Ten Years of the National Socialist State"). In this piece he sang the praises of Hitler and his medical henchmen for supporting racial hygiene in word and deed. He did not neglect to mention what the SS achieved through its demands that its men marry their racial equals. What is unusual, even for Rüdin, was the vitriolic rhetoric he employed when discussing "enemies of the state." Jews and Gypsies received a linguistic lashing as "parasitic foreign-blooded races." Rüdin applauded the Nuremberg Laws as "preventing further invasion of Jewish blood into the German gene pool."

As we have seen, Rüdin previously expressed his belief that war was inevitable, if dysgenic. In his article, the Genealogy Department head went on to discuss the present situation; in so doing, he correctly suggested that the conflict was a "racial war." Rüdin maintained that the recently enlarged German-*Völkisch* Reich now included German-blooded individuals who were formerly living as "an island [community] under foreign peoples; thanks to the work of the *Führer*, they were finally protected from languishing or disappearing entirely." The current war was inflicted upon the world by the "Jewish-Plutocratic-Bolshevistic-influenced powers," Rüdin continued, slavishly regurgitating commonplace Nazi rhetoric. It was not desired by Germany. His adopted homeland, Rüdin insisted, merely wanted to be free to establish a humane economy for its people and put its racial hygienic theories into practice. At the moment there was nothing to do "but fight until victory." The experience of the last ten years, the Munich director assured his readers, made him confident that Germany will redouble its racial hygienic efforts after its victory.[134]

This article would have serious consequences for the Munich director in the immediate postwar period. Although after 1945 Rüdin would later claim that he penned this piece against his better judgment to save his Institute from the SS, there were probably other factors at work.[135] It could have been more or less forced upon him as the price of freeing himself from the "political hornets' nest"[136] owing to the controversial article published in the *Archiv* a year earlier. But perhaps Rüdin had, by this time, accepted the worldview of his new masters. The publication might have represented his true feelings—sentiments radicalized by the long-term symbiotic relationship between Nazi racial policies and human heredity under the swastika. This interpretation is certainly conceivable in the light of the grim news from Stalingrad and reports of the brutalities on the Eastern Front. Moreover, at the time he wrote this vitriolic article, Rüdin could hardly have remained unaffected by his knowledge of, and "involvement" with, the project to exterminate "lives not worthy of living." Perhaps this, too, explains his hardened heart and heated rhetoric.

Epilogue: Rüdin, War, and "Euthanasia"

By the time Rüdin initiated his second "Munich pact," Nazi racial policy had already taken a radical turn. On August 31, 1939, the day prior to the outbreak of war, the Reich Ministry of the Interior decreed that the implementation of the Sterilization Law should be drastically cut to cover only the most "urgent cases." Only two weeks before, the Ministry mandated that doctors would have to report all children up to the age of three who were alleged to have certain diseases.[137] The transition from the Nazi state's ability to revoke an individual's right to parenthood to "euthanasia" had begun. Here again we see the radical symbiosis between National Socialist racial policy and German biomedical science leading researchers to the doorsteps of committing others to physical death—and themselves to moral collapse.

Although there had been discussions of the killing of "lives unworthy of life" as early as 1920—especially in the aftermath of the "sacrifice" made by the "best blood" in World War I—eugenicists initially rejected the idea and it was not pursued. It was seen as too radical and inhumane. There were some psychiatrists, however, who viewed the elimination of incurable mentally handicapped individuals as the flip side of a "reform" program in their medical specialty. This emphasized treating acute cases of mental illness through a combination of methods. Only those patients who could be brought back to "useful work" were candidates for such medical intervention. In order to rid psychiatry of its negative image within the medical sciences (after all, it had a low rate of curing psychiatric disorders), some practitioners became sympathetic to a "cure or kill" mentality. When the time was ripe for the implementation of a "euthanasia" program, enough German physicians were ready to participate. There were also new scientific frontiers to be conquered. Academics involved in the Nazi state's first planned foray into mass murder understood that such killings would yield "interesting material" for research purposes.[138]

For his part, Hitler had discussed the desirability of eliminating these "defectives" or this "ballast" as least as early as 1935. Like the mass extermination of the Jews for which this project served as a testing ground, the decision to launch a systematic and stealth "euthanasia" program was made possible under the cover of war. As we have already seen in the case of the Dahlem Institute's connection to Auschwitz, war makes actions wholly impermissible during normal times appear more acceptable—especially to physicians already inclined to view the interests of society as more important than those of the individual. It also opens up new windows of opportunity. Like the case of individuals slated to die at Auschwitz, "euthanasia" victims, stripped as they were of their rights, could be actively exploited by biomedical scientists with impunity. Decades of diatribes deriding the racial pollution and social cost of "defectives" combined with new wartime conditions help explain how such actions became a reality. The word "euthanasia"—a term that implies the notion of "mercy killing" to end the suffering of a terminally ill person—was a Nazi euphemism to describe the systematic slaughter of perhaps more than two hundred thousand individuals in Greater Germany and Poland by gas, lethal injections, medication, or outright planned starvation. Some were killed for eugenic reasons; others out of economic considerations or the medical contingencies of war (such as the need for hospital space). The perpetrators—both the well-known psychiatrists and their lesser-known male and female, medical and nonmedical, staff—undertook their tasks owing to a conjunction of eugenic, economic, scientific, and careerist motives. In terms of ideology, personnel, and methodology, the "euthanasia" action was in every way the antechamber of the Holocaust.[139]

Although the details of this human and moral tragedy are richly documented elsewhere, suffice it to say that there were three main systematic operations within the larger "euthanasia" project. They were organized by the *Führer's* Chancellery, Dr. Karl Brandt (1904–48) (Hitler's personal physician), and the Reich Ministry of the Interior. The first entailed the killing of infants and children; the second, the so-called T4 action, targeted adults ("T4" is shorthand for the address of the Berlin villa headquarters of the operation: *Tiergartenstrasse* 4); and, finally, operation 14 f 13, which dealt with "diseased" concentration camp inmates.

The most well-known mass murders took place simultaneously in six "killing centers" in Germany and Austria under the T4 program. Its cold bureaucratic efficiency was responsible for the gassing of approximately seventy thousand mentally handicapped adults between September 1939 and August 1941, although there is evidence that some asylum patients were killed by other means even prior to the start of the war.[140] So-called expert physicians were asked to "evaluate" the patients, most of whom they never met. A plus or a minus meant the difference between life and death; those unfortunates slated to die were usually brought by bus to one of the killing facilities. Reports of patients hiding or attempting to escape their captors were commonplace. After their murder, relatives were informed that they died owing to "natural causes"—although frequently mistakes were made in selecting a "disease" and their loved ones became suspicious. Some members of the clergy, such as Bishop Clemens August von Galen of Münster, spoke out against the measure from the pulpit. Most, unfortunately, did not.

When word about the operation became an open secret in some regions by 1941, Hitler halted its systematic form, and it became more clandestine. The *Führer* could not risk alienating a large segment of the nation during wartime. At this point, the so-called *wilde Euthanasie* (decentralized euthanasia) commenced in hospitals throughout Germany and Poland from 1941 until 1945. In addition, between 1942 and 1943 *Ostarbeiter* ("East Workers," a synonym for civilians from occupied Eastern Europe forced to work as slave laborers in the Reich) deemed to be mentally ill were also murdered in special euthanasia centers.[141]

Given Rüdin's privileged position as leader of the "coordinated" German professional organizations in psychiatry and as director of the Munich Institute we can hardly avoid asking about his involvement with the "euthanasia project." This calls for a careful weighing of the available evidence; one's answer to this question will also depend upon one's understanding of what it means to be "involved."

We know that in addition to his research and his commitment to the institutionalization of racial hygiene, the Munich director was a passionate defender of his profession—psychiatry. Long before he served as *Führer* of the forcefully amalgamated professional organizations in his field, he had been a member of numerous associations dealing with psychiatry and mental health. During the Third Reich, Rüdin was anxious to ensure that his field remained professional, and he took a stand against allowing nonphysicians to teach in the medical faculties of universities. Like others in his discipline, the Munich director was worried about the image of psychiatry as a medical specialty and sought to do all in his power to preserve its prestige.[142] We have observed Rüdin's role in organizing the important racial biology seminar for psychiatrists at his Institute in 1934; the Munich director was also called upon to give talks to a wide variety of audiences.[143] In addition, Rüdin held an influential position as a scientific evaluator at the German Research Council. Any research project having to do with psychiatry, or human heredity more broadly defined, needed his stamp of approval.[144] In sum, Rüdin took an active role in his profession both before and after 1933; he also wielded an extraordinary amount of power in shaping his discipline under the Nazis.

Among his close professional associates were several influential psychiatrists who later actively participated in the "euthanasia" program. Perhaps the Munich director's closest colleague, at least in terms of the volume of correspondence, was psychiatrist Paul Nitsche (1876–1948) whom he knew as far back as 1910. Serving as director of asylums for the mentally ill and as an advisor on health matters to the State of Saxony during the Weimar Republic, Nitsche falls into the above-mentioned category of psychiatric "reformers." He believed his profession would be best served by separating those who could be treated and brought back to useful work from the so-called hopeless cases that should be neglected or eliminated. During the Third Reich Nitsche was the secretary of the German Society of Neurologists and Psychiatrists—the "coordinated" organization headed by Rüdin. When the T4 project got underway, Nitsche served as its deputy and later as its head. Although letters from the adult "euthanasia" period are missing, the correspondence leaves little doubt that Rüdin knew about conscious cost-cutting measures in the asylums during the 1930s that negatively impacted the welfare of patients, as well as the official killings themselves.[145] The Munich director was well aware of the "euthanasia" action.

Nitsche as well as other professional associates of Rüdin such as Maximinian de Crinis (1889–1945), Kurt Pohlisch, Hans Heinze (1895–1983), and Werner Heyde (1902–64) were active participants in the "euthanasia" program. They belong to those physicians who either directly ran the T4 operation, headed asylums where patients were exterminated, served as evaluators in determining the life or death of patients, or were otherwise involved in the physical killing of the handicapped. From extensive research on the subject we know that Rüdin does not belong to this category of victimizers, his professional contacts with many of them notwithstanding.

However, this group of several hundred individuals does not incorporate a larger number of professionals who might also be implicated in this deadly project. They include people who, both before and during the killings, provided the scientific legitimation for "euthanasia" as well as those who served to professionally profit from the murdered victims' data. Rüdin certainly appears to fall into this category, thus further implicating him in the crimes associated with the T4 project.[146]

At this point we need to analyze a wide array of evidence: Rüdin's public statements on the question of "euthanasia," the language he used to describe the untreatable mentally handicapped, his reaction to colleagues who wished to take a stand against the killing, his relationship to "reform" psychiatrists, especially to those involved in the "euthanasia" action, and postwar testimonials. We must also consider how the troubles he experienced with the SS in his Institute might have impacted his connection to "euthanasia." Assessing Rüdin's "involvement" with these medical crimes clearly demonstrates that history is not merely about assembling facts, but interpreting these facts.

If we examine the writings of Rüdin prior to 1933, we can find nothing that suggests that he openly supported the killing of the mentally handicapped. Even during the prewar years of the Third Reich, the Munich director openly opposed such action. In two interviews given in 1935 to correspondents working for nonspecialist Nazi publications, Rüdin reported that the "destruction of unworthy life in the harsh form in which it occurs in the plant and animal kingdom is something that we cannot accept. For this reason we [attempt] to solve the problem through germ plasma selection."[147] Yet what is interesting to note in the Munich director's language is his use of the phrase "destruction of unworthy life"—an expression uncannily close to the title of the infamous pamphlet first advocating "euthanasia" in Germany in 1920, *Die Freigabe der Vernichtung des lebensunwerten Lebens* (*Permission for the Destruction of Life Not Worth Living*). Elsewhere he characterized the mentally handicapped as "ballast," a term also originally used in this same publication.[148] Moreover, he clearly believed that "inferior people" played no useful role in the state. We will recall that one of the participants in the psychiatrists' seminar hosted by Rüdin in 1934 had queried whether the nation needed a certain number of feebleminded to perform menial tasks. Luxenburger responded that even with the Sterilization Law there would be enough "coolies" for such jobs. In an article for a welfare publication, Rüdin went further than his colleague. From a cost-benefit perspective, the Munich director argued, an efficient regulation of "the work process . . . makes the inferior portions of the population superfluous."[149] Rüdin obviously viewed the worth of individuals in terms of their contribution to the "national body." For the Munich director, the severely mentally handicapped were nonproductive and thus void of value. It would not be unfair to say that prior to the war, Rüdin's rhetoric contradicts his public stance against killing the "defective."

We have already mentioned Rüdin's close professional connections to "euthanasia" physicians. Almost all of these individuals can be grouped as belonging to the "cure or kill" circle of "reformers" in psychiatry. Although it is difficult to assess whether the Munich director also belonged in this category during the prewar years, we know that he was very interested in active therapies to treat those who might be brought back to useful work. In 1938 Rüdin received a letter from an asylum physician at Eglfing-Haar near Munich, Anton von Braunmühl, describing the positive effects of the insulin shock therapy he had been testing on schizophrenic patients. The merits of this treatment were not merely medical. Insulin shock therapy fit well into the "reform" program of reducing the length of a patient's required stay in a hospital, hence cutting costs. The treatment, however, did not always work; Braunmühl killed two of his patients. From his letter to the Munich director, however, we can surmise that Rüdin, like some of his other colleagues, followed the progress of this "therapeutic" measure. Braunmühl was not a "euthanasia" physician; however, he was a "reformer" anxious to improve his profession's status and help economize German asylums. It would appear that Rüdin shared his concerns.[150]

That not all "reform" psychiatrists were willing to accept the killing of the mentally handicapped is demonstrated by the actions of Hans Roemer, the director of the Illenau Asylum in Baden, a colleague of Rüdin, and a member of the Advisory Board of the German Society of Neurologists and Psychiatrists. Although Roemer was an active eugenicist and a supporter of measures like insulin shock therapy to treat the curable mentally ill, he was opposed to the "euthanasia" action. Roemer tried to convince Rüdin to take an open stand against it in December 1939; other psychiatrists wishing to professionally protest the killing attempted the same. Rüdin refused to use the professional society of psychiatrists as a platform for protest, despite his promise to his former coworker Luxenburger to the contrary. Roemer resigned as director of the Illenau Asylum; he later also gave up his position on the Advisory Board of the German Society of Neurologists and Psychiatrists.[151] Unlike his colleague, the Munich director never took a public stand against the formal "euthanasia action," although he appears to have expressed his misgivings about the killings, at least at this stage, in private.[152]

As was mentioned, the killing of the handicapped continued after the decision to halt the formal action was made in August 1941. It is during this so-called period of "decentralized euthanasia," which lasted until the end of the war, that the Munich director appears most closely connected to these crimes. What was the Munich director's "involvement" with this development?

Rüdin's longtime colleague Carl Schneider (1891–1946) was a professor of psychiatry and neurology at the venerable University of Heidelberg. He was also an active "euthanasia" physician who worked closely with the central T4 Berlin office; later he was in charge of selecting patients for the killing operation. It was during the "decentralized euthanasia" period, however, that he developed a research program to establish criteria separating traumatic or infectious (nonhereditary) forms of mental deficiency and epilepsy from those that were genetic. As a "cure or kill reformer," Schneider desired this information to separate out those who would be treated and brought back to useful work from those who would be euthanized. In the case of mentally handicapped children, he could use this diagnosis to encourage families with genetically feebleminded progeny to forgo having more children; those with offspring whose mental illness was not hereditary would be encouraged to increase the size of their families, as the "struggle for survival of the German people" during the war demanded. Schneider undertook his research by bringing patients, primarily children from a nearby asylum, to undergo a wide variety of tests at his University clinic—some of which were quite painful and which in one case led to death. The children were then brought to the Eichberg Asylum near Wiesbaden where at least twenty-one were killed through an overdose of medication. The histopathological dissections of many of the children examined, while alive, in Heidelberg, were brought back and compared to earlier clinical studies. An assistant at the Rüdin Institute, the psychiatrist Julius Deussen, became actively involved in all facets of Schneider's research. His main task was to examine the children's alleged genetic traits; he even toured neighboring asylums to identify potential candidates for the project headed by Schneider.[153]

A member of Rüdin's Department since 1939, Deussen had first trained under von Verschuer at his Institute for Hereditary Biology and Racial Hygiene in Frankfurt.[154] An active Nazi Party member since 1933, Deussen researched "vagabond asocials" while in Munich. Although Deussen, like other young coworkers at the Munich Institute, was conscripted for military duty, he obtained a leave of absence from the army to pursue his science. In November 1943 he joined Schneider's research team. We will recall that Deussen was one of Rüdin's coworkers who defended his boss against the charges leveled against him by Riedel in 1940. Rüdin returned the favor by providing financial support for him during his stay in Heidelberg. The Munich director also recommended him for a professorship there. Deussen's research division, Rüdin believed, could become the center of an institute for genetics and racial hygiene, which did not yet exist at the University.[155]

The Rüdin-Deussen-Schneider connection is not the only evidence of the Munich director's "involvement" with "euthanasia." In 1942 Rüdin was asked by Conti to outline the research priorities in his field, since such work had to be efficiently organized because of the contingencies of war. After listing numerous topics he mentioned the eugenic importance of "distinguishing which children could, already as children, be clearly categorized as so valueless and worthy of elimination that . . . they could be recommended for euthanasia in their own interest and that of the German people." The Munich director viewed this action as a "humane and certain countermeasure to the negative selective processes facing the German national body." "We have no interest," Rüdin continued, "in preserving the life of incurable and ruined victims of hereditary." On the other hand, we should "save what can be saved," at least to preserve their social utility.[156]

In the following year the Munich director, along with the "euthanasia" physicians Nitsche, de Crinis, and Heinze, joined Schneider in crafting a memorandum on the future of psychiatry for Conti and Hitler's personal physician, Brandt. Here we find the "cure or kill" program clearly articulated: "the measures of the euthanasia program will meet more understanding and approval if it is guaranteed and publicly known that in each case of mental illness all possibilities are utilized to cure the patients or at least to improve their condition to such an extent that they . . . are directed into activities that are of value to the national economy."[157] The plan for the future of psychiatry not only emphasized its importance for the well-being of the state. Psychiatry would be the discipline "responsible to educate and train the profession of medicine in general." "Everything must be done," the memorandum continued "to counteract the discrediting of the psychiatric profession which nowadays has occurred so frequently; instead, emphasis has to be placed on pointing out everywhere the importance . . . of psychiatric work."[158] Several months later, in early 1944, Rüdin wrote his friend Nitsche a letter. Among other things, he asked why their colleague Heinze had not yet written an article for Rüdin's *Archiv* legitimizing euthanasia based on his "thoroughly investigated children." Rüdin queried whether material like this could be published.[159]

The facts presented here—some of them seemingly contradictory— raise a host of questions regarding Rüdin's involvement with "euthanasia." His first official endorsement of these crimes comes in 1942, during the period of "decentralized euthanasia," although his unwillingness to take a public stand against the killings earlier would suggest that he at least tolerated them from the outset. The historian, as well as the reader, is left pondering significant queries. For example, was Rüdin part of a larger discipline-wide movement to implement a "cure or kill policy" in psychiatry, or did he align himself with the "reform" "euthanasia" physicians for other reasons? Did the troubles that he faced in his Institute as a result of the SS play a role in his more active support of "euthanasia," as it might have done in his decision to write the 1943 *Archiv* article, "Ten Years of the National Socialist State"? Or, alternatively, did the more radical policy in his Institute predispose him to take eugenics to its logical next step? Was he attempting to preserve his diminished power and prestige in the field of Nazi racial policy? Or, finally, was the Munich director motivated primarily by the additional research opportunities that the "euthanasia" action could provide? After all, as he stated many years earlier, data were "a matter of survival." Whether he was bothered by the fact that "subjects" might not survive the process of procuring this valuable data is not known. In short, Rüdin's connection with the "euthanasia" project cannot be doubted; his precise "involvement" with it and the motivations for his actions are open-ended questions with which one must wrestle.

We cannot end the discussion of Rüdin and "euthanasia" without mentioning how other Departments of the German Research Institute for Psychiatry profited from these medical crimes. The Department of Neuropathology, headed by Willibald Scholz since Spielmeyer's death, used hundreds of brains from "euthanasia" victims for research. There is evidence that the "diseased brains" that found their way to the *Kraepelinstrasse* were not sent entirely randomly. Although it is unlikely that Scholz or any of his coworkers themselves targeted individuals to be killed for research purposes, asylum directors at "euthanasia" facilities certainly knew the research interests of the various neuropathological institutes. Apparently few, if any, of the scientists at the Munich Institute receiving the organic remains of these victims had any qualms about using them.[160]

We have seen how von Harnack's hope that the renowned German Research Institute for Psychiatry would serve as "the starting point of a new epoch in healing" the ill led, after numerous historical twists and turns, to a biomedical center connected to and profiting from a notorious era in killing hospital patients. The Munich Institute, as envisioned by its founder, Emil Kraepelin, was meant to be the vanguard research center in Germany for the scientific study of psychiatric disorders. The Institute's patron, Loeb, had faith that the funds he invested in this new Kaiser Wilhelm Institute would be well spent. Boasting such world-famous biomedical scientists such as the neuropathologist Spielmeyer and the serologist Plaut, the German Research Institute certainly enjoyed an international reputation second to none.

By the late 1920s, the Institute already displayed elements of the symbiotic relationship between human heredity and politics that later became the hallmark of its legacy. Rüdin's investigations into psychiatric genetics were essential scientific resources for a financially strapped government interested in eugenic solutions to social problems. Like his Berlin colleague Fischer, Rüdin had a clear sense of his value for a country already preoccupied with the role of genetics in social disorders; the Kaiser Wilhelm Society was forced to drive a hard bargain to secure his return from Basel in 1928. Rüdin's price: the acquisition of a new Institute building made possible through Rockefeller Foundation support, a well-financed Department of Genealogy, and a handsome salary. What no one was able to guarantee, however, was that his "genetics research factory" would continue to receive the fiscal support that the head of the Genealogy Department believed was his due. By the time he assumed the directorship of the Institute in 1931, Rüdin already faced the fiscal problems that would continue to plague him for the next fourteen years.

With the coming of the "national revolution," the conservative mandarin Rüdin made the first of his two "Munich Pacts" with the Nazi government. This first Faustian bargain between the Munich director and the Nazi state medical bureaucracy in the person of Gütt could not have been more natural or profitable for both parties. Never having to face the political suspicion that Fischer did during the opening years of the Third Reich, Rüdin was easily led to believe that he was a key intellectual resource for the "racial state"; during the prewar years, he never doubted—nor was given any reason to doubt—that he was Germany's academic "court" advisor for Nazi racial policy. His self-fashioning as Nazi Germany's most important racial hygiene expert colored his arrogant, almost brazen, discourse with organizations like the German Research Council and the Kaiser Wilhelm Society. Even more interested in shaping Nazi Germany's genetic future than his rival in Dahlem, he served the medical masters of the swastika in a wide variety of ways, most notably by writing the medical commentary to the notorious Sterilization Law. Evidence suggests that the Munich director ruled like a *Führer* over his fiefdom. He was supportive of those who were useful to him; those he viewed as competitors—either in the scientific or racial policy domain—were treated harshly whenever he had the opportunity to do so. Unlike Fischer, he was not beloved by his coworkers.

Despite his public image as the most pro-Nazi human geneticist and his position as director of an Institute devoted to laying the biological underpinning of Nazi racial policy, he had to become increasingly creative to secure the money for his Department's expensive genealogical-demographic research. This was especially true after the Rockefeller Foundation and the Loeb estate terminated their relationship with the Munich Institute in large measure owing to Rüdin's prominent role in advancing Nazi racial policy. Prior to the war, he could always count on his *Führer* to help him and his Department stay afloat financially. Warnings from all sides to rein in his fiscal appetite counted for little for this man. Rüdin, at the height of his power in early 1939, would soon learn the hard way.

Driven by financial necessity, the Munich director turned to the SS Ahnenerbe just prior to the outbreak of war in the hope of securing badly needed funding. This second Faustian bargain—more so than the first one—was initiated solely by Rüdin. Although the Munich director extended his hand to Wüst and Himmler out of shear desperation, he certainly had no qualms about their cultural and political projects. It was only after they insisted on installing a fifth column of loyal SS researchers hostile to Rüdin and his older coworkers in the Genealogy Department as the price for support that he probably had doubts about the usefulness of this second "Munich Pact" for him. His Institute, never very harmonious from the outset, was now divisive beyond repair. Rüdin was forced to invest much time and political kowtowing to preserving his Institute from a complete takeover by the SS. Sadly for him, his partner in this second Faustian bargain was far more demanding than Fischer's patron, Conti. In fact, it might well have been to counter Conti's growing influence in the health sector that Himmler was so interested in the Munich research center.

Just as in the case of the Dahlem Institute, war had a reciprocally radicalizing impact on the symbiotic relationship between human heredity and politics in Munich. War-related budgetary shortages kept Rüdin tied to the SS even after its deleterious effects on the Institute became obvious. Rüdin's connection to the Nazi "euthanasia" project demonstrates this nefarious radicalized symbiosis most clearly. Comparisons with the Dahlem Institute after its turn to *Phänogenetik* are instructive. As we have seen in the last chapter, von Verschuer and Magnussen were involved in wholly unethical research during the late war years designed to provide a more scientific racial diagnosis to separate out "racial enemies" from members of the *Volksgemeinschaft*. A prerequisite for this work was the availability of a large number of "subjects" incarcerated in Auschwitz who—either dead or alive—could advance the research interests of the Dahlem scientists. In the field of psychiatry, "reform euthanasia" physicians were also interested in research that could distinguish "valuable" from "valueless" individuals—in this case between people suffering from curable and incurable forms of mental disorders. Just as an up-to-date racial diagnosis pursued by the Dahlem team required that clinical observation of the living be combined with information procured from dead inmates at the extermination camp, a determination of treatable or nontreatable (usually viewed as genetic) mental disorders also required that clinical observation of living patients be compared to histopathological findings acquired by dissections. This naturally presupposed the death of the subject. And just as the Dahlem researchers had involuntary subjects at their disposal, so too did the "euthanasia" physicians during the mid-to-late war years.[161] The data obtained from the dead and living subjects of both the Munich and Dahlem scientists would be used against these and other "racial undesirables" to physically eliminate them from the *Volksgemeinschaft*.

We have noted Rüdin's own position on this "cure or kill reform" psychiatry and "euthanasia" beginning in 1942. Although he was not an active "euthanasia" physician like many of his colleagues, the Munich director certainly accepted the measure, as many of his actions demonstrate. Scores of brains removed from murdered "euthanasia" victims found their way to the Institute and were used by the Munich scientists for research. The mutually beneficial relationship between human heredity and politics during the Third Reich reached its ethical low point through unbridled research on subjects without rights, which was accepted and legitimized by Rüdin and practiced by several Dahlem scientists.

We have now examined the unholy symbiosis between human genetics and Nazi racial policy in two institutional settings. By their very nature, research institutes concern themselves with the production of knowledge. In the cases already analyzed, the two most important German research centers for human heredity produced scientific knowledge that became an indispensable intellectual resource for the Nazi regime. The scientists working at these institutes in turn derived clear benefits from selling their "commodity" to state and party medical officials. In addition to the production of "useful" biomedical knowledge, the leaders of the Third Reich were equally interested in its dissemination. Perhaps the act of disseminating human heredity research is nowhere better demonstrated than at professional conferences. In chapter 4, the focus shifts from the production of biomedical knowledge to its diffusion by following Germany's leading human geneticists as they legitimized their regime's racial policies and advanced their own professional interests on the national and international stage.

FOUR

The Politics of
Professional Talk

About a year prior to an important population policy congress scheduled for Paris in 1937, von Verschuer seemed to have a premonition of the ideological challenge he and his German colleagues would have to confront there and at other international conferences. In an article entitled "Rassenhygiene als Wissenschaft und Staatsaufgabe" ("Racial Hygiene as Science and National Duty"), von Verschuer voiced his concerns that he and other German human geneticists might meet with scientific opposition outside of their country. He lost little time articulating a strategy to deal with such a challenge: "The struggle over opinions regarding genetics and race hygiene in the international arena is especially intense," the director of the Frankfurt Institute for Human Heredity and Racial Hygiene reminded his readers. "There are many scientific attempts afloat designed to attack [efforts at] hereditary and racial care in National Socialist Germany. For this reason, the sword of our science must be well sharpened and well guided."[1]

In the past two chapters we examined the symbiotic relationship between human heredity and politics during the Third Reich in two of the most prestigious institutes in Germany, the Kaiser Wilhelm Institute for Anthropology and the German Research Institute/Kaiser Wilhelm Institute for Psychiatry; we saw how the biomedical sciences and Nazi political goals in the "racial state" became mutually beneficial resources. Given the nature of these institutes and their tasks under the Nazi state, their allegiance to the regime's policy goals is hardly surprising. Yet at first glance, there is one important service to the state that might appear unusual: the German human geneticists' use of the "sword of [their] science" as a political weapon at both national and international professional talks.

Ironically, in doing what comes naturally to all scientists—participating in conferences and delivering professional lectures—German human geneticists of all stripes served as the standard-bearers for the regime's political interests in a myriad of subtle ways, in addition to the obvious one revealed in von Verschuer's quote: bestowing international professional legitimacy for Nazi racial policies. We must not forget, however, the dilemma they faced in serving this function. On the one hand, German human geneticists, like their counterparts in other countries, were anxious to travel both at home and abroad, chat with German and international colleagues, keep up with trends in their field, share their research findings, and exploit the prestige that attending such meetings confer. After all, participating in professional conferences is an indispensable part of doing scientific research. On the other hand, while there were undeniable political constraints placed on them, German human geneticists consciously desired to legitimize the political interests of their country, in this case Nazi racial policy. Prestigious scientific organizations like the KWS also endeavored to capitalize upon their members' research in this field. And lest we forget that these researchers were hardly passive pawns in this activity but rather conscious actors anxious to use such meetings for their own professional gain, we need to also consider the extent to which German human geneticists profited from this while simultaneously aiding the national and foreign policy goals of their government.

In exploring these thorny issues, the primary focus will again be on the human geneticists affiliated with the KWS. We will follow them as they offer lectures in Berlin as well as in other German cities with ties to the Society. We will examine the politics involved in selecting or turning down a particular speaker or a researcher's lecture. What did the world-renowned KWS hope to gain from sponsoring such talks? Who attended these lectures, and how were decisions made on who would be invited to such talks? Turning from these KWS-sponsored lectures, we need to ask about the role of biomedical scientists in the international arena—both before and during the war. Prior to the outbreak of hostilities, these men presented papers at numerous internationally respected conferences throughout the capitals of Europe. After 1939, with the absence of any truly international conferences, biomedical professionals engaged in delivering talks at arranged lecture series in occupied, allied, or "friendly" countries. How can we evaluate their activities? National

Socialist Germany was a dictatorship; there were numerous restrictions placed on who could attend such coveted international meetings. To be sure, there were also political expectations made clear to those permitted to attend. But there is unequivocal evidence that most of our German human geneticists went beyond a mere obligation to write a report for the government on the meeting in question; they sought to legitimize the Nazi regime through their professional activities abroad. What is certain, however, is that scientific talks, whether national or international, were overtly politicized during the Third Reich. Similar to the institutional studies previously examined, human heredity and politics became indispensable intellectual, rhetorical, and financial resources for each other at what most imagine is an apolitical venue: professional meetings and public scientific talks.

KWS-Sponsored National Lectures

Long before the advent of the Third Reich, the Kaiser Wilhelm Society spotlighted the research of its scientific members by sponsoring public talks. The opening of a new Kaiser Wilhelm Institute was almost always accompanied by its director holding a lecture to a large, respectable group of citizens and friends of the Society. Such eminent KWS scientists as the chemists Emil Fischer and Paul Ehrlich, the physicists Max Planck and Otto Hahn, the geneticists Carl Correns and Erwin Baur, and the aeronautical engineer Ludwig Prandtl held talks during the Society's early years in the German Empire and throughout the Weimar Republic.[2] This task was made more convenient with the opening of the Harnack-Haus (Harnack House) in Dahlem on May 9, 1929. Named after the first president of the Society, Adolf von Harnack, it was designed not only as a place where the Society could offer lectures to an intellectually hungry elite Berlin audience, but as an international meeting place for prominent German and foreign scientists. It was also a center where scientific members of the Society, especially those whose Institutes were located in this plush district on the outskirts of Berlin, could meet over dinner to discuss their work. One would have been hard pressed to find a place in the Reich capital where so many Nobel Prize winners, industrialists, and top government officials congregated in one building.[3] Although the stately new structure became an intellectual focal point for the KWS, the Society continued to showcase the newest findings of its august body of researchers in other cities as well. Not only could prominent individuals become members of the Society; entire cities could, and did. KWS-sponsored lectures were held in cities that were either Society members or might become such. Hence, from the outset, there was a symbiotic relationship between the professional talks of the Society's scientific members and the financial interests and prestige of the KWS.

The KWS's human geneticists were prominent among the speakers at Society-sponsored meetings. Even before the opening of his Institute in 1927, Fischer held a talk on "Constitution and Race" in the Berlin Palace, the elegant former residence of German emperors and Prussian kings. On December 2, 1932, von Verschuer delivered a lecture accompanied by illustrations in the charming old city of Münster on the hotly debated topic of "The Heredity-Environment Problem in Humans." And Hermann Muckermann, the tireless crusader for eugenics at the KWIA, spoke to a packed audience in the large, predominantly Catholic city of Cologne (four thousand invitations were extended to notable citizens) on January 26, 1933, just four days before Hitler was appointed Chancellor. The title of his talk: "Eugenics in the Service of National Welfare." Even the city's lord mayor and future Chancellor of postwar West Germany, Adenauer, attended, hearing Muckermann speak on the dangers of degeneration and the need to do everything possible to boost the number of "genetically healthy" families. After all, Adenauer was the chair of the Prussian Upper House at the time deliberations for the ill-fated draft voluntary sterilization law reached it in 1932; he and his Catholic Center Party were supportive. The KWS eagerly publicized Muckermann's upcoming talk, since it undoubtedly placed the Society and the research undertaken at the new KWIA in the limelight. Such positive advertising was good for the Society and its scholars.[4]

On the whole, the KWS prospered under National Socialism, even though certain Nazi measures like the expulsion of its Jewish scientists were not always to the KWS administration's liking. Although the KWS did not applaud all the Nazi state demanded of it, there was enough overlap between the Society's largely conservative administration and the new regime to sustain a good working relationship.[5] The Society certainly hoped and expected that the Nazi state would be better positioned to fund the KWS than the former Weimar Republic, especially during the lean depression years. In this the KWS was not mistaken. As Max Lukas von Cranach, an important administrator of the KWS, related to a senator and Nazi Party member of the Society as early as 1934, the money to the KWS "has, as expected, happily increased after the National Socialist takeover of power." He contributed this improved financial situation to the Society's international reputation. Its international prestige, von Cranach continued, contributed in no small measure to making "the

Kaiser Wilhelm Society also so popular in the Third Reich." Nobody could accuse the KWS of mere intellectual masturbation—"operating in thin air"—as might be said of other scientific institutes, von Cranach added. "Our members build a bridge between the German people and research. In this way we have always succeeded in keeping the Kaiser Wilhelm Institutes tied to the *Volk*."[6] Indeed, funding for the Society as a whole increased dramatically under the swastika. What perhaps started out as a pragmatic acceptance of National Socialism by the KWS during the early years of the regime turned into a mutually beneficial partnership later on, especially during the war years.[7]

KWS-sponsored talks continued throughout the Third Reich. We have already examined Fischer's controversial lecture given at the Harnack House on February 1, 1933, and how its political fallout resulted in a denunciation campaign directed against the Dahlem director. This was perhaps not the best publicity for the KWS, and the next time the Society sponsored one of its human geneticists' talks, it made sure that both the topic and its presentation would at least be less controversial for Germany's new political masters. The subject matter needed to hold large public interest—perhaps enough to secure the KWS new financial backers—since the Society was dependent not only on the state purse but private monies. Possibly as a means of throwing an olive branch to indignant Nazi Party dignitaries still reeling over Fischer's ambiguous pronouncements on the race question, the KWS chose von Verschuer to hold a talk. The venue: the medieval city of Nuremberg, site of the Nazi Party's largest rallies. It was scheduled for February 16, 1934. Von Verschuer presented a talk entitled "Paths Leading to the Hereditary Health of the German People."[8]

The groundwork for this lecture was carefully laid. The Duke of Saxony-Coburg and Gotha, a high nobleman and a senator of the KWS, was asked to perform the formalities at the upcoming talk, since Max Planck, the Society's president, could not attend. Months before the lecture, von Cranach was in contact with Nuremberg city councilman Dr. Dürr, a longtime member of the KWS. Von Cranach was appreciative that Dürr would not only find an appropriate site for von Verschuer's talk, but also generate the "necessary propaganda" to secure a large and respectable audience.[9] The raw material for this propaganda was provided in the form of a report on the KWIA's research in general and von Verschuer's in particular. It was approved, if not actually written, by the KWS administration and distributed to the press. As the press release noted, "the *völkisch* state has begun a great task: the biological renewal of the German people. It hasn't limited itself to the health care of the living generation.

It begins from the standpoint of human heredity and racial research and attempts to influence the genetic substrate of a people. The Law for the Prevention of Genetically Diseased Offspring and other laws demonstrate that this task will be carried out vigorously and successfully." Nor did the report omit the defining role of von Verschuer's Institute in its propaganda statement. "The research of the Kaiser Wilhelm Institute for Anthropology, Human Heredity, and Eugenics is of the greatest importance for the scientific underpinning of hereditary and racial care policies," the press release emphasized.[10]

A special prelecture dinner for "leading personalities" was planned for February 15, the evening prior to von Verschuer's talk. Dürr promised the KWS that he would come up with a special invitation list that could be altered as needed. Von Cranach urged Dürr to include distinguished guests "who could be important for us." As an example, von Cranach named the "President of the Reich Railroad and other such people"—potential contributors to the financial well-being of the KWS. Von Cranach also informed the Nuremberg councilman that he planned to invite the German press for tea at his hotel prior to the dinner for dignitaries. This would allow the press's representatives to become better informed about the Society.[11] Through such a strategy, the KWS stood to gain potential new donors and extend its national influence.

There seemed to be a general agreement between Dürr and the KWS administration that in addition to key representatives of German industry, important state and Nazi Party officials must be invited. Among the party elite requested to attend was Julius Streicher, as well as police president and SA Group Leader von Obernitz. Representatives of the German army were also welcomed. Indeed von Cranach specifically wrote Colonel Otto, commander of the Twenty-First Bavarian Infantry Regiment, requesting that he extend his invitation to his entire battalion.[12] Disagreements arose, however, over which human geneticists should be allowed to attend von Verschuer's talk. Rüdin's Institute in Munich was only a few hours from Nuremberg by train. On Dürr's original invitation list was not only Rüdin's name but also those of his coworkers Plaut and Spielmeyer. At the time both were still members of the GRIP. However, the latter two names appear to have been crossed off the list, perhaps by von Cranach, perhaps by someone else. As we saw in our last chapter, Plaut was Jewish; Spielmeyer was married to a woman with "Jewish blood" and was known to be unsympathetic to the new order. By early 1934, German Jewish human geneticists or those with spouses tainted by "Jewish blood" who could not be expected to toe the party line were not welcome guests at a KWS-sponsored lecture—even if they were still scientific members of the Society. The KWS, if not the instigator of the removal of Plaut and Spielmeyer from the invitation list, did nothing to try to keep them on it.[13]

Unlike Fischer a year earlier, von Verschuer hardly compromised the KWS by his public talk. He began his speech by emphasizing that although there had formerly been a one-sided emphasis on population quantity, the situation was now totally different. In today's state, von Verschuer reminded his audience, the "care of the national body through the preservation of its genetic health and its racial characteristics are the content [of population policy]." "With a remarkable sense of its mission," the speaker added, the state was on its way to accomplishing its genetic goals through the Sterilization Law and the Law against Dangerous Habitual Criminals. It is solving its racial problems through the 1933 Law for the Reestablishment of the Professional German Civil Service, which removed Jews and political undesirables from all civil service positions and revoked German citizenship for newly arrived unwanted foreign elements. Von Verschuer stressed that the newest research he and his colleagues were undertaking at the KWIA would serve as "an important basis for all hereditary health measures." "Every person," von Verschuer concluded, "can contribute his part such that the biological renewal of our people becomes a reality."[14] Von Verschuer's talk quite unambiguously supported the new racial order. However, it simultaneously served a rhetorical and financial resource for the KWS.

As planned, the Duke of Saxony-Coburg and Gotha gave the closing remarks following von Verschuer's talk. He reminded the audience that the KWS's existence was made possible by the goodwill of its members and the entire German people. In his last statement he thanked the audience for supporting the "quiet academic work" of its members, such as von Verschuer. Indeed, in a sentence crafted with the prestige of the KWS in mind, he assured all in attendance that "German science plays a leading role in the reconstruction of our Fatherland under the direction of our People's Chancellor, Adolf Hitler."[15]

On February 15, 1935, Rüdin delivered a KWS-sponsored lecture in the city of Weimar. The National Assembly that drew up the Weimar Republic's constitution also convened in this popular city of Germany's most illustrious poets, Goethe and Schiller. As was the case with von Verschuer's talk, the Society used it as an opportunity to gain recognition for itself. Dr. Karl Astel, a fanatical Nazi and head of the Landesamt für Rassenwesen (State Office for Racial Policy) in Thuringia, was scheduled to give an address that evening. His Office was cosponsor of the event. With his unwavering concern that the KWS's interests were not neglected, von Cranach wrote Astel asking if the Duke of Saxony-Coburg and Gotha could give a three- to-five-minute speech prior to Rüdin's talk. He also requested that Astel provide details on the size of the room in which Rüdin would deliver his lecture. This time, however, von Cranach did not say that the size was important for attracting potential KWS donors; instead he was eager that the event "place great emphasis on familiarizing the broadest possible number of people with [the Society's] efforts." Avoiding any semblance of elitism or class prejudice—a tactic pursued by the Nazi Party in order to create the illusion of a unified *Volk* devoid of all class divisions—von Cranach emphasized the special nature of the KWS to Astel, as he did to Nazi Party leader and KWS senator Otto a year earlier. As von Cranach reported, since the founding of the Society in 1911, the Society has sought to "build a bridge between pure research and working people."[16] It was critical that Nazi Party officials who had an important say in whether a Society member's lecture could be held have no reason to prevent the KWS from spotlighting its work for the nation.

Apparently this tactic worked, since the Duke delivered an introductory talk prior to Rüdin's lecture. There was a decidedly *völkisch* flavor to the Duke's speech. He was not above mentioning that KWS scientists were following the hope of the future leader of the German Air Force, Hermann Göring, when the latter stated that "it is the most important prerequisite for [the KWS'] effectiveness that it is completely one with the feeling and thinking of its people." The Duke and KWS senator once again reminded his audience that "for the reconstruction of our Fatherland, we need science and research more than ever." "German comrades," he continued, "to make you more familiar with the goals of the Society, is the goal of this evening. We first, however, wish to thank our *Führer* and Chancellor as well as the government for the support of German science, whose international standing should not only remain stable but advance. Heil Hitler!"[17] Obviously, Nazi rhetoric had become an effective linguistic resource for the KWS to advance its goals. Its prestige, which von Cranach unabashedly flaunted in a letter to Astel, also lent an air of scientific respectability to the Thuringian Racial Policy Office president's aims.

For Rüdin, who lectured on "Predictions of Hereditarily Diseased and Normal Children," the evening was a chance to once again demonstrate the usefulness of his Institute and assert his personal loyalty to the regime and its leader, as well as legitimize its racial policies, especially in the area of sterilization. "Our *Führer*, as an ingenious political pioneer, has a watchword," Rüdin proclaimed, "help bring the healthy, talented and pure race to victory." "Government, people and science are turning their attention to the welfare of the healthy of the coming generation," the

Munich psychiatric geneticist continued. Since it is difficult to discern who is truly of "good race," the speaker argued, his own research tool, the empirical hereditary prognosis, is a critical instrument serving the state's policy needs. Children of the mentally ill are 10 to 60 percent more likely to become sick than those of a normal population, Rüdin maintained. Moreover, "according to the newest research," 40 percent of the children of the feebleminded are abnormal. On the basis of 250 epileptic twins, the GRIP director emphasized, the hereditability of epilepsy in their progeny could be demonstrated almost 100 percent of the time. As such, "the Law for the Prevention of Genetically Diseased Offspring is well-supported with trustworthy numbers [based upon] genetic prognoses." Although Germany still had a way to go before reaching a final consensus, Rüdin admitted, "science is on the way to establishing a usable racial hygienic step ladder of the genetic worth of individuals." "The socialist organization of the support for the individual depends upon how, through racial hygiene, a people eliminate its unhappy ballasts through the prevention of the reproduction of its genetic defectives," Rüdin concluded.[18] Whatever nonideological motivations might have moved the Munich director to give such a talk, there is little doubt that his language was harsh. Like his talks examined in the previous chapter, this one may also have paved the rhetorical path leading to something far more sinister than Nazi Germany's draconian Sterilization Law: its "euthanasia" project. As we have seen, Rüdin was not detached from the "euthanasia" initiative after it was initiated by the regime.

Fritz Lenz, head of the Eugenics Division of the KWIA after Muckermann's removal, was also invited by the KWS to present a public lecture at the Harnack House. Interestingly, Lenz appeared unsure about how scholarly his talk should be. Fischer, his boss, urged him not to tackle a difficult scientific problem. Instead, he should offer a topic of current concern. Remarking to President Planck that his research field offered a good opportunity for such a nationally relevant talk, Lenz held his lecture on February 20, 1935. The title: "Problems of Practical Race Hygiene."[19] Like Rüdin after him, Lenz stressed the importance of the Nazi Sterilization Law as a negative eugenic measure. He argued that if the Law for the Prevention of Genetically Diseased Offspring were vigorously applied, the number of feebleminded could be cut in half in a matter of years. But Lenz's real interest was in positive eugenics: the increase of the so-called fitter elements of the population. In his lecture he offered a number of suggestions for how this might be accomplished. Among the several points he raised in his talk, Lenz championed the Nazi law that limited the number of people who could pursue advanced degrees (in reality, impacting Jews, "Aryan women," and politically unreliable individuals). Not only would it reverse the dangerous trend that resulted in "fitter" people having fewer children; it would also prevent the decline of talent in academia. Rhetorically capitalizing on the National Socialist view that character was more important than dry intelligence, this most intellectually elitist of human geneticists conceded that although scientific talent appeared to have recently declined in the universities, the "selection of character" has become better. This could be gleaned, Lenz reminded his audience, in "the achievements of the young academic generation at Langemarck [a village in the Flemish part of Belgium where German troops attempted to take back ground from the British and French in 1917 during World War I], in the quelling of the communist revolt in the post–[World War I period in Weimar Germany] and in the struggle for the National Socialist idea."[20] Lenz ended his talk with an interesting, if not uncontroversial, point: "selection of the highest grade racial elements cannot be achieved with reference to external traits. One champions the culturally productive elements most effectively," Lenz asserted, by making "efficiency and actual accomplishments in the service of the entire people and culture" the measure of fitness.[21] Here Lenz is taking issue with the crude racial-anthropological views of some leading Nazis who believed that only those who looked "Nordic" were worth selecting. He was opposed to creating a hierarchy of "Aryan Germans" based on physiognomy. This would be divisive and would not strengthen the people's community. Hence, Lenz's public speech at the Harnack House deftly incorporated party language and sentiment while simultaneously deviating from it; his talk was shaded gray. Whether the speaker felt he was upholding his intellectual integrity by offering this gray-toned speech is unknown.

Although Lenz's 1935 public lecture was an important event for him, the KWS, and the Nazi state, it paled in comparison to his Hamburg talk delivered a little more than a year later. A Hanseatic port on the North Sea, Hamburg had traditionally been a city of merchants made rich through trade. The KWS appears to have had good connections in this old city, and undoubtedly it hoped that a talk focusing on whether heredity or environment dominated in human traits might draw a large crowd, hence bringing the Society new financial backers and additional party support. Corresponding with Planck, Lenz suggested this particular topic rather than a more popular one. He believed it might reflect better on the work undertaken in the KWIA. Planck agreed with him. He also added, however, that it was important for the lecture to offer something for the specialist.[22] Perhaps the president hoped to offset some of the wilder notions of racial hygiene held by certain less academic Nazi officials; maybe he wished to uphold the scientific integrity of the Society. Or possibly he believed that those most likely to financially support the Society would prefer a scholarly talk in the long-standing tradition of the KWS. At any rate, this public lecture was important enough for the aging Planck to preside over it. The talk came on the heels of the twenty-fifth anniversary of the Society's founding. Nothing about the lecture was left to chance.

Like other KWS-sponsored public lectures, von Cranach asked the speaker to write something about his Institute and himself that would be rewritten into a press release for the talk and the Society.[23] What is particularly interesting about Lenz's speech is that we have his own appraisal of his Institute as well as his impact on Nazi racial policy along with its reformulation for the press by von Cranach. These two documents demonstrate how language was renegotiated for political purposes.

Lenz was absolutely candid in his draft to von Cranach. After explaining the nature of the KWIA and his own research, including the new clinical work that he assumed after von Verschuer's departure to head his own institute in Frankfurt, he touched upon his duties for the state. Lenz reminded von Cranach that he was a member of the Reich Ministry of the Interior's Expert Committee for Racial and Population Policy. But the eugenicist lamented that although he penned numerous testimonials on practical racial hygiene measures, "few have found their way into the legislation" and none have been openly discussed. He admitted that he was "all but uninvolved" in any racial laws presently on the books. Lenz was, however, politically astute enough to recognize that it would probably be unwise to emphasize this in any press release.[24]

If we look at the reformulated press release written by von Cranach we find an even greater emphasis on the role of the KWIA for Nazi racial policy than in Lenz's draft. Indeed, parts of von Cranach's version appear to be little more than a KWS attempt to demonstrate the KWIA's usefulness for the biological policies of the regime. After explaining when and why the KWIA was established, von Cranach commented that when it opened its doors, "no person could have imagined that the research direction of the new institute would, in so short a time, become the center of interest of the entire *Volk*." Clearly, he continued, no one could have anticipated that it would be used so heavily as an intellectual resource by the state to formulate its population policy. In his final paragraph—one completely absent in Lenz's draft—von Cranach stressed that "owing to the enormous growth in the importance of genetic and racial teaching in the new Reich," teaching duties for physicians and pastors along with public talks have been added to the KWIA's "real task": research. "As such, this Institute of the KWS stands in a very special close relationship with the *Volk* and State," he concluded.[25] When one compares the two versions of the press release, there can be little doubt that the Society, perhaps more than Lenz himself, was interested in milking this upcoming event for all it was worth politically.

As in earlier KWS-sponsored talks, the invitation list for the lecture, the specific venue, and the honored guests invited to dinner before the talk were negotiated affairs. Both the KWS administration and state and party officials in Hamburg wanted to make sure everything went off without a hitch. An important city councilman, Dr. Lindemann, had the largest hand in drawing up the list of notables who would dine in the exquisite restaurant in Hamburg's finest hotel, the Vier Jahreszeiten (the Four Seasons). A high-ranking official for scientific and educational affairs in the city, Dr. Witt, suggested to von Cranach that the original invitation list, with its representatives from the army, SS, SA (*Sturmabteilung* or Storm Troopers), party, and Hamburg government, lacked an appropriate number of important representatives of industry—something not irrelevant to the KWS. He also queried whether teachers and members of the university community should be included. Von Cranach, of course, insisted that members of the Society in Hamburg and the surrounding areas receive an invitation. By the time all interested parties finished their deliberations, over two thousand invitations were extended to Lenz's talk.[26] The dinner preceding the lecture must have been deemed exceptionally politically and socially important for the KWS, since detailed seating arrangements were drawn up. Indeed von Cranach was so concerned about who sat next to whom that he requested Lindemann to inform him if the mayor of Hamburg was planning to attend the dinner. In the event of the mayor's absence, his representative had to be placed on the right of side of President Planck. The highest ranking general of the army in attendance, von Cranach stated, should be seated on Planck's left side. If he was unable to attend, it was imperative that the next oldest general sit to the left of the Society president.[27] Such formalities were not just a matter of civility; they were an exercise in politics and power.

Judging from the numerous newspaper reports, Lenz's talk, though scientific rather than popular, was a big success. Indeed, so many people accepted an invitation to attend the public lecture that it was necessary to use the city music hall to accommodate all the guests, rather than two large halls at the Hamburg University, as originally planned. According to one newspaper report, even the large music hall auditorium could barely hold all who chose to attend the nearly two-hour lecture. Every seat was taken. President Planck reminded the packed hall of the "tremendous importance of pure research," the kind of research that the KWS had always been renowned for. Although it often did not immediately lead to practical applications, Planck continued, without such research, "any progress and any development in our culture would be unthinkable."[28] With those words he introduced the evening's speaker, Lenz.

If one examines the language used by Lenz in his talk, one finds that he was far more reticent than Rüdin had been to exaggerate the exactitude of genetic knowledge that scientists acquired from human beings. He spoke at great length on how difficult it was to predict with scientific certainty whether a trait was really genetic—certainly not something that was music to the regime's ears. This is because one can only deduce hereditability of traits from experiments on plants and animals; such experiments on humans could never be carried out, Lenz maintained. Perhaps even more politically controversial was his statement that human genetics could not count on the records of the medically staffed Hereditary Health Courts to implement the Nazi Sterilization Law. Unfortunately, such records were "too full of holes" and did not constitute a representative sample of all defective people in the nation.[29] Fischer, Rüdin, and von Verschuer, it will be recalled, had been given special access to such records for their research by Arthur Gütt. And Rüdin, in his own KWS-sponsored talk, had all but announced that enough genetic knowledge was available to support all Nazi racial policy measures. But Lenz's words of caution should not be construed as a critique of sterilization. As the Eugenics Division head put it, whether a disease like "feeblemindedness" is genetic or acquired is really immaterial. "In either case," Lenz reminded his audience, such a defective person was "unfit for parenthood," and it was certainly not in the interest of society that he or she reproduces. At least one of the papers covering Lenz's talk ended its column by noting that achieving an ever more reliable assessment of the heritability of traits would be accomplished through the ideological push provided by the Nazi worldview.[30] Again, the nationalistic and elitist Lenz offered his listeners a talk that served the interests of his employer, the KWS, but could only be seen as ambivalent from the standpoint of Nazi enthusiasts. On the one hand, he was indirectly advocating that a larger number of people be robbed of the right to reproduce; the Law for the Prevention of Genetically Diseased Offspring, as draconian on paper as it was in practice, only denied parenthood to those suffering from certain so-called genetic disorders. On the other hand, Lenz indirectly reminded his audience again that the scientific basis of much of racial policy might be on shakier ground than most thought. Perhaps he believed that if he emphasized how much work in human genetics was left to be done, more money might flow into the coffers of his Institute. At any rate, since there is no indication that he suffered any political repercussions from this lecture, we must assume that the good points of the talk outweighed the bad. We do not know whether his critical statements on the state of knowledge of his field bothered other German human geneticists such as Fischer and Rüdin.

The KWS's attempt to organize public lectures for the Society's human geneticists was not without its amusing episodes. The politics of professional talk took on an almost comical air in the case of a talk scheduled for von Verschuer in 1936. As we have seen, by February 1936 von Verschuer already headed his own human genetics and racial hygiene institute in Frankfurt. It was viewed as a model, and its existence and work were reported internationally. Von Verschuer remained, however, a scientific member of the KWS. As such, it was no problem for the Society to tap his growing reputation in the field of human heredity and medical genetics. Von Verschuer anticipated holding a talk in Saarbrücken, capital of the Saarland, entitled "Hereditary Talent and Genetic Taint."[31] The Saarland, taken from Germany and put under control of the League of Nations by the Treaty of Versailles, was returned to Germany on January 13, 1935, following a plebiscite. In the wake of a heavy propaganda campaign conducted by the Nazi state, over 90 percent of those living in this industrial region bordering France and Luxembourg voted to once again become part of the Reich. One might imagine that the newly reincorporated region would welcome the KWS and one of its most renowned researchers in the field of human geneticists with open arms. Yet sometime between September 1935 and early February 1936, the ambitious District Leader and Reich Commissioner of the Saarland, Josef Bürckel, decided to refuse to allow the lecture to take place. Although there was nothing the KWS could do about this decision, it used its influence to try to find out the reason for this refusal. After writing to its contacts in the Reichskanzlei (Reich Chancellery)—a testimony to the Society's good connections in the highest places—von Cranach received word on July 31, 1936, from a representative of the Ministry of the Interior containing an explanation from Bürckel.[32] Apparently, he forbade the talk because when the request for it reached his desk, a popular article from "reactionary" circles honoring the former kaiser, Wilhelm II, caused trouble for the Nazis in the region and angered party members. Bürckel admitted that he failed to recognize that the lecture was sponsored by the "scientific institute" known as the KWS. He assumed that von Verschuer was someone wishing to speak for some promonarchy "Kaiser Wilhelm Organization." This Bürckel could naturally not allow. Moreover, the District Leader emphasized, the request for permission for von Verschuer's talk came not long after the Saarland was reunited with the Reich. There were so many people and organizations wishing to capitalize on this occasion—the first new territory gained by the Nazi government after the hated Versailles Treaty—that even party organizations were refused permission to hold rallies there, Bürckel claimed.[33]

This comedy of errors notwithstanding, there are several interesting things to note. First, by 1936 the party obviously had the last word over whether any national talk could be held. That having been said, the KWS was obviously important enough to expect and receive an explanation from high-ranking political organizations and personalities for this seemingly arbitrary decision. To make sure that party members recognized the name of the Kaiser Wilhelm Society in the Saarland in the future, von Cranach sent literature on the KWS to an official of the Saarbrücken city government. Von Cranach told him that he would be obliged if the official could enlighten the entire Saar region to the nature of the Society.[34] And finally, by the time it learned the reason for the refusal, the KWS had already decided to hold von Verschuer's talk in a different industrial region—again in the hope of gaining financial backing. In a letter to the treasurer of the KWS and Ruhr steel magnate Dr. Albert Vögler (in 1941 Vögler would be appointed president of the Society), von Cranach suggested that von Verschuer's lecture be delivered in Wuppertal, in the steel and coal-producing Ruhr region. The city of Wuppertal was not yet a member of the KWS, von Cranach admitted, but "maybe it could be wooed on this occasion." Surprisingly, there were not many cities in the Ruhr that were presently members of the Society, von Cranach mentioned to Vögler. The KWS treasurer thought holding the lecture there was a good idea, although he admitted that he didn't have many connections to the city. Ultimately, von Verschuer's public lecture was held in Remscheid, a Ruhr city that was already a Society member.[35]

Throughout the Nazi period, the KWS continued to sponsor lectures delivered by its scientists in the Harnack House in Berlin as well as in other cities. These professional talks did not cease with the outbreak of war, even if the Reich Minister of Interior did remind the KWS that in sponsoring lectures dealing with population in the East, great caution was in order. "Backlashes and misunderstandings" could result.[36] That highly contentious talks were held under the Society's banner during wartime can be gleaned from a memo from the SS dated October 22, 1941. It lists six KWS-sponsored lectures for which it requested invitations in winter 1941–42. Among the talks of interest to the SS was Fischer's lecture on "Problems and Tasks in White Africa" scheduled for January 13, 1942.[37]

26 The Ninth Meeting of the German Society for Physical Anthropology, Tübingen, 1937. During the conference, the participants voted to change the name of their organization to the German Society for Racial Research. The young Josef Mengele is pictured in the second row on the extreme left. In the middle of that row we find the elderly, white-bearded Alfred Ploetz. Standing second and third to the right of Ploetz are Fischer and von Verschuer. Photo courtesy of the Archiv der Max-Planck-Gesellschaft, Berlin Dahlem.

Fischer responded enthusiastically to a request by the general secretary of the KWS, Telschow, to hold this talk. He replied that "considering the future colonial problems, the topic was . . . especially relevant."[38] If the Society did not merely wish to "survive" the Third Reich but to flourish within it, the role played by KWS-sponsored talks to confer prestige on the organization and to help fill its coffers was significant. It goes without saying that such national talks were also professionally useful to the researchers who delivered them.

German human geneticists not only spoke at KWS-sponsored lectures. Prior to the Third Reich, they also presented papers at national professional conferences. To give but one example, in 1930, Rüdin held a talk in Munich entitled "Means and Goals for the Biological Investigation of Criminals with Special Consideration of the Role of Heredity," at the German meeting of the Gesellschaft für Kriminalbiologie (Criminal-Biological Society).[39] Such activities continued after the Nazi takeover. We know that after 1933, Fischer gave talks at such established venues as the Deutsche Gesellschaft für Innere Medizin (German Society for Internal Medicine), the Berliner Akademie für medizinische Fortbildung (Berlin Academy for Continuing Medical Education), and even the esteemed Preußische Akademie der Wissenschaften (Prussian Academy of

Sciences).[40] His colleagues did the same. At a meeting of the Deutsche Gesellschaft für physische Anthropologie (German Society for Physical Anthropology) held in Tübingen in September 1937, von Verschuer, Fischer, and Alfred Ploetz made their presence felt. According to a report in von Verschuer's journal, *Der Erbarzt*, by the then young and largely unknown human geneticist Josef Mengele, these men and their colleagues decided halfway into the conference to change the organization's name. The new name, the Deutsche Gesellschaft für Rassenforschung (German Society for Racial Research), was certainly a professional title that reflected both the research of these men and the tenor of the times.[41] Yet it is in the international arena, at foreign-held professional conferences and lecture series sponsored in neutral, occupied or "friendly" countries during the war, that one can best see the symbiotic relationship between human heredity and politics under the swastika.

International Professional Conferences and Lecture Series

As was the case for KWS-sponsored and national conferences, Germany's community of human geneticists frequently attended international meetings in their field. As we have seen, Fischer, von Verschuer, Rüdin, and others took an active role in international IFEO conferences during the 1920s and early 1930s. Recall that prior to the Nazi takeover, Fischer, with his American colleague Davenport, was anxious to influence Mussolini on population policy. They eyed the 1929 Rome meeting of the IFEO as a unique opportunity to have a concrete political impact on the new fascist government in Italy.

Although German human geneticists, like their non-German colleagues, used international conferences to push their eugenics agendas, the rules for attending such meetings changed for the Germans dramatically after 1933. Under the Nazi regime the mere desire to participate in an international meeting was no guarantee that a scientist could do so. There were numerous political and economic restraints placed on those who wished to make their mark in the international arena. As strange as it may sound to us today, foreign professional meetings during the Third Reich were viewed as a venue for airing National Socialist foreign policy concerns. German human geneticists knew this and still gladly participated in them.

During the prewar years, there were initially three state organs that were involved in the politically and fiscally delicate operations of coordinating German scientists' attendance at professional conferences abroad:

Konstantin von Neurath's Auswärtiges Amt (Foreign Office), Bernhard Rust's Reichsministerium für Wissenschaft, Erziehung und Volksbildung (Reich Ministry of Education), and Josef Goebbels's Reichsministerium für Volksaufklärung und Propaganda (Reich Ministry for Popular Enlightenment and Propaganda). Naturally, there were conflicts of interest among the various state and party organs that believed international conferences fell under their bailiwick.[42] As we have mentioned, the Third Reich was not a monolithic entity where Hitler made all the decisions. Yet there does not seem to have been any appreciable difference in the attitude of the major state organs toward international scientific conferences, although Goebbels's ministry seems to have been most vocal in its demands on those attending.

The Deutsche Kongress-Zentrale (DKZ) (German Congress Center) was established as a division of Goebbels's Ministry of Popular Enlightenment and Propaganda in 1934—probably owing to Goebbels's feeling that his voice was not being heard regarding the propaganda value of international conferences. Beginning in 1936, those seeking to attend international conferences needed the DKZ's approval. From then on, it was responsible for questions of hard currency.[43] Hence all applications made by individual scientists or institutions in the name of their researchers, like the Kaiser Wilhelm Society, were dependent on this office to get the needed money to attend an international conference. From the earliest days of the Third Reich, Germans attending such meetings were organized as delegations with a "delegation leader" at the helm. In addition, delegation leaders and the researchers under them were expected to meet with official German representatives in the country where the conference was being held. And finally, all scientists who traveled abroad were required to submit a report upon their return home.

The DKZ pulled no punches about its view of such international conferences and its demands of delegation leaders at such meetings. As the DKZ's *Richtlinien für die Leiter Deutscher Abordnungen* (*Guidelines for Delegation Leaders*) points out, a delegation leader must understand that his task is not merely a professional one toward his scientific specialty. Rather, he must view it as "political or cultural-propagandistic pioneer work in the sense of German world prestige." "Our present view of international congresses," the *Guidelines* continued, "differs decidedly from earlier, more traditional views." Moreover, the DKZ emphasized, "congresses are one of the most effective weapons in the struggle against poisoning the minds of people; in this manner we can, through efforts and personal impressions, eliminate prejudices and hateful lies without recourse to direct political propaganda." Complaining that about 75 percent of all international conferences were held in Paris or Brussels, the *Guidelines* argued that Germany should take its cue from France and consciously recognize that such meetings served as a form of cultural propaganda that "in the hand of the statesman can be used as an unrivalled political weapon." Declaring that Germany play "a leading role, if not *the* leading role" at these international meetings, the head of the German delegation and the scientists under his leadership must do all they can to bring this about. Among other things, this included that the delegation leader bring those under him to a "unified group with one will."[44] Moreover, special attention must also be given to questions at conferences touching such politically sensitive issues as "racial hygiene, sterilization [and the] Jewish Problem." Delegation leaders were instructed to answer these queries in an objective manner and directly rebuke any attempt at a critique of National Socialist racial policies. And finally, the heads of the German delegation had to recognize that permission to speak at such conferences was an internal affair. The Nazi regime would decide who could represent Germany internationally.[45]

Under a section of the *Guidelines* entitled "It must not happen that . . .," the DKZ clearly articulated several taboos for international conference etiquette, including the significant point that a German scientist should never contradict another in matters of Nazi ideology. Members of the German delegation found to be a political liability, despite having passed the political litmus test for attending such conferences, were to be sent home immediately. In addition, the DKZ stressed the need for the heads of German delegations to international conferences to deposit invitations, memoranda, and anything distributed there in a DKZ archive created specifically for this purpose.[46] And finally, as a last word of advice to all those traveling abroad, the DKZ offered the following: "remember, that in the eyes of foreigners, you represent Germany; for you there is no longer a 'private sphere.'" They must recognize that "your main task is to represent the interests of Germany in a worthy manner and through your presence to acquire prestige for your Fatherland."[47] The language of these commandments could not have been plainer.

Although none of the other state offices involved with foreign scientific conferences were as explicit as the DKZ, the reports submitted by Fischer, von Verschuer, Rüdin, and other German human geneticists suggest that these and other official regulations were heeded. We know that the content of the reports must have been fairly accurate since frequently important members of the Nazi Party, such as Walter Gross of the Racial Policy Office, attended international genetics conferences to keep a watchful eye over the behavior of German biomedical scientists.

One of the first international meetings relevant to German human geneticists—as well as to Nazi racial policy—was the first Internationaler Kongress für Anthropologie und Ethnologie (International Anthropological and Ethnological Congress) held in London between July 20 and August 4, 1934.[48] Despite the serious economic difficulties facing the Reich, the Foreign Office appointed Fischer as head of the delegation; travel for German scientists was granted.[49] What is especially noteworthy about this meeting is that it was held during the interparty and state denunciation campaign against Fischer. We will recall that both jealous colleagues and important Nazi Party officials felt that this "non-Nazi" racial scientist wielded far too much power. As it might have caused international repercussions to remove him as delegation leader, Gross did not insist on Fischer's dismissal. He did, however, accompany the delegation leader.[50]

In his obligatory report, Fischer stressed that over a thousand scientists from forty-nine countries attended the Congress. He also dutifully reported that there were a small number of Jewish immigrants at the meeting, but "they did not make a display." According to the KWIA director, one Englishman had the audacity to make a "tactless" remark. He claimed that a famous philologist named Müller had demonstrated that the term "Aryan" did not designate a race but a linguistic group. Moreover, according to this same expert, the original "Aryan culture" was mixed with Near Eastern, Semitic, and other ethnic groups. Although the Englishman did not specifically attack Nazi racial policy, Fischer complained to the general secretary of the Congress about this ostensibly impertinent comment. Fischer stated in his report that he would see to it that this "tactless" speech be shortened in the meeting's official report.[51] Most interesting, however, was Fischer's emphasis on the honors bestowed upon him at this prestigious meeting. As Fischer stated in his official report to Reich Minister of Education Rust: "I received the assignment to greet all of the universities and societies from every country in the name of all the representatives of every nation [attending the Congress], to thank Prince George and the President [of the Congress], and to wish it success. There were no other speakers. In addition, I was invited to dinner at the House of Lords by the Earl of Onslow," with only twenty other important individuals invited to attend the dinner. Fischer was also proud that "several Germans took turns chairing sessions with [delegates from other countries]." It would have been a grave mistake, Fischer concluded, had the German delegation not been so heavily represented.[52]

From Fischer's report and his actions at this conference it is clear that he served his own interests as well as those of the state by attending this

meeting. He also demonstrated his loyalty to the regime during his denunciation campaign and pointed to his value as a scientific resource for
the Third Reich. The KWIA director clearly proved that it was advantageous for the National Socialist state to retain him as head of the German
delegation. Moreover, it was probably not accidental that the denunciation campaign against Fischer stopped soon afterward. As Fischer later
stated in his unpublished autobiography with regard to his need to "walk
on eggshells" at this particular gathering: "I remained scientifically objective and all went splendidly; there were no disturbances, and I received
applause."[53] We must question, however, whether a scientist suspected
by Nazi officials would behave any differently or write a report that would
significantly vary from the one he submitted. That Walter Gross kept an
eye over the delegation in London may have further induced Fischer to
compliance at a time when his professional stature was in jeopardy.

In 1935, the International Union for the Scientific Investigation of
Population Problems (IUSIPP) hosted its World Population Conference
in the capital of the "new Germany." Fischer was appointed acting scientific president of the Conference by the government. Although some
members of the IUSIPP had reservations about holding the conference
in Berlin after the "Nazi seizure of power" (but before the passing of
the Nuremberg Laws), individuals of the stature of American geneticist
Raymond Pearl believed that a "proven and broadminded scientist like
Eugen Fischer could guarantee the scientific neutrality of the conference."[54] Suffice it to say that the Conference was truly a propaganda
showcase for Nazi racial policy in every sense of the word. The accomplishments of the young National Socialist state were lauded in virtually
every German scientific paper, as well as in Fischer's opening remarks.
Fischer spared no words slavishly praising the achievements of Hitler for
all his great work in the field of racial hygiene:

We are full of prideful joy in the knowledge . . . that our government, especially our
Führer and Reich Chancellor Adolf Hitler, has recognized this deep and far-reaching
meaning of the science of population policy, and that he has the will to take the consequences [from this knowledge]. As such, we should begin our work today by honoring the man whose strong hand has the desire, and God-willing, the energy, to turn
the German people around from the population policy fate that led past cultures and
people to their demise. I hope and wish for the same for all state leaders and governments of all other nations and people. In this wish for all—as we gather together on
German soil and in the Reich capital—let us respectfully pay homage to the *Führer* and
Reich Chancellor of the German people. . . . *Heil, Hitler.*[55]

Von Verschuer, Fischer's protégé, also reiterated the importance of human heredity for National Socialist racial policy in his own conference talk, "Genetics as the Foundation of Population Policy." Numerous other German human geneticists held lectures at the conference, including Rüdin. As such, the Berlin meeting also shed a positive light on the KWS since so much of the new regime's racial science was undertaken in the Fischer and Rüdin institutes.

Perhaps less obviously propagandistic is the way important state officials like Reich Minister of the Interior and Honorary Congress President Wilhelm Frick used the conference for Nazi foreign policy ends. To be sure, Frick, like most of the German researchers in attendance, laid bare the biological vision of Nazi politics. He stressed the importance of a specifically racial population policy. "What sense does foreign, financial or economic policy make," Frick remarked, "if people are destroyed racially?" But then he went further. Frick not only tried to deflect from the negative image of Germany's sterilization policy abroad, insisting that it was merely "an emergency measure . . . to banish the acute danger for the time being." He also argued that following the logic of race hygiene and population policy, National Socialists must be enemies of war. "The German people," Frick continued, "wish for nothing more than to maintain their own population within the framework of other nations and to contribute their share to the further development of human culture and civilization."[56] When one considers that by this time Germany was already in the throes of its illegal rearmament program and that it was one of Hitler's conscious foreign policy strategies to present himself as a man of peace, one realizes how such a conference could, and did, serve Nazi goals in the international arena. Perhaps Frick was influenced by the views of Rüdin. The Munich psychiatric geneticist gave an interview to the Nazi press with the title "What Is Racial Hygiene?" just several weeks prior to the Berlin Congress. Like Frick, he also stressed that German eugenicists desired nothing more than peace on earth. War, Rüdin declared, only destroys the "fittest" people. This report came about a year after one appearing in the *Völkischer Beobachter*, which also championed racial hygienists as opponents of war. It was based on a resolution made at the IFEO conference held in Zurich in July 1934—a meeting presided by Rüdin and attended by numerous German human geneticists.[57]

Fischer's success at hosting the World Population Conference at home ensured his appointment as delegation leader when it was held in Paris two years later. Recognizing his scientific value to the regime, he requested a large sum of money for forty scientists, including junior people and spouses of some of the researchers. Fischer stressed the cultural-political importance of a large German delegation to attend a conference in Paris—where the political situation would be far more delicate than the Conference he hosted at home.[58] Indeed, according to the delegation leader, "even women could be used to promote the German cause at such occasions"—a suggestion that was endorsed by the DKZ when it wrote its *Guidelines* a year later.[59] In a letter to his trusted friend von Verschuer, Fischer strongly encouraged him to make the journey to Paris; it was all well and good that Nazi officials appear, but it was also necessary that "prestigious representatives of science attend."[60] A conflict with those foreign human geneticists who did not see eye to eye with the German delegation was expected in Paris.

Owing to illness, Fischer was unable to attend, although he had made all the necessary preparations for this important meeting. In his place, Rüdin was appointed substitute German delegation leader.[61] The expected ideological conflict in Paris—where the Germans were unable to control events as they did in Berlin—was not slow in coming. Although the details cannot concern us here, suffice it to say that three Jewish scientists, including the renowned cultural anthropologist Franz Boas (who emigrated from Germany to the United States in the late nineteenth century) questioned the importance of genetics as the determining factor in such traits as intelligence and denied that a country's intellectual development was dependent upon its inhabitants' race. Moreover, Boas and his like-minded colleagues argued that the individual's or group's environment largely shapes so-called racial traits. Boas subsequently presented his views in a French publication entitled *Races et Racisme (Races and Racism)*. He could hardly have expected to get a hearing in a German professional journal at this time.[62]

In their reports of this conference, both von Verschuer and Rüdin emphasized how they brandished "the sword of [their] science" to refute the claims of the Jewish participants. Von Verschuer accomplished this by stressing that these speakers were not in step with the newest hereditary research. Ernst Rodenwaldt, another member of the German delegation, labeled the criticism of these Jewish scientists "rabbinical." They had nothing to do with customary scientific discussions found in European science, he added. Moreover, von Verschuer pointed out that German racial legislation did not aim to assign a "value" to individual races. Allegedly, Germany was only interested in protecting "its own people from an infusion of completely alien racial elements."[63] Von Verschuer and Fischer appeared to agree among themselves on this and other rhetorically slippery means of negotiating the politically sensitive "Jewish question" in order to protect their own scientific reputations abroad while simultaneously trying not to anger the Nazi government. For example, both scientists never used the phrase "less valuable race" to depict Jews. Instead they were labeled "different."[64] Von Verschuer concluded this portion of his report by stating that all attempts of the Jewish partici-pants to discredit his colleagues' research faltered on "German scientific thoroughness."[65]

Delegation leader Rüdin stressed his Institute members' scientific contributions to combat the opposition. Hans Luxenburger, Friedrich Stumpfl, and Klaus Conrad of the GRIP gave talks that lent support for the large role of heredity in all sorts of psychiatric disorders. The Munich director presented a paper on "The Eugenics of Insanity," which, as he claimed, particularly offended the worldview of the Jewish scientists at the conference. In his talk Rüdin stressed the need for both negative eu-genic measures (sterilization) as well as positive racial hygiene (increasing the number of the "fit"). According to Rüdin, this is merely what nature does, only more brutally. It actually allows the unfit to die, not some-thing aimed at by eugenicists, Rüdin asserted. Although he specifically stated that his intention was not to threaten the lives of these hereditarily defective individuals, one wonders how his language would have come across to his audience: "The push to increase the birth rate of the geneti-cally healthy and hereditarily talented must be the first concern of all cultured nations. And this will be made easier if the available help now funneled to the mass of genetically diseased and hereditarily defective, especially the genetic insane . . . and feebleminded . . . whose unhappy, if primarily parasitic, existence is a burden for the working population, is no longer wasted. Overcoming the birth rate [problem] will not be made more difficult by an eliminationist eugenics, but easier." Rüdin had long since been an advocate of mandatory sterilization. He also held the view that "the good of the whole comes before the good of the individual" even before it became an ubiquitous Nazi slogan.[66]

In his report to the authorities, Rüdin proudly echoed von Verschuer's observations by stating that "the German position was defended in a worthy manner and undoubtedly won an intellectual and moral victory" at the conference. However, he further argued that it was necessary to go to international meetings even when such unpleasant instances occur, in order to know what the other side thought about Germany's science and politics and to immediately report any incidents that occur. Almost two years later, only months before the outbreak of the war, Rüdin showed his willingness to do the regime's bidding in a report that was sent to the general director of the KWS. Asked if he thought new international scientific congresses were necessary, Rüdin replied that what was important was not to create new conferences, but to ensure that "Germany's interests are secured at the ones already in existence." It was most important, however, "to bring the existing ones to Germany. I have worked towards this end," he asserted.[67]

The GRIP director had the good fortune of serving as delegation head for another conference held in Paris during the same month as the controversial Population Policy meeting. This was the Second International Kongress für psychische Hygiene (Congress for Mental Health). As was customary for delegation leaders, Rüdin prepared a list of participants requiring both permission and money to attend. In his letter to the DKZ, he, like Fischer, specifically requested that money be given for the spouses of the researchers. As Rüdin explained, it was important that women attend as they were could effectively disseminate propaganda for the Reich. The men naturally had to focus upon conference papers and had less time for such matters. Such money for spouses, Rüdin argued, "would be very usefully spent."[68]

Rüdin's talk at the conference, "The Conditions and Role of Eugenics in the Prophylaxis of Mental Illness," appears to have been generally well received, except by the "non-Aryan" participants, as he noted in his report. At least this time, the Munich director remarked, the pope avoided issuing any anti-eugenic remarks, as he did at the 1935 meeting in Rome. Rüdin was also anxious to host the next Congress meeting in Munich, demonstrating in word and deed what he had reported to the KWS administration. Holding it in his home would undoubtedly popularize the "practical advances in psychiatry" to a larger audience, Rüdin argued. More to the point: it would serve to showcase what Germany has accomplished in the field of genetic and racial care. This was something, the Munich director remarked, the country could be proud of.[69]

Rüdin's paper at the Mental Hygiene Congress in Paris contained the same well-worn arguments in favor of mandatory sterilization that colored all of his national and international professional talks. After all, the unfit, Rüdin contended, could not be counted on to forsake parenthood voluntarily. And for the well-being of society, people are subjected to all sorts of regulations. Why should eugenic sterilization be different? What was different, if not odd, in his speech was his reference to German Jews. Apparently, eugenics strove for an "ideal physical and mental strengthening of our Jews, even our Jewesses." This would train the younger generation of German Jews in the virtues of "community spirit, sacrifice, selflessness and especially responsibility for the preservation of the bodily and mental health of our *Volk*." One wonders why Rüdin went out of his way to say this. There were undoubtedly numerous "non-Aryan" scientists at this particular conference. Whether he believed he could curry favor with "Aryan" opponents of his views through such a remark is difficult to imagine. He certainly gave no indication elsewhere that he viewed Germany's Jews as part of the German national community.[70]

Although the symbiotic relationship between science and Nazi politics is best observed through the lens of biomedical researchers in the field of human heredity, there were those working in other genetics subspecialties who clearly served the Third Reich. They, too, profited from their willingness to aid their country in the international arena. Perhaps one of the best examples is the renowned KWS plant geneticist Fritz von Wettstein (1895–1945). Unlike almost all the major German human geneticists under discussion, the Austrian-born von Wettstein never became a party member. Indeed, he is frequently viewed as having been anything but a Nazi.

In September 1937, von Wettstein, codirector of the prestigious KWI for Biology in Dahlem, received an invitation from the Genetics Department of the Carnegie Institution of Washington at Cold Spring Harbor, New York, to spend two to three months there and hold a series of lectures. He was also requested to give a talk at the prestigious American Genetics Society Meetings in Indianapolis in December of that year. Von Wettstein was one of the most generously funded German plant geneticists during the Third Reich and a scientist with extensive international connections. Who would prevent him from taking part in a professional activity that stood to aid his country as much as it would enhance his own reputation? As the report of his trip makes clear, von Wettstein was indeed given permission to go; it also sheds much light on the degree to which he was willing to use his time at Cold Spring Harbor to gather information on the institutional structure of scientific, especially genetic, research in the United States. Moreover, it reveals that he was eager to inform his German patrons about what needed to be done to ensure that German science, particularly genetics, remained competitive on the world stage. Von Wettstein apparently felt that the international standing of Germany in the field of genetics was threatened by the work done in the United States.

In a section of his report entitled "Scientific Life" von Wettstein outlined what he believed helped account for the strength of American science.

The strength and impressive aspect of American scientific life in its competition with other nations is the large number of working scientists and the existing institutes. This

was the first important impression [I had as early as] Indianapolis, and it was continually confirmed later. Not only is the number of universities and institutions of higher
learning very large; one also finds a large number of different researchers for individual
subspecialties in each institute. The *quality* of American scientists is not better than it
is here—in many cases it is certainly worse. But if a leading light somehow discovers a
new problem, a large number of older and younger scientists are immediately on the
scene to follow it through theoretically, and, most especially, experimentally.[71]

Von Wettstein remarked that every American institute had numerous researchers for each subdiscipline—a situation that simultaneously strongly
encouraged new work and fostered "deep divisions and *one-sidedness.*"
Although von Wettstein did not hold back his criticism of individual
aspects of American research life, he stressed the importance of the newly
founded Rockefeller institutes. He felt that, although Germany did not
yet lag behind the United States in science, the trend in North America
was such "that we must make *every* effort to hold the current position
of equality."[72] In particular, von Wettstein praised the "general education" given at American institutions of higher learning. "The social life
in the dormitories results in a good *esprit de corps,*" the KWI for Biology
codirector continued. He recommended "a tougher education" in his
own homeland, something that would not only improve the state of German science but result "in an avoidance of the [moral] decline of [our]
youth." Von Wettstein closed this portion of his report with what he felt
Germany should strive for in academic life: "the correct [path] is the sum-
mation of a large number of specialized personnel under the direction of
a leading [scientific] dignitary."[73]

It is clear that von Wettstein's professional visit to the United States
served the National Socialist regime well, insofar as the KWI for Biology
codirector went out of his way to describe the advantages and disadvantages of the structure of American scientific research and offered a
concrete suggestion for how Germany could remain competitive in light
of the great strides being made in the United States. Von Wettstein was
particularly interested in doing all he could to modernize his field and
give German genetics a competitive edge in the international arena. This
was not just a matter of insuring that his Institute remained in the vanguard of plant and agricultural genetics; for von Wettstein, it was also "a
matter of patriotic pride."[74]

But von Wettstein did not limit himself to a discussion of the pros and
cons of American research structures. He spent an exceptionally large
amount of time discussing Germany's image in the United States and
what could be done to improve the less-than-positive view of the Third

Reich there. In a special section of his report entitled "Our Propaganda," von Wettstein discussed what he believed was Nazi Germany's most serious problem in the United States: "One of the worst impressions that any person acquires who has lived in [the United States] for any length of time is the anti-German propaganda, especially that which originates from the hate press. This is so bad that a remedy is absolutely essential." The plant geneticist suggested that "even if it requires considerable financial resources" an "independent newspaper should be established that can simply bring clear, true news without inopportune propaganda. Most [Americans] are subjected to this hate press, as nothing else exists."[75] It is clear from the rest of his report that von Wettstein's "hate press" is synonymous with the so-called Jewish press. He appears to have accepted the National Socialist view that Jews were in control of the media in the United States, and that Germany would have to actively combat this if anything like an "objective" view of the Third Reich could reach non-Jewish Americans. Von Wettstein reported that "anti-Semitism is increasing heavily in many regions." Considering this trend, he suggested that "a wise, unobtrusive propaganda [campaign] demonstrating our true development would, especially now, fall on fertile ground. This newspaper propaganda should be supplemented by clever films that can offset the hate films."[76] Like many national-conservative German mandarins,[77] von Wettstein might have become more open to anti-Semitic propaganda over time, since his report leaves open the possibility that he believed the Jews were responsible for America's negative view of the "new Germany."

In addition to combating the "hate press," von Wettstein suggested that Germany consider an exchange program with young American academics. These people would study in German institutions of higher learning and experience firsthand the truth about the Third Reich. "A long stay in our country is the best propaganda. I have observed that in connection to the English and American students in my Institute," von Wettstein assured Minister Rust.[78] And finally, the codirector of the KWI for Biology even contemplated implementing "increased cultural propaganda by sending artists, scientists, and poets to conferences." These efforts could even begin on ships carrying young American academics to Germany.[79] Von Wettstein concluded his six-page report to Rust by reiterating the danger of not taking anti-German propaganda seriously. "I believe that we cannot pay enough attention to this anti-propaganda. Even if we need to spend a certain amount of money, we must neutralize this hate—both generally and especially in institutions of higher learning."[80]

In fact, von Wettstein's interest in combating anti-German propaganda in the United States was part of a much larger project to establish Germany's world hegemony in his field, control Eastern Europe's stock of agricultural resources, and secure his country's dominant position in the future European "new order."[81] As we can see, even nonhuman genetics and Nazi politics were mutually reinforcing during the Third Reich.

The war, at least the British involvement in it, did hamper some German geneticists' professional hopes. Von Verschuer, for example, gave a high-profile, high-prestige talk at the Royal Society in London on twin studies just months before its outbreak. He had hoped to secure an exchange of junior researchers between his Frankfurt Institute and the Francis Galton Laboratory. Allegedly, he viewed it as a way to quiet things down in the internationally and politically contested field of human genetics. In a private letter, Fischer congratulated his colleague on his achievements and mentioned that he spoke to the general secretary of the KWS, Telschow, about his accomplishments. It was clear to both, Fischer added, that von Verschuer would be his successor at the KWIA. Von Verschuer's scientific and foreign policy successes could thus be used to promote him to the position of director of one of the most prestigious KWIs for human genetics—a plan long since forged by both men.[82]

Turning to the state and party organs involved in these matters, one finds that even before the war there were important changes in the Foreign Office. Nazi Party official Joachim von Ribbentrop replaced conservative von Neurath, and the SS presence there became increasingly obvious.[83] By this time, Ribbentrop's office had also acquired its own "cultural-political department" responsible for overseeing international scientific conferences. The Foreign Organization of the NSDAP took on a new importance during the war years; Rust's Reich Ministry of Education showed itself more aggressive as well. Indeed, in 1939, shortly after the outbreak of hostilities, a memo was sent to the KWS entitled "German International Cultural Propaganda" categorizing its members into those useful for "purely professional talks" and those with the ability to speak on more general scientific topics. Although Fischer, von Verschuer, von Wettstein, and Rüdin fell into the latter category, the subject matter of their science was so abjectly political that any meaningful distinction between scientific and political lectures fell by the wayside. The memo also stressed that the KWS, owing to the "completely apolitical manner in which it was viewed abroad," would be perfect for the kind of cultural-political work the regime now had in mind. It requested that the KWS encourage its members to invite scientists to hold talks at the Harnack House. It also desired KWS scientists to hold talks in appropriate foreign countries. To facilitate matters, Fischer, von Verschuer, von Wettstein, and other important KWS scientists were required to fill out a form pertaining to their foreign scientific contacts in neutral countries and requesting information on their ability to hold talks in foreign languages.[84]

With the beginning of the war, other high-level changes were made in the way the state dealt with international scientific conferences. On November 12, 1940, a meeting was organized to discuss all existing international scientific organizations and how they could be used or discarded to advance Germany's interest. Fischer was part of the commission legislated to make this important decision.[85] In 1941, the Reich Ministry of Education circulated a secret memo stating that German scientists were to have as little to do with their Polish counterparts as possible.[86] And when, in 1942, von Verschuer was invited to give a talk at the new Reich University of Posen on "Twin Studies as a Basis of Contemporary Racial Hygiene," the poster announcing the talk specifically stated that "the *German* population is welcome to attend" (author's emphasis). This exclusion of non-Germans paralleled the experience surrounding the lecture of the internationally famous author of the "uncertainty principle" and director of the KWI for Physics, Werner Heisenberg, when he held a talk in Krakow in 1943.[87]

Let us examine three sets of conferences where Fischer, von Verschuer, and Rüdin held talks during the war. Since normal international scientific conferences ended with the outbreak of hostilities, German geneticists were reduced to holding talks in friendly or occupied countries, frequently as part of a lecture series sponsored by the German embassy (through the German Institute). These talks were perhaps more important to the regime from a cultural-political point of view than those held at respected international meetings prior to the war.

In 1940, von Verschuer and Rüdin were scheduled to give talks related to their field of human genetics at the KWI für Kunst-und Kulturwissenschaften (KWI for Art and the Cultural Sciences) in Rome (the Society founded institutes in foreign countries as well). That a series of talks on such a subject would be held in what was, until 1934, known simply as the Bibliotheca Hertziana (BH)—the first humanistic institute opened by the KWS in 1913, dedicated largely to Italian art—requires an explanation.[88] In 1934, the new Nazi Party director of the BH, Werner Hoppenstedt, wrote a memorandum in which he argued—allegedly with the good wishes of the *Führer*—that a new cultural institute should be established in the Eternal City. The KWS agreed to have it appended to the BH and to extend its area of competence to a study of the relationship between

Italian and German culture. As Hoppenstedt explained, "it is the hope that with [the founding of such an Institute] a place in Rome could be secured that would advertise the German position and German politics in a meaningful and clear way without the word 'propaganda' having to be written at the entrance."[89] To accomplish this, scientists from both inside and outside the Institute would deliver seminars and colloquia that would attract the Italian public, especially the youth.

According to von Verschuer's report, both he and Rüdin did indeed journey to Rome to present lectures on human genetics. In addition, Nazi Party and state officials responsible for racial policy such as Walter Gross and Leonardo Conti were also slated to present papers. Although the last two individuals definitely did not attend, von Verschuer could be pleased that his lecture, held in German, went over well with a large audience of Italian scientists and physicians. He intimated that the Italians had a lot of catching up to do in medical genetics and racial hygiene, but fortunately they were eager to learn. An Italian journal entitled *La Difesa della Razza* (*The Defense of the Race*) (with a circulation of 150,000) agreed to carry a special issue dedicated to the subject.[90]

Fischer and von Verschuer held numerous talks throughout Europe in 1941 and 1942. From October 23 through November 8, 1941, Fischer delivered a series of lectures in Romania. He also gave one talk in Hungary. What is revealing about his travel report is the amount of political information it contained, especially on the tensions between Romanians and Hungarians, as well as between Romanians and the ethnic Germans living among them. Fischer stressed the positive role played by the Deutsches Institut (German Institute) in Bucharest as a mediator between academic and political circles in Romania, as well as between the latter and Germans. Interestingly, he warned against having scientists lecture to Romanians and the ethnic Germans at the same time; separate events would strengthen the ties between Reich Germans and Romanian academics. Fischer also viewed it as a mistake to combine a trip to the city of Klausenburg (Cluj) in Hungary (now located in Romania) with Romania, since the Romanians still viewed the city's university as their own. Nonetheless, declared Fischer, "the present foreign policy situation has never been as favorable as it is now" for Germany. We would do well, he added, to invite Romanian academics to Germany not because of any scientific talent, "as, in general, they cannot bring us too much," but for cultural-political reasons. Fischer closed his report with a positive assessment of his effectiveness in handling the delicate question of "race" resulting from his status and age. After all, the Romanians knew

27 Scientific member and director of the KWIA, Eugen Fischer, as well as other dignitaries, at
an anthropological conference in Budapest during the war. Fischer is seated second to the
left. Directly behind him is von Verschuer. Photo courtesy of the Archiv der Max-Planck-
Gesellschaft, Berlin Dahlem.

his scientific position on race even before 1933.[91] In this series of lectures
Fischer assumed the position of cultural ambassador not unlike the one
played by Heisenberg.[92] However, unlike those of the physicist, Fischer's
scientific talks served to spread the political taint of Nazi racial policy
beyond the borders of the Reich.

In late 1941 and early 1942, the German Institute[93] in occupied Paris
initiated a series of lectures dealing with issues of health and racial hygiene, presumably at the most respected institution of higher learning in
France, the Sorbonne. Von Verschuer held a talk entitled "Human Genetics."[94] Fischer, by now a party member, decided to speak on the critical
topic of "Race and German Legislation."[95] The lectures were delivered
to a group of elite French scientists in their native tongue. As should be
obvious, here the goal was no longer to legitimize Nazi racial policy in
the abstract; it was to win approval for its implementation in occupied
France. Indeed, Fischer's discussion of the "Jewish problem" was held
only weeks before the infamous Wannsee Conference, held in Berlin's ex-
clusive southwest lake district on January 20, 1942. Headed by Reinhard
Heydrich and attended by leading state and Nazi Party bureaucrats, it

officially slated European Jewry for extermination, 165,000 of whom lived in occupied France.[96] As such, one could argue that the KWIA director not only served Nazi foreign policy objectives but also its genocidal goals as well. There can be no doubt that Fischer realized something terrible was happening to the Jews, at least those in Eastern Europe. In 1940 he sent two students to the Lodz ghetto to find pictures of "typical Jews" to use for his blatantly anti-Semitic book published with the German theologian, Gerhard Kittel, entitled *World Jewry in Antiquity*.[97]

Although the original papers no longer appear to exist, because the talks in this series were later published as a booklet entitled *Etat et Santé* (*State and Health*), we know exactly what Fischer said at the meeting.[98] After offering his definition of race, Fischer used studies from American "mainline eugenicists" such as Davenport to support the idea that important intellectual differences existed among various races. Not surprisingly, he played down differences among the so-called European races, not merely to avoid offending his audience, but because he genuinely believed that some racial mixture among allegedly "closely related races" was not harmful. Matters were entirely different with regard to Jews, however. Although Fischer noted that there were isolated Jews who made remarkable achievements, they nonetheless had a very marked racial mentality and character that separated them from Europeans. "The moral tendency and all of the actions of the Bolshevik Jews lay bare such a monstrous mentality that we can only speak of inferiority and [the Jews representing] a species different from our own." If a people wish to preserve the culture of their ancestors, it is imperative that they exclude those races whose character traits are so alien from their own, Fischer concluded.[99] We see here that the Dahlem director moved away from his earlier, so-called objective terminology of racial "difference" to racial "inferiority" when he spoke about Jews to his audience.

Perhaps even more disconcerting than Fischer's talk was his appraisal of it in his official report. After praising the German Institute for the wise decision to hold the talk at a French university, he mentioned that it was both well attended and well reported in the local newspapers. He explained that his "extremely open, but purely scientific manner" of discussing the "Negro problem" and "Jewish problem" in France was accepted without rebuttal. Indeed, individual French men of science, Fischer claimed, "acknowledged that I discussed the topic honestly and courageously." Unfortunately, one could not trust collaborating with most of these people. The majority of the anthropological institutes in Paris were anti-German, at least regarding Nazi Germany's concept of race, Fischer reported. "That is not unimportant for [our] policy as a whole." According to Fischer, only "scientifically-modern and German-friendly" researchers should be allowed continued influence in the field.[100]

_____

As we have seen, German human geneticists certainly used "the sword of [their] science" as a weapon to support the racial policy goals of the Nazi state by attending professional conferences at home as well as international meetings abroad. But this was only the tip of the iceberg. We can summarize some of the less obvious ways in which these meetings intersected with the national and foreign policy concerns of the Third Reich in the prewar period as follows: First, KWS-sponsored meetings as well as international professional conferences bolstered the prestige of German science. In light of the pariah status Germany experienced during the early Weimar years when its scientists were excluded from foreign conferences, this was no small matter. Second, insofar as international conferences were moved to Germany or to "countries well-disposed toward Germany," they furthered Nazi aims since they were controlled by human geneticists sympathetic to National Socialism. Moreover, we can look at the roles of renowned German geneticists like Fischer, von Verschuer, von Wettstein, and Rüdin and view how they directly used their influence to advance Nazi national and foreign policy interests prior to 1939. On the home front, they accomplished this task by disseminating their racial science research to an elite public (although this was not relevant in the case of plant geneticist von Wettstein). Abroad, German geneticists undertook an important service to their country through their attempts to change the shape of international conferences to reflect German interests as well as their efforts, through indirect channels and personal relations with foreign scientists, to influence their colleagues in a pro-German direction. And finally, through the reports these German scientists were forced to write, they gave the regime valuable information about the political and scientific state of the host country of such conferences.

During the war, Fischer, von Verschuer, and Rüdin continued to support Nazi foreign policy goals by legitimating the execution of Nazi racial policy in occupied and "friendly" countries. Moreover, through their support of Nazi cultural policy, they helped win the hearts and minds of professionals in neutral countries. And lastly, these scientists helped prepare the way for the "new order" in Europe by showing the virtues of German science, in general, and human genetics, in particular.

How did the KWS and its genetic researchers profit from the talks the Society sponsored? And what can we say about the relationship between these scientists and the National Socialist state if we examine their activities in the international arena? We can draw two conclusions here. First, the KWS certainly enriched itself financially and politically by hosting its talks and spotlighting the human geneticists on its payroll. These lectures bestowed prestige on both the Society and the researchers who agreed to lecture. Indeed professional conferences, especially those held abroad, conferred influence at home; the KWS and its human geneticists' international reputations were scientific capital for the Nazi regime and they knew it. The scientists in particular could exploit it for their own ends: to terminate denunciation campaigns, to secure a directorship of the KWIA (in von Verschuer's case), and to obtain more money for their institutes. This last activity should not be underestimated, since more financial backing for their institutes and their research directly served the racial policy needs of the Nazi state. Second, attending international conferences gave German geneticists a chance to meet with their peers and exchange ideas that could enhance their own work—something that, given the parameters of the Nazi state and its racial policy, also served the interest of the regime.

The most that one can say for certain is that the politics of professional talk lays bare the radical symbiotic relationship made between the junior and senior partners to the "Faustian bargain"—human geneticists and the Nazi state, respectively—at the outset of the Third Reich. As in the case of the production of biomedical knowledge discussed in chapters 2 and 3, the dissemination of this knowledge at professional meetings both at home and abroad reminds us how human genetics and politics interfaced under National Socialism in a variety of subtle and not-so-subtle ways. As we will see in our next chapter, this knowledge was popularized even more broadly throughout Germany's secondary schools.

FIVE

Politicized Pedagogy

In late January 1934, approximately a year after the beginning of the "Thousand Year-Long Reich," at least two male
students in the graduating class of the Bismarck-Gymnasium,
a college preparatory school in the middle-class district of
Berlin-Wilmersdorf, chose a biological topic for one of their
Abituraufsätze (college preparatory exit exams). Because the
school did not specialize in the natural sciences, not all
of the eighteen- and nineteen-year-old boys who received
their *Abitur* (diploma needed for university study) from this
tradition-bound institution that year were required to take
an exit exam in biology. The Bismarck-Gymnasium stressed
ancient and modern languages and culture, including German. Interestingly, the teacher of German language offered
a relevant biological/racial theme as one of several possible
topics to graduating students.

That he—or, far less likely, she—could do so was made
possible by another teacher at the school. Dr. F. not only
taught botany and zoology at the Bismarck-Gymnasium
but also "racial science." This was the biology instructor's
mandate. Less than nine months after the "Nazi seizure of
power," then Prussian Minister of Education and former elementary school teacher Bernhard Rust issued a decree or-
dering the addition of new subjects to the final grades of the
Prussian secondary and primary schools. Two years later,
in 1935, it was extended to all German schools. The aim of
this new meta-field was to safeguard the racial and genetic
substrate of the *Volk*. The pillars of racial improvement that
fell under the rubric of "racial science" included genetics,
eugenics/racial hygiene, population policy, genealogy, and

Rassenkunde (ethnology).[1] The two students in question had obviously learned the lessons of this biology teacher and were enthusiastic enough about the topic to select it instead of more usual themes offered for a German language exit exam. The two wrote on "The Biological Foundations of *Völkisch* Racial Care." Dr. F., a PhD in biology, took part in evaluating the students' work. This biology teacher not only played a role in giving them a final grade; he also commented on the content of the essays. Although there might only have been two teenagers who chose to select a biological topic for their exams at this institution that year, all the boys graduating from Bismarck-Gymnasium in 1934 had to undergo an oral examination on some facet of "racial science."[2]

In our last chapter we saw how biomedical researchers disseminated human heredity knowledge at national and international professional conferences. It was clear that the politics of professional talk benefited the scientists themselves, the KWS, and the Nazi regime. The biomedical researchers' newest findings, however, reached a relatively limited audience: other scientists like themselves and a select interested lay public. It is certainly fair to say that higher secondary school biology instructors (those employed in college preparatory schools) who taught human heredity in their classrooms throughout Germany during the Nazi era—although they taught less than 10 percent of all school age children[3]—influenced the lives of more impressionable individuals than did those working within the confines of the KWS. And similar to biomedical scientists such as Fischer, von Verschuer, and Rüdin, these secondary school biology teachers also served their own professional interests as well as those of the state. Here, too, we find a symbiotic relationship between human heredity and politics during the Third Reich. As will become evident, biology education was of great import to National Socialist pedagogues and party members. After all, what better way to impress upon Germany's youth the so-called scientific foundations of the Nazi worldview than through biology instruction? Hans Schemm, a prominent Nazi pedagogue and head of the Nationalsozialistischer Lehrerbund (National Socialist Teachers League) (NSLB), declared that "National Socialism is applied biology." Two higher secondary school biology *Studienräte* (teachers at college preparatory schools) reiterated this view in their handbook, *Biologie, Nationalsozialismus und die neue Erziehung* (*Biology, National Socialism and the New Education*), by asserting that the new movement is based on an ideology "that is grounded in biology and approaches the world biologically." And if one reads the turgid prose of Hitler's *Mein Kampf,* the Bible of Nazism, one finds that it also reeks of biological metaphors as well as references to racial struggle and so-called eugenic practices in nature.[4] Although biology teachers were mandated by the Ministry of Education to emphasize various aspects of human heredity, their desire for a larger piece of the curriculum pie and increased status for their discipline in the hierarchy of school subjects often made them "willing executioners" of this new decree and those that followed later. One Hamburg secondary school teacher spoke for many of his colleagues when he proclaimed "[we] biologists are unbelievably happy that a place has been created for these important [biological] questions in the new Reich and that our *Führer* Adolf Hitler and his helpers have such an interest in them." This will ensure, he continued "that our youth will begin to think biologically early and view the future of our state as being dependent upon noble biological laws."[5]

Schools have always done more than disseminate knowledge. They served, and continue to serve, a social function: to integrate their charges into society. They also undertake the unarticulated task of teaching children society's norms and cultivating national identities. We see this every day in American schools and in those of other democratic countries. Think of the function of the Pledge of Allegiance or the role of decorating an American school classroom for Thanksgiving or Presidents' Day. Yet if we listen to the oral and written testimonies of individuals schooled during the Third Reich as well as to scholars who studied education under the swastika, we are reminded that the Nazis went far beyond any legitimate attempt to integrate the youth into German society. Indeed, instruction in virtually all subjects under National Socialism consciously attempted to inculcate young Germans with all the unsavory components of Nazi ideology, especially racial hatred. Formal education in the schools desensitized German pupils to the oppression all around them; it also reinforced the notion of the separation between "national comrades" who were part of the *Volk* and "community aliens" who were not. The latter—Jews, Roma and Sinti, Afro-Germans, the handicapped, homosexuals, and asocials—became national, if not international, pariahs. These "degenerates" or "racially alien" individuals were no longer part of the pupils' ethical universe, or so they were taught. Considering the significance of biology for the Nazi worldview, it has always been singled out as one of the most ideological subjects in the Nazi schools.[6] Although human heredity instruction (broadly defined to include "racial science") was not the only ideological tool at the biology instructor's disposal, it is the part of the biology curriculum usually viewed as having the most direct connection in engendering racial enmity toward "the other."

While even a cursory glance of the numerous curricular guidelines for secondary school biology and handbooks for biology teachers would support this assertion, we must ask whether the Nazi pedagogical ideal for this subject actually reflected the reality of classroom practice, and how classroom teachers went about presenting their subject to their students. Like members of the state-funded KWIA, biology teachers were civil servants, and as such they may have felt a special obligation to the state that employed them or an ethical imperative to do its bidding. Moreover, secondary school biologists, even if they held PhDs, did not normally do original research in human heredity. The existence of one elementary school teacher's racial study of the head size of nine hundred East European-born Jews living in Germany and their German-born children—a project undertaken at the KWIA—is certainly a rare exception.[7] As such, biology instructors had to rely on the knowledge of others who did such scholarly work.

In order to address the issue of secondary school biology instruction in the Third Reich it is necessary to first focus on the ideal of human genetics education (again, broadly defined to include the five subfields of "racial science") by analyzing the proclamations on the subject issued by prominent Nazi pedagogues and university-based human geneticists. Although the emphasis will be on these college-preparatory institutions, passing reference to other types of German schools will be made, especially since they educated far more children than did the three major traditional forms of the higher secondary schools: the *humanistisches Gymnasium* (emphasizing ancient languages and culture), the *Realgymnasium* (stressing modern European languages and culture), and the *Oberrealschule* (emphasizing the natural sciences) that existed until the reform of 1938 made changes in the structure of these institutions. Biology textbooks designed for pupils will also be surveyed. And finally, it will be made clear what ideologically committed secondary school biology teachers did to promote their subject outside the classroom.

After exploring the Nazi ideal of human heredity instruction, we will have the opportunity to examine its practice with an eye toward comparing the ideal to the real. This latter task—ascertaining what was actually taught in the classroom—is far more daunting than studying the racial science ideal. One of the most difficult problems is accumulating sources or documents that show how the subject was actually taught, not merely how it was supposed to be taught. Fortunately, we have a rare archival find to aid us: topics for biology exit exams along with the students' responses from a few Berlin higher secondary schools (some even from the Weimar period), and oral testimonials from individuals schooled during the Third Reich. These sources can stand as either a corrective to or corroborate—however imperfect because of their small number and concentration in one city—the content of official school textbooks and curricular guidelines. We can also utilize so-called *Stimmungsberichte* (morale reports) communicated by Nazi Party school biology teachers to their superiors to help refine the historical picture. These speak to the political attitude of biology instructors in the *Gaue,* or various districts, of the NSLB, an organization that represented some 97 percent of all German school instructors during the Third Reich. (Nazi Party auxiliary organizations were structured, like the party itself, into a hierarchy of larger and smaller regions and districts).[8] Yet before turning to our subject during the Nazi era, it is worth examining the role of biology and human heredity in the schools prior to the Third Reich as well as the attitude of biology teachers toward the position their subject occupied in the school curriculum. This will address the issue of continuity and discontinuity between pre-1933 and Nazi traditions in biology education.

Biology Instruction before the Third Reich

Although school subjects were never as abjectly politicized as they were under National Socialism, it would be a mistake to think that the Nazis were the first to see the political relevance of controlling the traditional disciplines taught to pupils. Even during the Second Empire, school subjects were used for political ends, especially history education. Wilhelm II (reigning from 1888 to 1918) and his ministers made an urgent plea that school instruction become a more effective weapon in the fight against German Social Democracy, a political movement then viewed by almost everyone, even its advocates, the German working class, as a threat to the established authoritarian state. The Emperor demanded that more emphasis be placed on modern and contemporary history in the German classroom. Such instruction would demonstrate that "only state authority could protect the individual's family, freedom and rights." Wilhelm II went so far as to attack an entire school type for its political irrelevance. His pet peeve: the *humanistisches Gymnasium.* "We should be educating young national Germans and not ancient Greeks and Romans," the German monarch insisted.[9] Interestingly, even the potential social relevance of biology instruction was not ignored. It, too, was instrumentalized for political purposes under the Empire, although in this case, the initiative did not come from the monarch. In the 1880s, pedagogical reformer Friedrich Junge (1832–1905) proposed changing the biology curriculum (albeit in the elementary schools) from an emphasis on dry taxonomy to one stressing the *Lebensgemeinschaft* (biotic community). This shift would present nature as a harmonious community and hence combat the political divisiveness of social democracy and stress moral values in the school biology classroom. It also promised to boost the professional aspirations of elementary school teachers. Ultimately, the idea caught on and the future political implications of this biological paradigm for the Nazi worldview—where all organisms, including humans, were understood as part of a larger, transindividual whole—were not forgotten.[10]

The German Revolution of 1918–19 and the creation of the fragile Weimar Republic that followed not only transformed German politics. It also signaled important curricular changes in the German schools. Paragraph 148 of the new Weimar Constitution mandated the teaching of civics throughout the school curriculum. Perhaps surprisingly, biology teachers in all schools, especially the higher secondary schools, saw an important role for their subject in this new mandate. Among other things, biology instruction could focus on "health education" as a part of this new educational task. According to Walter Schoenichen, director of the Pedagogical Department of the Preußisches Zentralinstitut für Erziehung und Unterricht (Prussian Central Institute for Education and Classroom Instruction) and Weimar Germany's leading biology didactician, children needed to understand that individual health was not merely a private but also a civic and political affair: students had to learn that it was their duty to be healthy and to combat anything detrimental to the "efficiency of Germany's human resources." In order to do this, human beings must become the major object of biology education. This was a prerequisite for discussing the problems of alcoholism and infectious diseases, especially sexually transmitted ones. To complete the civic function of health education, genetics and eugenics should occupy a prominent place in the biology school curriculum, especially in the higher secondary schools, he concluded.[11] This fitted comfortably into the eugenic outlook of the Weimar Republic; we will recall that it was the Weimar state and federal governments that supported the establishment of the eugenically related KWIs in Munich and Berlin.

The slow-turning wheels of the German state bureaucracies notwithstanding (education during the Weimar Republic was the prerogative of individual German state governments, not the federal government), many of the civic and political aims involving the teaching of heredity and eugenics found their way into biology curricular guidelines in Prussia and elsewhere. They were also somewhat integrated into school practice. For example, in 1930, an official meeting regulating the teaching of biology in the upper grades of one Hamburg *Oberrealschule* for girls mandated that, in addition to evolution, pupils must be taught Mendelism, chromosome theory, human heredity, sex-linked genetics, the relationship between genes and the environment, and eugenics. About three years earlier, a biology teacher in another Hamburg higher secondary girls' school offered an experiential team-based class on genetics and eugenics. Although it was voluntary, ten girls were interested enough in the subject to take part. Where they were taught, such biological classes appear to have been quite popular. A 1928 report from the Helene-Lange-Oberrealschule in Hamburg noted that its older female pupils were so interested in biological subjects demonstrating "immediate relevance to contemporary life," that they themselves set up an *Arbeitsgemeinschaft* (team-based class) in a semester where one was not originally planned.[12]

In order to satisfy the demand of school teachers who lacked this new socially relevant knowledge in their field, biology didacticians wrote handbooks for higher secondary school biology instructors to acquaint them with the newest scientific developments in genetics and eugenics. Suggestions were often provided on how to teach such material effectively in the classroom. Perhaps one of the most extensive treatments of the subject was written by a former secondary school biology teacher, Dr. Jakob Graf, a man sympathetic to the Nazis during the Weimar Republic (he joined the NSDAP in 1932). After 1933, he became both an important biology textbook author and a leading figure in the biology division of the NSLB during the Third Reich. His 263-page handbook *Vererbungslehre und Erbgesundheitspflege* (*Genetics and Hereditary Health Care*), published in 1930, was exhaustive. He discussed all the recent scientific findings on genetics, human heredity, genealogy, and population policy; he also supplemented his text with an extensive bibliography of the works of KWS biomedical scientists such as Baur, Lenz, Fischer, and Rüdin. However, there was no treatment of what became central to Nazi pedagogues: *Rassenkunde* (ethnology). Yet the conclusion to this important handbook left little doubt that Graf was a thoroughgoing eugenicist and an advocate of the biological needs of a genetically healthy *Volk:* "The teaching of heredity and selection has demonstrated which way leads to [cultural] ascent and which to decline. Let us hope that this knowledge is quickly and broadly disseminated . . . and that the lessons to be learned take root in our youth such that they are aware of the lofty task they have towards their heredity. In addition to being responsible for themselves, they [the youth] must develop a responsibility to their *Volk* and the coming generation, if civilization is not to become a biological tragedy for our people."[13] Philipp Depdolla, another important biology pedagogue during the late Weimar years, tirelessly emphasized the need to teach heredity and eugenics in the schools. He wrote numerous articles emphasizing the cultural importance of these subjects in anthologies, journals for teachers, and professional eugenics publications. Günther Just, a renowned human geneticist and eugenicist at the University of Greifswald, included Depdolla's plea for more eugenics education in the secondary schools in his widely read anthology, *Vererbungslehre und natur-wissenschaftliche Erziehung* (*Heredity and Science Education*). According to Depdolla, eugenics had an important civic function: "the spreading of a consciousness of ethical responsibility toward the entire nation and race." Just was merely one of many professionals in heredity and eugenics who himself took part in a conference sponsored by the Prussian Central Institute for Education and Instruction. The purpose of this seminar was to bring teachers from all types of Prussian schools together to hear what experts, such as KWIA Eugenics Division Head Muckermann, had to say about the newest developments in these fields and the necessity that these subjects be introduced into the biology classroom as quickly as possible.[14]

Not only handbooks and professional articles but also biology text-books from the Weimar period suggest that genetics and eugenics were taught in some higher secondary schools. For example, Cäsar Schäffer's popular 1930 *Leitfaden der Biologie* (*Themes in Biology*) provided an over-view of Mendelian genetics, genealogy, and eugenics for the middle grades of the Prussian higher secondary schools. The author treated these subjects more extensively in a book published a year earlier. Designed for graduating classes of these schools and for "self-instruction," the author not only went into great detail about heredity and eugenics, but even included a section on *Rassenkunde* (ethnology) written by the *völkisch* Leipzig anthropologist, Otto Reche. This, however, was probably the only biology school textbook written prior to the Third Reich that included a section on ethnology. Yet it neither contained a value judgment of the various "human races" standard in such books after 1933 nor did the author mention the Jews. An overtly racist ethnology would not have been an appropriate linguistic resource for a school book in the Weimar Republic.[15]

The best evidence that human heredity and eugenics was actually taught in at least some of Germany's higher secondary schools during the Weimar Republic is provided by biology exit exams. In 1932, Dr. M., a biology teacher at the Leibniz-Oberrealschule, suggested three topics for his teenage boys' final examination. Ultimately, however, the six pupils graduating from this institution in the upper-middle-class section of Berlin-Charlottenburg selected the most relevant of the three themes originally suggested as a possible exam topic: "Eugenics: An Overview of the Efforts to Improve Hereditary Quality." The work of these six seniors were corrected, graded, and assessed in writing by the same biology teacher who formulated the test question.

So what did the pupils at the Leibniz-Oberrealschule learn about human heredity and eugenics? Did their biology teacher, a man who held a doctorate, attempt to educate his charges in a responsible manner, following the available scientific literature on the subject? First, these teenage boys were taught that eugenics was concerned with "the hygiene of human reproduction." Its goal, pupil J.W. wrote, was "to elevate the hereditary substrate of a people." Virtually all noted that "eugenics was a science" or "part of the science of biology with its own methodology" and mentioned that its German name was *"Rassenhygiene"* (racial hygiene). This suggests that the teacher used the terms eugenics and racial hygiene interchangeably, unlike many *völkisch* right-wing eugenic researchers, especially those who would later remain at the KWS after 1933. These scientists, such as Lenz and Rüdin, consciously employed the term "racial hygiene" because it left open the possibility that anthropological race was also a criterion of "fitness." Dr. M. had not taught this to his pupils; he apparently viewed the moderate form of eugenics popularized by political Centrists and Social Democrats as scientific. Dr. M. also stressed (or used a textbook that stressed) the "theoretical" and a "practical" side of eugenics. The former, according to student G.N., focused on the "relationship between heredity and procreation." "We know from genetics," he continued, "that people do not have the same hereditary make-up; rather, they have healthy and diseased traits." It is of utmost importance whether individuals with healthy or unhealthy genes are more fecund, G.N. argued. All of the exams emphasized this particular point, and they provided numerous examples of how the "unfit" posed a danger for the nation owing to their allegedly larger than average number of children. According to one study noted frequently by the boys writing on this topic, "feebleminded" mothers in Rostock were said to give birth to 6.4 children on average. Although these children died more frequently in childbirth, 4.7 lived to marriageable age—a disaster for national efficiency. Following a Munich statistic, G.H. reported, genetically healthy parents produced only 1.87 children on average while families with "feebleminded children" attending special schools had approximately two siblings each. Pupil B.S. added the ostensible fact that members of the lower classes tended to have a higher birth rate than those belonging to the higher social orders, another eugenic danger. "The biological efficiency" of a *Volk* would increase with each generation, G.N. continued, "if those who are, hereditarily speaking, completely worthy" only chose to have more offspring. And G.H. explained that we now know the cause of the fall of the ancient Greeks and Romans (at which point the teacher added, in the margins, "in ancient cultures," viewing the student's omission of the Babylonians and Assyrians as an error). Today, however, with the help of racial hygiene, we can avoid their common fate, G.H. asserted with relief. Dr. M.'s boys had obviously learned the standard moderate eugenic line on population quality.

In their discussion of "theoretical eugenics," the pupils examined diseases and cultural practices from the standpoint of the genetic health of future generations; they were labeled either "selective" or "counterselective." War, it was noted, while at one time selective, was now counterselective. The best and the brightest died on the battlefield; the inferior in mind and body stayed at home and reproduced. Nearsightedness, at one time a serious obstacle in the "struggle for survival" as well as in a person's ability to find a mate ("no one would marry a nearsighted person" in days of old, J.W. argued), no longer played a role in modern civilization. Glasses have offset this form of "natural selection."

Biology educators, as we have seen, believed they served an important civic function in their treatment of national health in the classroom. The boys' discussion of the eugenic danger of venereal disease, especially gonorrhea, must have reflected Dr. M.'s particular concerns (or preoccupation), since every child writing an exam discussed this point in great detail. As G.H. expressed it, "the eugenic impact of gonorrhea can be seen in the fact that up to 90% of men with a double testicle infection [the result of gonorrhea] become sterile. On average, Germany looses 200,000 births to this form of venereal disease. . . . [Unfortunately], different occupational groups are affected differently. The longer the training period and, hence, the [need] to remain single, the more frequent is the disease. This is proved by statistics: in Berlin, 9% of workers, 16% of businessmen and 25% of [university] students are afflicted." The point, of course, is that university students (usually from the middle and upper classes) were most likely to suffer from gonorrhea, and their allegedly better than average genetic makeup would be lost to posterity. J.W. demonstrated to his teacher his knowledge that the unhappy fate of approximately one-third of all childless German couples can be traced to this infectious disease. Syphilis, he added, led to spontaneous abortions and infected infants in otherwise genetically sound women.

Alcoholism and tuberculosis were also included among diseases that were overwhelmingly counterselective, a clear indication that eugenics during the Weimar period was not taught as a field opposed to social hygiene measures. Indeed, some of the graduates indicated that poverty and an unfavorable environment were largely responsible for the higher rate of TB in the poorer classes, and Dr. M. did not tell them they were wrong. Indeed, he complained at the end of one exam that the pupil neglected to mention the improvement of living quarters as a possible positive eugenic measure. Interestingly, total alcohol abstinence, while accepted for youth, was not universally viewed as beneficial for adults. In particular, the prohibition movement in the United States that criminalized alcohol production was sharply criticized by two students; it led to nothing but "bootlegging and crime," B.S. insisted. Instead, alcohol should be taxed more heavily, according to A.H. Diabetes and idiocy, however, were deemed selective diseases by all pupils, since individuals with these afflictions rarely reached reproductive age or were otherwise unable to produce offspring.

The best grades on this question were given to those boys who also dealt with the "practical" dimensions of eugenics in some detail. Indeed, on several of the exams, Dr. M. complained that this important aspect was not sufficiently treated by his students. Among the practical measures stressed by all taking the exam was "popular education." This "should be supported as strongly as possible," since "[through such measures] a tremendous amount can already be accomplished," argued A.H. The boys with the highest test results discussed practical measures like marriage counseling, the exchange of health certificates before marriage, and tax reform based on the number of children a healthy couple produces. The two pupils receiving a "good" and a "very good" on their exams—grades not given out lightly from the strict *Studienräte* during the Weimar Republic—also broached the question of eugenic sterilization. That they mentioned it at all demonstrates that it was discussed in class. The exams reveal that they knew what it was and how it differed from castration. Moreover, at least A.H. was aware of mandatory sterilization laws in the United States. Both he and B.S. understood that such an operation was more dangerous for women than for men, although they both had their qualms. While it was not a bad idea, sterilization was only useful if applied liberally. Even in the United States, where eugenic sterilization was tested more than elsewhere, the alleged "3–4,000 procedures [*sic*]" mentioned in A.H.'s essay did not, he maintained, produce the desired effect; there were also complicated legal issues (surprisingly, Dr. M. did not correct this serious underestimation of the number of Americans forcibly sterilized). Indeed, A.H. knew that even voluntary sterilization was not yet legalized in Germany. All forms of mandatory state eugenic measures, he remarked, lay in the distant future! Yet the country that first manages to take better account of the principles of eugenics, one of the less talented pupils concluded, "will have a huge leg up over all others."[16]

These exams certainly tell us something important about how human heredity was presented in the classroom and about the mind-set of the biology instructor who taught it. Yet the clear-cut case of genetics and eugenics education in the Leibniz-Oberrealschule notwithstanding, most secondary school biology teachers bemoaned that not enough was done to improve the status of their subject, especially in those college-preparatory higher schools lacking a natural science concentration. Indeed, German biology teachers had long complained that their subject was grossly undervalued by those formulating school curricula. Generally viewed by nonbiology educators as being little more than a descriptive science possessing little pedagogical value, biology was unable to compete in the battle for school hours with the "cultural disciplines" of religion, German, and history. Although the details cannot concern us here, suffice it to say that whereas the position of biology as a school subject had improved by the Weimar period, biology teachers were still far from their stated goal of two hours of classroom instruction per week in all secondary school grades. As we have seen, only in the *Oberrealschulen* was biology taught in the graduating class.

Even in these schools, however, biology was often considered the stepchild of the other natural sciences. Biology teachers employed in the higher secondary schools lamented this sad state of affairs in Germany's numerous pedagogical journals. They were particularly outraged over the teaching of biology by nonbiologists—indeed by nonscientists. This was especially irksome since many biology teachers trained for higher secondary schools, as we have seen, had PhDs in their subject. Though perhaps an extreme example, a man who allegedly could not tell the difference between a spider and a beetle had a position as a secondary school biology teacher![17]

During the financially lean final years of the Weimar Republic, secondary school biology teachers faced the same obstacles as other academically trained professionals: budgetary cuts impacting employment chances. We have seen how frustrated Fischer became—despite his prestigious position as head of the KWIA—when the KWS could not give him the material resources he needed to keep his Institute's research on track. It certainly contributed to his favorable reception of the new Nationalist-Nazi government. Similarly, biology pedagogues and biology instructors felt they were in an equally untenable situation as the final curtain was drawn on the unloved Republic. They sought to promote their interest by joining hands with university biologists in the establishment of the Der Deutsche Biologenverband (the German Association of Biologists) in 1931. Their professional goals were boldly pronounced in the organization's journal, *Der Biologe*. In an article published in 1932 entitled "Guidelines for Secondary School Biology Education," one can find the perennially voiced demands to introduce heredity and eugenics as well as the broader political and civic goals of biology education into the curriculum. "The eugenically endangered future of the German people," the Guidelines argued, "makes a thoroughgoing instruction in genetics and eugenics imperative." The last of the ten points in the Guidelines ended by stating that "biology is to be accorded a place among the central *Kernfächer* [central subjects] in all schools." Little did instructors teaching biology know that in less than one year, many of their professional aspirations would be met by a new regime anxious to demonstrate that their area of expertise was the intellectual kernel of its worldview. This development was welcomed among most German higher secondary school biology teachers, many of whom were nationalist in sentiment and harbored no great love for the political vicissitudes of democracy.[18] Given their professional predicament and their political outlook it comes as little surprise that many secondary school biology teachers looked to the new regime with fond anticipation.

The Racial Science Ideal

Long before the Nazis secured the power they desperately sought, Hitler had laid out his views on education in his political testament, *Mein Kampf*. In addition to stressing the importance of both physical prowess and character over intellect, the future *Führer*—who himself had not excelled in the classroom and had nothing but contempt for the traditional school masters' emphasis on book learning—made racial consciousness the cornerstone of instruction in his future *Reich*. "The crown of the *völkisch* state's entire work in education and training must be to burn the racial sense and racial feeling into the instinct and the intellect, the heart and the brain of the youth entrusted to it. No boy and no girl must leave school without having been led to an ultimate realization of the necessity and essence of blood purity."[19]Although neither Hitler nor the Nazi Party ever formulated a concrete plan to transform the schools to meet this goal, there was unspoken recognition among German pedagogues, even before Rust's 1933 decree mandating the teaching of "racial science" in the graduating classes of all schools, that the biologically grounded concept of "race" and related subjects would be examined in the nation's biology classrooms. Reich Interior Minister Frick made this somewhat more official when he stated, just months after Hitler's appointment as Chancellor, that biology instruction must be centered on ethnology (albeit in its racist form), racial hygiene, and genealogy. This was a position for which biologically trained *völkisch* educators—especially at teacher training academies and universities—had argued even prior to the Third Reich. Also noteworthy is the fact that biologically trained school teachers played a major role constructing the final form of Frick's proclamation and Rust's decree, since they developed curricular guidelines that were considered in the wording of the official 1933 racial science mandate.[20]

Rust's 1933 decree not only gave biology teachers an extra two to three hours a week of valuable classroom time to execute their mandate "if necessary, at the expense of mathematics and foreign languages";[21] it also generated a veritable flood of pedagogical articles, teacher handbooks, and student textbooks on the subject of "racial science." These educational projects were deemed necessary for many reasons. Since biology instructors in the higher secondary schools had attended university during the Empire and were unacquainted with the new material, such advanced training was unavoidable. In addition, there were still many older school texts containing little, if anything, on genetics and eugenics. Moreover, nothing about the racist form of ethnology was available in biology school books prior to 1933. And finally, important Nazi pedagogues such as Rudolf Benze demanded that biology teachers play a leading role in schools—one demanding proper training. They were not merely in charge of offering the scientific foundations of racial science in their classrooms; biology teachers were also primarily responsible for the success of *völkisch* education as a whole. They would accomplish this by serving as experts when racial science was discussed by instructors in other subjects such as history or German. After all, biology was now officially valued as its school practitioners had always wished: as a "central subject" and the queen of natural sciences.[22] These considerations aside, it should be obvious that since biology educators were the authors of the required texts, writing them to meet the new educational demands was professionally self-serving. It also simultaneously supported the goals of the Nazi state.

It should come as no surprise that renowned university-based geneticists such as Fischer, Lenz, Just, and Russian-born Nikolai Timofeeff-Ressovsky (1900–81) came to the aid of their fellow biologists in secondary schools. They did so by publishing pedagogical articles stressing the importance both of the biologically minded Nazi leadership and the new subjects under the banner of racial science. Fischer, for example, wrote an essay in 1933 entitled "People and Race as a National Question" in the *Monatsschrift für die höhere Schule* (*Monthly Journal for the Secondary School*), a periodical designed for instructors at the college-preparatory schools. Here he unabashedly praised the Nazis for being the first political party in history to take the racial future of its people into its own hands. One year later, in an openly Nazi mouthpiece dealing with "political education," the KWIA director answered questions on the topic "Why Hereditary and Racial Research?" At a time when his denunciation campaign was still in full swing, Fischer told his readers that what his science accomplished through tedious research, the *Führer's* "ingenious foresight" enabled him to understand instinctively. The Berlin University Rector ended his remarks by proclaiming that "the domestic and international fate of a people is dependant upon heredity and race." In a 1934 issue of the *Nationalsozialistische Korrespondenz* (*National Socialist Correspondence*), the same paper in which Fischer praised Hitler's biological wisdom, Lenz explained "that the only way to lead our race to a higher level of health and efficiency is through racial hygienic selection." It is a task, he concluded, that "our grandchildren will thank us for." Human geneticist Just took his message directly to secondary school biology teachers. At a 1933 meeting for middle school teachers in Kiel, the Greifswald professor held a talk entitled "Eugenics and Education." In the published version of his lecture, Just stressed the need for eugenics education, especially in middle school biology classes—the type of school that educated the kind of pupils who formed an "indispensable energy reservoir of our national body." Timofeeff-Ressovsky, the brilliant geneticist employed in the Berlin KWI for Brain Research, directed his attention to university-bound teenagers. He wrote an article aimed at biology educators on the importance of teaching genetics and "higher Mendelism" for older students in the team-based classes for biology of Germany's college-preparatory schools. And lest we forget that school biology pedagogues were singing the praises of the new racial science present in their curricula, an article entitled "Genealogy from the Perspective of Higher Secondary School Biology" by Jakob Graf offered Germany's biology instructors practical advice on how to teach genealogy and help students construct their family trees. Such instruction would lead children to an "understanding of the genetic and racial composition of the family and the *Volk*."[23]

More interesting than the handbooks written for higher secondary school biology teachers were the countless new textbooks for the graduating classes of the various schools as well as the supplements to older biology school books that made their debut in the wake of the Rust decree.[24] From these latter two sources we can glean what the students were expected to learn about racial science. Most texts treated all five components of this meta-complex: genetics, genealogy, population policy, racial hygiene/eugenics, and racist ethnology. Many asked pupils to undertake studies and perform experiments to make the subject more real to them.

Almost all textbooks began with the basics of genetics. There were ample diagrams of fertilization, mitosis, meiosis, and Mendel's laws (including the use of Punnet squares to demonstrate the results of Mendelian crosses) applied to both plants and animals. Most books included an explanation of dominant and recessive traits. But even here, biology pedagogues deftly integrated the prejudices in favor of Nordic racism, large families, and healthy peasant stock into a discussion of this allegedly neutral subject. In one biology textbook discussing dominant and recessive genes—in this case eye color—the desirability of the Nazi ideal was subtly communicated to pupils. Children were told that brown-eyed peasant mother Frau Berghof loved the blue eyes of her husband and desperately wanted to see them in her five children. But as diagram number 9 in the textbook demonstrated for those who completed the exercise, Frau Berghof would be disappointed. Her brood all had brown eyes. One of her children, however, eventually married a peasant neighbor blessed with blue eyes. Fortunately, as students who did their assignment conscientiously noticed, Frau Berghof lived to see the precious blue eyes of her husband in some of her grandchildren. Recessive traits, these pupils learned, happily did not disappear from the worthy Berghof family forever.[25]

This ideological injection into genetics education notwithstanding, school biology textbooks provided the most up-to-date information on the subject. Even crossing-over, sex linkage, and the origins of mutations were sometimes discussed. Frequently, authors emphasized the importance of the fruit fly as the object of genetic research par excellence or the significance of twin studies for human heredity. Some texts not only specified the number of human chromosomes, but also offered illustrations of their relative size and shape. One book even offered a chromosome map of the fruit fly (taken, if somewhat ironically, from the Jewish geneticist Curt Stern). However, if we compare the presentation of genetic knowledge in these books to what was communicated just a few years before during the Weimar Republic, there are two major differences: German words replaced internationally accepted terms stemming from Latin (e.g., *Kernschleife* instead of *Chromosom*); and the power of heredity

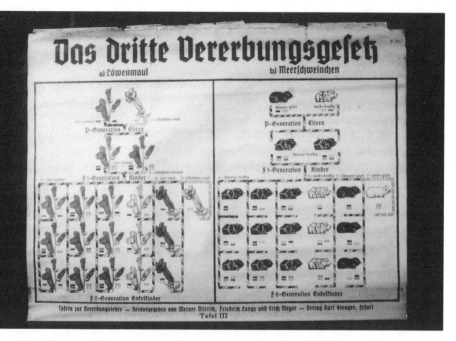

28 A high school biology classroom wall hanging demonstrating Mendel's Third Law: when two alleles come together during fertilization, one will be dominant and one recessive. The particular trait in question will have the appearance of the dominant allele. Photo courtesy of the Universitätsbibliothek Duisburg-Essen.

was emphasized to the virtual exclusion of the environment. The latter was now viewed as virtually impotent. Either it was not stressed at all or mentioned dismissively: "Character traits acquired through the environment," one textbook flatly asserted, "were not inherited."[26] Moreover, what is obvious in the books under discussion is that the laws of genetics and the newest findings in heredity were not generally presented as ends in and of themselves—as knowledge useful for its own sake. Most textbooks available to students included modern genetics to serve as the biological background for the far more politicized constellation of subjects now taught in the biology classroom.

Although biology pedagogues prior to the Third Reich had occasionally emphasized the usefulness of genealogy, it neither loomed large in the demands of school teachers for revisions in the biology curriculum nor in their classroom practice. This changed in 1933. Genealogy, especially the pupils' construction of family trees and the interpretation of famous German lineages like the Bach family, played both an ideological and practical role in Nazi racial policy. Ideologically, it helped make Germany's boys and girls aware of their connection to the larger "organic community." Nazi pedagogue Benze and his coauthor Alfred Pudelko bluntly remarked that the National Socialist "organic ideal" was the cornerstone of genealogy. Careful attention to this subject placed the individual "historically, socially, biologically, and racially in an intricate relationship to a transcendental whole." It also apprised the child of the role of heredity in his or her own family, the nucleus of the National Socialist state. As one biology workbook put it, tracing one's family lineage had "deep meaning." Indeed, "we are dealing with the life unit out of which a people are formed." "The individual person is not its foundation stone; whoever thinks so is making a dangerous mistake. The significance of the family for the rise (or fall) of our *Volk* must be learned anew," the workbook maintained. This "dangerous mistake" should be interpreted as both an incorrect assumption as well as a politically nefarious inference. Although attacks against excessive individualism were common in the biology classrooms of the Republic, during the Third Reich, the individual was always supposed to take a back seat to some larger biological unit. The standard Nazi phrases, "you are nothing; your *Volk* is everything" and "the individual is nothing more than a simple link in a chain," would have been inappropriate rhetoric in the classroom prior to 1933.[27]

Showing how traits such as head shape and nose size were passed down through the generations in a child's own family also concretely demonstrated the power of human heredity. Since pupils were also asked to trace the history of alleged genetic illnesses in their lineage, such practical genealogical exercises simultaneously made them more eugenically aware. In addition, it informed them about their family's fecundity, knowledge that was important for the regime's population policy. Practically, these homework assignments—along with family tree charts frequently appended to biology texts and supplements to be filled out by pupils with the help of parents and relatives—were a means to separate out "Aryans" from Jews and other "racial aliens," especially in the years prior to the notorious 1935 Nuremberg Laws.[28] There was no easier way to ascertain if a child was a "pure Aryan" than to record the ethnological information that students included in these assignments. One can well imagine that pupils often innocently mentioned the unfortunate fact that one or more of their relatives was Jewish, not realizing the consequences of including such information in their homework. In some areas, biomedical professionals outside the classroom encouraged school instructors to have children construct their family trees. It was hoped that pupils would thereby inadvertently ensure that genetically unfit members of their own family were rendered sterile.[29]

As eugenics, racial hygiene, and population policy were usually treated together in school biology texts during the Weimar years, it is appropriate to examine how these two components of racial science were communicated to pupils under the swastika. We have noted that some college-preparatory biology instructors were keen to discuss these issues as part of biology education's civic contribution to the state prior to 1933. This was far more the norm under the Nazis. Indeed, biology textbooks for the graduating classes of all the primary and secondary schools probably devoted more time to these subjects than to any other, in some cases even more than Nazi-specific ethnology. Although the issues of unfavorable population trends among "valuable" citizens, the dangers of degeneration, and the duty of all healthy Germans to have large families were the same both before and after 1933, their treatment under the Nazis was imbued with images and a language of contempt for the "less valuable" that were muted or absent during the Weimar years. This was frequently accomplished by comparisons to what animals allegedly did to preserve their species in nature. "As soon as an animal is seriously wounded or sick, or when, owing to illness, his efficiency is impaired, others recoil from it and leave it to its fate. Such animals frequently separate themselves from the herd on their own and await their end in some hideout. That appears hard to us, but it is necessary for the health of the [species] as a whole."[30] The lesson to be learned is that humans need to adopt "hard" measures to ensure the preservation of the *Volk*.

These texts also favorably reviewed the draconian Sterilization Law of 1933. Pupils sometimes learned of other eugenically progressive countries that also saw the biological wisdom of mandatory sterilization. A German "must be willing to accept the infringement of his personal rights," biology pedagogue Otto Steche told his readers, "when the needs of the national community demand it." In what was certainly an attempt to offer scientific support for involuntary sterilization, biology textbook authors went out of their way to demonstrate the enormous cost of Germany's "degenerates." As Hans Feldkamp explained, these "hereditary defectives" not only strapped the public purse to the tune of 350 million RM per year; he also claimed it would be "inappropriate compassion" not to improve the genetic efficiency of the national community and take the financial strain off national comrades by reducing the number of the "unfit" through mandatory sterilization. After drumming home to his young readers the cost of supporting a special needs child, a mentally ill

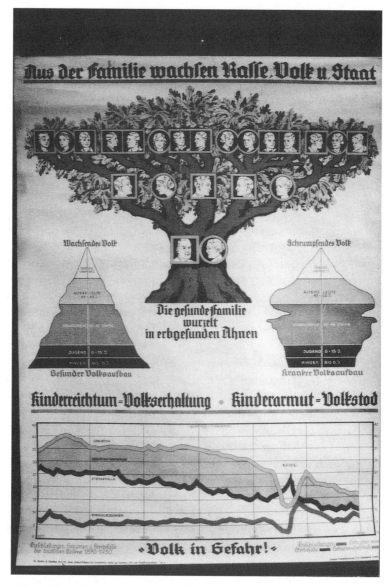

29 "Race, *Volk*, and State Grow out of the Family" and "The *Volk* Is in Danger." These are the two messages of this high school biology classroom wall hanging. Notice the top picture: all the faces appear to look "Nordic." The bottom picture depicts negative population trends from 1870 to 1930. Photo courtesy of the Universitätsbibliothek Duisburg-Essen.

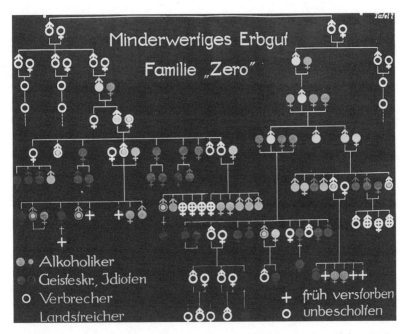

30 A high school biology classroom wall hanging demonstrating the inheritance of degeneracy: the so-called Family Zero. Photo courtesy of the Universitätsbibliothek Duisburg-Essen.

person, and a blind or deaf student, Feldkamp appeared to backtrack by asserting that "certainly the ill deserve acceptable and caring aid." After all, he argued, the Germans were a civilized people. However, the author reminded his readers, the hereditarily fit "are the pillars of the future *Volk;* only healthy parents can expect healthy children." As scientific evidence for his position he included literature from university human geneticists and eugenicists such as Lenz, Fischer, and Hamburg University anthropologist Walter Scheidt.[31]

If there was any modicum of compassion left for the "genetically unfit" during the early years of the Third Reich, this changed by the start of the war. By this time, four sets of official school biology textbooks were in place for use in college-preparatory schools, one of which was authored by the early Nazi sympathizer Jakob Graf. The use of language to discuss this topic in the newly approved books reveals the symbiotic relationship between Nazi racial policy and human heredity that existed from the start of the new regime. However, the pictorial images of the mentally infirm that were specially altered and designed to scare pupils—to render these unfortunates outside the sphere of children's moral obligation—suggest a radicalization of this relationship. These intentionally distorted illustrations of the mentally handicapped were not only available in school texts; they were shown in "documentary" movies such as *Erbkrank* (*Genetically Ill*) throughout Germany. After 1939, it appears that no sympathy at all remained for the genetically "burdensome." A more derogatory rhetorical image of the "unfit" went hand in hand with the visual ones. As one official biology textbook for college-bound students put it, "today it is the case that the state invests more money to drag around a more or less worthless member of the community than a father often has for the sustenance and education of one of his healthy children." The handicapped were henceforth devoid of all value. By using these textbooks, were biology teachers, however unwittingly, preparing their charges for the acceptance of the secret/not-so-secret "euthanasia" project—the "opening act" to the "final solution"?[32] We will recall that Nazi Germany's first foray into state-sponsored murder was directed against handicapped children and adults beginning in late 1939–40. There is certainly no easy answer to this question, but raising it does force one to think critically about the nature of complicity in a dictatorial state, especially of those individuals standing at a distance from the center of political influence and power. The extent of biology instructors' responsibility for the moral indifference of an entire generation of youth toward the fate of Germany's most vulnerable citizens, the handicapped (Jews, Roma, and Sinti were no longer citizens by this time), thus must form a central query of any examination of biology secondary school instruction. It appears that the youth did indeed become particularly insensitive to the plight of these victims of Nazi biological politics.[33]

And what was the official textbook line on ethnology—the subfield of racial science without any prerequisite in the biology classes of the Republic yet closest to Hitler's heart? As in their treatment of other components of racial science, biology textbook authors handled this new form of racist ethnology in a uniform manner. Ethnology in school texts was, by and large, based on the writings of the Jena professor Hans F. K. Günther, a man sometimes referred to as "Race Günther" or "Race Pope" owing to his obsession with his subject.

In his most widely read treatise, *Rassenkunde des deutschen Volkes* (*Racial Science of the German People*), Günther postulated the existence of six so-called European races. Although there were no pure races left anywhere in Europe, some people and nations were blessed with being predominately "Nordic"—the crown of human creation, according to the author. Everything aesthetically, spiritually, and physically positive was embodied in, and attributed to, this so-called race. Nordic blood was responsible for the high level of civilization in both ancient Greece and Rome as well as in contemporary Europe. While the other five European races also had their virtues, especially in combination with the Nordic, none were as lofty as this pinnacle of human evolution. More important for subsequent Nazi anti-Semitism, Jews, Günther argued, were not biological members of the European community. They were an ostensible mixture of two non-European races, the Near Eastern and Oriental. Different physical and mental traits separated Jews from the "Aryan" population. Günther discussed his views on the racial inferiority of Jews in more detail in a separate book, *Rassenkunde des jüdischen Volkes* (*The Racial Science of the Jews*), which, like his earlier treatise, was also published during the Weimar Republic. If, we recall, Social Democrats and Centrists were concerned that the KWIA—with its research emphasis on racial genetics—might undermine the Republic when it first opened its doors in 1927, this was largely owing to the writings of people like Günther. It would take the Third Reich, however, to elevate the Jena professor's views from the intellectual mainstay of a handful of *völkisch* thinkers to the highpoint of secondary school biology education. His popularity among influential members of the Nazi Party elite notwithstanding, it is doubtful that Günther, a man not trained in the natural sciences, would have fueled the fires of racial hatred among the youth to such an extent had he not received some intellectual support from those academic human geneticists who made the Faustian bargain with the new regime.[34] As we have seen, respected non-German biologists also accepted Günther's Nordic ideology and anti-Semitism, even if their views were not grounded in his writings.

How did Günther's teachings on ethnology and that of the Nordic enthusiasts among the German human genetics community reveal themselves in Nazi school biology texts? The section of these books dealing with this subject generally began with the six European races followed by a physical description of them. Almost all biology texts, both those for students as well as those designed for the teachers, included representative pictures of individuals (sometimes drawn, sometimes photographed) who allegedly were prototypes of these races. Sometimes textbook authors chose illustrations of prominent German politicians and generals such as Otto von Bismarck or Frederick the Great to showcase representatives of the most desirable European races, the Nordic and Westphalian; frequently the Nordic physical ideal was communicated to pupils by showing it depicted in the art of ancient Greece and Rome as well as in sculptures of the German Middle Ages and Renaissance. Occasionally, books would explain to pupils how to mathematically derive the so-called skull index, an ostensibly important physical racial trait.[35]

Die Raſſen des deutſchen Volkes

nordiſch

fäliſch

oſtiſch

oſtbaltiſch

dinariſch

weſtiſch

31 A high school biology classroom wall hanging representing Hans Günther's ubiquitous "Pictures of the German Races." According to Günther, there were six so-called races that made up the German population. Jews were allegedly a mixture of two non-European races that were "alien" to those that comprised "Germans." For Günther, the most noble of the racial strains found among Germans was the so-called Nordic. Photo courtesy of the Universitäts-bibliothek Duisburg-Essen.

Although it was clear that Hitler and many of his henchmen believed that Nordics led humanity, there was a potential problem in overstressing the physical appearance of this so-called race. Nazi Party officials and pedagogues were fearful that an emphasis on the so-called physical traits of the Nordic—fair skin and hair, blue eyes, and a long head, etc.—would lead to divisions in the classroom between students who matched this ideal and those who did not. In other words, it would serve to pit one "national comrade" against the other, as class and occupational divisions had done in former times. Hence biology school teachers were instructed by NSLB biology experts and Education Minister Rust to stress that all Aryan Germans had a significant portion of Nordic blood running through their veins.[36] Moreover, following the opinion of Fritz Lenz, racial science textbooks downplayed the physical attributes of European races—which were seen as being of little intrinsic importance—but rather emphasized their alleged mental and character traits. As one biology textbook author put it, "the human races are not only different from each other physically, but also mentally. If there were only physical differences, the racial question would in essence be meaningless."[37] Lenz's position on the subject was precisely the same.

So what were the spiritual traits of the Nordics, allegedly the highest "mental" and physical race? Nordics, biology textbook author Hans Feldkamp reported, embodied "confidence, strength of will and military preparedness." Quoting Günther but also following Lenz's tendency to read the desirable mental traits of the educated middle classes into the "Nordic race," Feldkamp continued his description of Nordic virtues. "The Nordic individual is typified," the author continued, "by his pronounced sense of truth." Nordics are also known for their "objectivity" and their trend toward a "calm rationality." "They are the most creative of all the races," Feldkamp added. Indeed, "through their creativity, rationality and spirit of sacrifice they have mastered nature. As such, one can correctly state that with respect to leadership and mental talent, the Nordic race stands above all others."[38]

However, since one could not tell a true Nordic by outward appearance but through "achievement," as one teacher handbook put it,[39] even a brown-haired and round-faced German child could represent the best specimen of humanity. The precondition was that he or she demonstrated the desired Nazi virtues—more or less the same ones lauded by the traditional, pre-1933 eugenic understanding of "fitness." Feldkamp expressed this point most clearly when, including a quotation from Günther, he argued that "even if the Nordic ideal is not outwardly visible, the mental attitude always reveals Nordic genes. As such, no racial divisions exist in the German people. Rather, every one of us faces the decision of being 'for or against the prototype of the Nordic individual.' "[40] One can imagine how such an emphasis on the "mental" characteristics of race promoted conformity to Nazi principles among pupils. Real Nordic children followed Nazi principles in word and deed. Hence, to be a true "Nordic" it was necessary to be a committed Nazi. If "Aryan" children conformed to Nazi principles—if they joined the Hitler Jugend (Hitler Youth) or the Bund deutscher Mädel (Association of German Girls)—the party leadership was satisfied.

As long as these children were "national comrades," there was no problem in equating race with efficiency and conformity to Nazi ideals. But what held true for "Aryan" pupils did not apply to those considered by the regime to be "racial aliens." The major purpose of ethnology was not to have children memorize the six European races. It was not even primarily to fill them with Nordic pride. Ethnology, like eugenics and racial hygiene education, was a vehicle to pontificate on human inequality. "It is a mistake," one school biology textbook author proclaimed, "to accept that 'anything with a human face' is equally valuable." Another textbook writer continued in the same vain: "It is one of the greatest lies of the French Revolution to maintain that all humans are equal. Nature knows no equality." The most important reason for mandating this subject was certainly to warn against miscegenation—to emphasize the alleged racial danger of "Aryans" mixing with Jews.[41]

Although Roma and Sinti as well as Afro-Germans were also viewed as racial outsiders, early Nazi secondary school biology texts almost exclusively emphasized the physical and mental differences between "Aryans" and Jews. Since they were far more integrated into German society than other "racial enemies," Jews allegedly posed a greater danger than any other "non-Aryan" group. The racial inferiority and threat of the Jews, of course, was also one of Hitler's major obsessions. Not a single biology textbook existed after 1933 that failed to stress the obligation of "Aryan Germans" to avoid physical contact with Jews. Even though such works varied in the amount of detail they included on the "Jewish question" and whether they actually stated that Jews were racially inferior to or "merely" "racially different" from Aryans—as Fischer and von Verschuer along with some biology pedagogues put it—the message was that "the Jewish race is alien to the European races and any mixing must have a negative impact." Hence, any German conscious of the importance of heredity should avoid such miscegenation "for genetic reasons." Indeed, for some biology textbook authors, it was not enough merely to forbid that Germans and Jews have physical contact with each other. Any mixture between these two alien groups should also be viewed by national comrades "as a defilement of their own species." Those texts written in the aftermath of the Nuremberg Laws frequently quoted them. Some included pictures designed to acquaint children with the arcane and demeaning Nazi terminology of "full Jews," "half Jews," "quarter Jews," etc., and remind them of who could marry whom.[42]

Although the worst diatribes against the Jews in the early years of the Third Reich were found in elementary school texts (or in books for elementary school teachers), with the 1938 school reform, and especially with the outbreak of war, there was a radicalization of the rhetoric against and images of all racial aliens, just as there was for the handicapped. Attacks against the Russians and even the Americans could be found after the former began to beat back the Germans and the latter entered the war. While the 1940 edition of Graf's *Biologie für Oberschulen und Gymnasien* (*Biology for Secondary and College Preparatory Schools*) avoided any discussion of the Russians (while the infamous Hitler-Stalin Pact still held), the 1943 edition (after the debacle for the Germans in Stalingrad) described them as having destroyed their Nordic blood in the 1917 Revolution and replaced it with the "defective hereditary material in the guise of Bolshevik sub-humanity." It would hence not be difficult for future German soldiers on the Eastern Front to carry out their "racial war" knowing that they were dealing with subhumans.

Interestingly, the same author (albeit in another volume of his book) also accused the United States of hypocrisy in racial matters. America, Graf argued, was the first to have sterilization and miscegenation laws. Now this country had the audacity to point its finger at Germany. He offered German students a seemingly viable explanation: Were it not for the "enormous Jewish influence in the United States," such a state of affairs would be impossible. On the same page, he presented German secondary school teenagers with a picture entitled, "Racial Mixture in a Harlem School; Harlem, the Nigger City of New York." Yet earlier in the same text, German soldiers were portrayed as gleefully talking to young "racially related" boys from occupied Holland.[43] It is clear that such textbooks lay bare the symbiotic relationship between human heredity and Nazi politics, which was radicalized by the contingencies of preparing German youth for a protracted war in the East against "subhumans" and for continued fighting on the Western front against a country allegedly in the hands of the Jews.

In addition to their classroom duties, many secondary biology teachers also served their own professional interests as well as the policies of the regime through a variety of extracurricular activities. In particular, they helped organize, held lectures at, or merely attended biology seminars on various facets of racial science. These seminars, sometimes sponsored by state offices, sometimes by the Racial Policy Office of the NSDAP or the NSLB, were numerous. The teachers were expected to pay for their own expenses. Such courses were naturally led by individuals able to teach the new subjects from the National Socialist viewpoint. Only Nazi biology educators, usually university professors or higher school biology teachers approved by the NSLB, were permitted to conduct them. Examining three such seminars will give us an idea of the additional subject matter presented to secondary school biology instructors.

On August 6, 1934, a two-day seminar for biology teachers of the higher secondary schools was held in Posen, a region in the eastern portion of "Greater Germany." More specifically, it was given at an asylum for the mentally ill in the small town of Oberwalde. In addition to having the advantage to see firsthand the tragic results for the *Volk* of the alleged increase in "degenerates," the teachers started their first day at 8:30 a.m. with a talk on "Hereditary and Racial Biology and Their Importance for the National Socialist Worldview"; from 10:00 to 11:30 a.m. instructors were enlightened by a state medical official on "The Genetic Foundations of Racial Hygiene." After a common lunch, teachers heard lectures on the diseases falling under the new Sterilization Law and the population-policy problem in Germany. The evening of the first day was reserved for a display of the newest literature and teaching aids on the subject.

The three-day school course held at the University of Münster in March 1934 was primarily devoted to ethnology. It, unlike the previous seminar, was organized by the NSLB. The thirty-three higher school teachers in attendance heard talks on "The Main Themes of Ethnology," "The Laws of Heredity," "Human Heredity," and the "Taxonomy of the German Races." On Saturday, the second day of the course, there were lectures on racial biology and racial mixtures, the prehistoric racial condition of the German people, as well as race and history. From the written report included on the event, this particular school course was a success.[44]

In 1938, the Racial Policy Office of the NSDAP felt compelled to hold a "racial science" symposium for biology teachers in Bavaria. Its title: "Racial Mixtures as Life-Detracting and Dangerous." A number of prominent secondary school biology textbook authors such as Werner Dittrich, Erich Stengel, and Karl Zimmermann held talks in the overfilled auditorium of the University of Munich. For good measure, Julius Streicher was asked to give the opening address at the event. In accordance with his and other high Nazi Party officials' depreciation of book learning, Streicher reminded the educators in his audience that, pedagogically speaking, "racial science should be presented to children in a simple and clear manner. No theoretical knowledge was necessary."[45]

Racial Science Practice

Despite the fact that biology instruction, and racial science education in particular, was deemed absolutely essential to the needs of the Nazi state, it took five years until new curricular guidelines and uniform textbooks were available for this "central subject." To be sure, all school disciplines had to wait until Rust managed to produce his *Erziehung und Unterricht in der Höheren Schule* (*Education and Instruction in the Secondary School*) in 1938. The lack of binding guidelines and official texts created an "orientation problem" for many school educators during the early years of the Third Reich. The famous left-wing German playwright Bertolt Brecht depicted the general dilemma of teachers in his play *Fear and Misery of the Third Reich* while in exile in the United States. In one particular scene, a *Studienrat* admits that he is ready to teach anything, if only he knew what "they" wanted him to teach. "What do I know about how Bismarck is supposed to be presented," the higher secondary school history instructor bemoaned in the play, "as long as they have not issued the new school texts!" Like the history instructor in Brecht's play, biology teachers anxious to do the bidding of the regime might have found it difficult to figure out the official line on their subject at the outset.[46]

When the new curricular guidelines were finally published, the twenty-four pages devoted specifically to biology instruction in the Reich reflected the subject's ideological importance for the Third Reich. More space was devoted to it than to physics and chemistry education combined. By the time Rust's reforms were instituted, any lingering "orientation problem" on the part of individual biology instructors notwithstanding, the community of interest between many secondary school biology teachers and party and state was already well developed. As was mentioned, college-preparatory biology instructors received what they had demanded since the turn of the century: two hours a week of biology instruction for each school year. Moreover, at least on paper, biology at most higher secondary schools now enjoyed more hours than physics and chemistry combined. And finally, all children attending such schools were officially required to take an oral examination in racial science. The party and state were also seemingly well served by secondary school biologists. As we have seen, in numerous books and pamphlets secondary school biology teachers went beyond the call of duty to demonstrate the compatibility of their subject matter, especially racial science, with Nazi doctrine. Ideologically committed biology pedagogues also authored the four sets of secondary school biology texts designed for classroom use after 1938.[47]

So much for the racial science ideal; now we must try to look behind the surface to see what biology teachers actually taught and what pupils really learned. We will also need to ask whether secondary school biology teachers were completely satisfied with the deal made with the Nazi state, and whether the regime was likewise content with the teachers' work. The mutually beneficial arrangement between the school instructors of human heredity and the National Socialist regime—while undoubtedly generally operative—might have been somewhat less than perfect in practice. As was mentioned earlier, assessing these matters is not straightforward.

One of the first steps toward discerning how completely symbiotic the relationship between secondary school biology teachers and the Nazi state was in practice is to try to assess teacher loyalty in the classrooms—again, a thorny enterprise. Although we lack statistics for biology instructors, only 3 percent of all *Studienräte* (355 secondary school teachers in all) were dismissed following the 1933 Law for the Restoration of the Professional Civil Service.[48] Since virtually all Jewish secondary school teachers were immediately forced into retirement as "racial aliens," this means that relatively few instructors were thrown out of the schools for being politically unreliable. Although many *Studienräte* were nationalists and not fundamentally opposed to the Nazis, certainly not all were. We can assume that democrats, political Catholics, and Social Democrats remained in their posts unless they were openly hostile to the new order. Those among them teaching biology—however much they wanted to improve the status of their subject within the school curriculum—were probably not anxious to instruct their students in the most overtly racist and scientifically controversial aspects of racial science, ethnology. They might have done so only under duress, especially if their principal was a hard-core Nazi. Such teachers might also have been sensitive to the plight of their Jewish pupils, who, at least until 1938, continued to attend secondary schools, albeit in ever-dwindling numbers. An overemphasis on the "Jewish question" and ethnology would obviously exacerbate the suffering of Jewish children in their midst. These teachers' duty as civil servants to the state might not have resulted in their carte blanche acceptance of the Nazi Party line on "race."

And even the nationalist-minded college-preparatory biology instructors—those who would have the most in common with the Nazi state—

were interested in preserving the integrity of their profession. This was true of all *Studienräte,* regardless of their political persuasion.[49] This personal identification with their profession might work against the symbiotic relationship between higher school instructors and the Nazi state. Let us not forget that Hitler and his paladins downplayed the importance of intellectual content in the schools. PhD biology teachers would have had to weigh their desire for increased prestige in the school curriculum and their own personal commitment to racial science with the threat of their authority being undermined by other Nazi "educational institutions" like the Hitler Youth (HJ)—the bane of many a *Studienrat*'s existence. HJ meetings were often used as an excuse by teenagers to miss class or avoid assignments. Even the NSLB, a Nazi Party organization, complained that the Hitler Youth movement lowered the level of academic work in the schools. In this conflict between the HJ and the NSLB we once again find an example of rivalries among state and party institutions that impacted governmental efficiency during the Third Reich. It was also generally known that members of the HJ often denounced teachers (and sometimes their own parents!) to their superiors. This, of course, could have consolidated ideological conformity in the classroom and worked in favor of the regime. Yet it might also have angered instructors such that they avoided going out of their way to teach their subject in an overtly ideological manner. They could at least have retreated into professional objectivity without much risk. HJ actions might also have led to resignation and apathy among secondary school instructors.[50] Why should even a pro-Nazi teacher have bothered to learn this new racial science material if his charges were allowed to attend a HJ sports event in lieu of doing their biology homework?

And then there was the question of implementing the appropriate textbooks to teach the new meta-field. Although, as we saw, there were literally dozens of newly approved textbooks for racial science instruction, relatively few of the Prussian secondary schools chose to adopt them outside of Berlin. Older biology textbooks, some originally written under the Empire, continued to be employed in many classrooms. This certainly exacerbated the "orientation problem" depicted by the *Studienrat* in Brecht's play. Even after Rust's 1938 reforms, the four state-mandated sets of secondary school biology textbooks written to reflect the new curriculum were not immediately used everywhere. And where they were adopted, these texts were only in use for a relatively short time. By 1943, allied bombing raids made school education more and more precarious, at least in the major cities. Children were frequently packed off to the countryside for their own safety. We should also note that after 1939, the additional time needed for physics and chemistry education—subjects more directly relevant to the war—spelled the end to biology exit exams in the natural science–oriented secondary schools. And even the oral examinations in racial science were not always given, although Rust never officially changed his 1935 mandate.[51] This might well have dampened the enthusiasm of some ideologically committed biology instructors. All of the above-mentioned ambiguities certainly ran counter to the convictions of Nazi pedagogues for whom racial education was central.

Morale reports of the NSLB district leaders for biology and "racial questions" suggest that at least some biology instructors, especially those teaching in the higher schools, were not completely satisfied with their situation. As one district leader from Mecklenburg bluntly put it, "the enthusiasm for frequent large events on the topic of race is, in general, lacking among educators." Another Nazi pedagogue, this time from Magdeburg, mentioned that given the ideological importance of biology, it would be good if this subject were taught by bona fide biology teachers. Although his report was written in 1937, there were complaints that "nonexperts" in college-preparatory schools still delivered instruction in this subject in his district. The morale of biology teachers in Merseburg-Halle also left much to be desired, since time once reserved for biology were now given over to sport. "The biology teachers," the district leader continued, "could not understand this." And finally, the report written by Jakob Graf, committed Nazi textbook author and district leader for racial questions among biology teachers in Hessen-Nassau, was especially damning. In his discussion on the morale of educators in his region, he revealed the stark reality: "There is a general apathetic and lukewarm attitude" among the teachers. The instructors at higher schools "justify their avoidance [of extracurricular schooling activities] by saying that they do not have the time, since they are dependent upon giving private lessons [in their field]." The other teachers legitimized their lack of enthusiasm for attending such retreats by stressing that they are "overburdened" with other Nazi Party service. The attendance at these extra-school activities is always larger, Graf remarked, "when the [teachers'] superiors [e.g., principals]" themselves show interest in various racial questions (suggesting, of course, that such was not always the case). Those responsible for organizing teacher seminars probably also attempted to cram too much into the instructors' heads at these retreats, Graf concluded.[52] Such reports not only reveal a certain amount of dissatisfaction from biology teachers; they were probably also not exactly music to Nazi diehards' ears.

If we turn from official reports to the written testimonies of individuals schooled during the Third Reich, we find that they also reveal the ambiguities of racial science practice under the swastika. Holger Börner, onetime prime minister of Hessen in West Germany, readily admitted that racial purity was the goal of biology education under the Nazis. On the basis of his own experience, however, he argued that racial science instruction had only a limited influence in college-preparatory schools during the Third Reich, although it was critical for the education of elementary school teachers. Hans Jung, a bishop of postwar West Germany's Protestant Church, thanked the biology education acquired under the Nazis for making him realize that racial theories are scientific rubbish. In Jung's opinion, the National Socialist leaders did not believe that schools were an efficient vehicle to communicate their ideology anyway. Another individual, a practicing judge in West Germany schooled in the Third Reich, commented on the way his particular biology instructor negotiated the problem of not believing in what he was expected to instruct. Naturally, the judge's *Studienrat* had to teach racial science. But he attempted to be objective in his instruction and, according to the judge's best recollection, informed the pupils that they shouldn't allow the "propagandistic racial teachings" to confuse them. "Hitler himself," the judge's teacher allegedly maintained, "doesn't understand anything about race." And finally, former president of West Germany Richard von Weizsäcker's testimonial confirms that the ideal and reality of racial science instruction might have differed. According to von Weizsäcker's account, many of his teachers were schooled in the pre-Hitler period and had nothing in common with National Socialist ideology. Since racial science "meant a lot in the Third Reich and should be conveyed in biology instruction," von Weizsäcker's higher secondary school biology teacher discussed the upcoming biology exit exam with his pupils. His biology teacher mentioned that he personally did not believe Nazi racial theory, but he had to offer some questions from this subject on the *Abitur* exam. He hinted to his pupils how they should answer these questions to receive a passing grade. "What this teacher did," von Weizsäcker admitted, "was courageous." "If one of the pupils had denounced him [the biology instructor], it would have been bad for him. But no pupil did so."[53] Although testimonials by prominent West Germans are certainly not representative, they suggest that racial science in the biology classroom was not always conveyed in a manner that would have won the approval of Nazi pedagogues. They also add a touch of gray to the classic attempt to paint biology education under the Nazis in simple black and white terms.

More typical Germans who attended schools during the Third Reich have also left their recollections of biology instruction under the swastika to posterity. H.P.H. recalls that his biology instructor Dr. Plantikow at the Kaiser-Wilhelm-Realgymnasium, located in a working-class district of Berlin, was a member of the Nazi Party. His membership, however, seems not to have had a real impact on his teaching. He appeared to H.P.H.—who as a "half Jew" would have been particularly sensitive to racial science diatribes—as "anything but a Nazi." Other former pupils who attended Plantikow's biology class confirmed H.P.H.'s assessment. On the other hand, Plantikow offered "The Nordic Race and Nordification" as a theme for graduating pupils in 1937. Whether he also provided hints on how to answer the question "correctly," as von Weizsäcker's biology teacher allegedly had done, is not known. What is clear is that not all biology instructors at this higher secondary school were members of the NSDAP. According to the testimony of his former pupils, *Studienrat* Dabel, never a party member, offered a livelier form of biology instruction at this same school than his colleague. Dabel allowed Eugen Fischer, "an absolute representative of racial science," to persuade his male students to do research. Fischer encouraged them to go to villages and measure the head shapes of the inhabitants. It is likely that the KWIA director made this suggestion at his Institute—the result of Dabel's biology class outing to Dahlem. Herr Dabel's teenage boys did indeed undertake ethnological head measurements in the countryside.[54] From this incident we can see Fischer's practical influence on secondary school biology education, and Herr Dabel may well have been proud to attract such a notable scientific researcher to speak in front of his class.

Exit exams, both oral and written, probably offer the best indication of what pupils actually learned in the biology classroom under the swastika. Although they might not have been tested everywhere after the start of the war, German pupils were grilled orally on racial science in the graduating classes of many higher secondary schools, not only in Berlin but as far away as the Bodensee, a picturesque lake district on the Swiss border.[55] And while written exit exams in biology were no longer mandatory after 1939, the few available to us before then provide us with an invaluable insight into the reality of higher secondary school biology education. Their relative scarcity is inversely proportional to their relevance for assessing everyday life in the biology classroom during the Third Reich. Given their small number, however, we unfortunately cannot consider them to be a representative sample of such exams. They can merely offer an impressionistic view of the reality of racial science education under the Nazis. These sources sometimes serve as a counterweight to, sometimes confirm, the general impression we obtain both from examining the words of Nazi pedagogues, official school texts, and curricular guidelines.

In the comfortable middle-class district of Berlin-Wilmersdorf, four-teen teenage girls, among them two Jews, took their oral exams in racial science in the Cecilienschule in 1934. Their biology teacher happened to be female—an exception among higher school educators, even in girls' schools. Frau M.-K. posed questions from all subfields of the racial science complex, but most were not directly related to ethnology. The two Jewish pupils, A.P. and O.S., were spared demeaning themes relating to their own alleged "racial difference" from the other graduates. We do not know the political orientation of Frau M.-K., but at least she appears to have been reluctant to torment her "racially alien" charges more than necessary. She asked them to discuss "The Genetics of Red-Green Colorblindness" and the meaning of the "Population Pyramid of 1925," respectively. One non-Jewish girl given the task of explaining, in ten minutes, "The Ger-man *Volk* in its Racial Composition" apparently had some difficulty with the topic. According to the protocol, she did not know the exact percent-age of any given race in the German population according to the views of Hans F. K. Günther.

Three years later the same instructor tested seven girls in biology, in-cluding two Jews. As was the case in 1934, the two Jewish teenagers were given more traditional topics to answer. One had to describe the cell nucleus; the other, the genetics of albinism. M.B., however, had fifteen minutes to tackle a more unusual question related to Germany's desire for economic self-sufficiency: "To What Extent Can Genetics and Its Practi-cal Application in Agriculture Help to Fulfill the Four-Year Plan?" Indeed, as late as 1943, biology teacher M.-K. was still testing her girls, although by this time there were no longer any Jewish pupils among them and less than one-third of the teenagers who completed their *Abitur* that year took an oral exam in racial science. Only one of the six young women who took such an oral exam had a question that was not on genetics, and even she was only asked to discuss a population policy theme.

The neighboring Viktoria-Luise-Schule, unlike the Cecilienschule, was a higher secondary school for girls stressing the natural sciences. The female pupils were still examined orally in racial science, this time by a male PhD biology instructor, Dr. O. Here, too, there were Jewish girls among the teenagers obtaining their *Abitur*. In 1934, Dr. O. allowed R.B., a Jew, to select her own topic. She chose to discuss some aspects of com-parative anatomy, a theme not directly relevant to racial science. The other Jewish girl, H.S., was asked about blood types. Dr. O. did, however, ask an "Aryan" pupil about the Nazi Sterilization Law; she was also ques-tioned about the position of other countries on rendering their "unfit" sterile. The other girls were asked a potpourri of questions on facets of racial science, whereby most of the topics were not directly related to ethnology. Among the oral exams in racial science held two years later, Dr. O. asked E.B. to explain the meaning of Reich Interior Minister Frick's 1933 address on population and racial policy. E.B. discussed—albeit with difficulty—the meaning of a Frick quote dictated by Dr. O.: "We must have the courage to divide the national body (*Volkskörper*) according to its genetic worth." She did better, however, in explaining measures designed to preserve racial purity. Here she mentioned the Nuremberg Laws. E.B. could also give examples of the "disharmony" that ostensibly resulted from racial mixing. Other questions that year varied. There were topics on the history of evolution and genetics (one even asked about turn of the nineteenth-century French biologist Georges Cuvier's "catastrophism"—a geological theory designed to explain the existence of fossils without postulating evolution), practical genetics exercises, questions on racial hygiene and population policy, and themes stemming from ethnology. Dr. O. made sure to represent all facets of racial science in his oral exams. Yet he, like his female colleague at the Cecilienschule, never forced a Jewish child to discuss ethnology. Moreover, the oral exams at these two schools provide no evidence of a gender-specific education in secondary school biology at the outset of the Third Reich. The situation appears to have changed, however, with the higher school reform in 1938.[56]

If we examine the topics presented for the written biology exit exam at various Berlin *Oberrealschulen* for 1934–35, we find that traditional themes not immediately relevant to racial science were given alongside those stressing pure genetics and some that were more obviously ideological. In other words, at one of Berlin's *Oberrealschulen* for girls, the Städtisches Oberlyzeum, graduating seniors had to write on "Fat and Carbohydrate Metabolism of the Human Body." Teenage girls attending the Agnes-Miegel-Schule, however, discussed "Plant Genetics." And the young women of the graduating class at the Städtisches Viktoria-Oberlyzeum were presented with the theme "Artificial Selection in Plants and Animals and Its Importance for the Autarchy of a Country." Given that there were thirty-six different *Oberrealschulen* for boys in Berlin (as opposed to a mere six for girls), there was a far greater variety of themes offered to them than to their female counterparts. Here, too, there were such traditional biology topics as "Animal Tissue," "The Developmental Stages of the Optic Nerve in Animals," as well as the "Active and Passive Immunization of the Human Body against Bacteria," offered at three such schools where biology was tested. There were also several exams dealing with themes in genetics such as "Hereditary Factors and Environmental Influences: The

Importance for the Phenotype," "A Genetics Experiment," and "Theory and Practice of Our Genetics Experiments in the School Garden."

But there were clearly numerous questions related to eugenics/racial hygiene and population policy at several institutions. Racial science themes that demonstrated the reality of ethnology instruction were also not lacking. At the Albrecht-Dürer-Oberrealschule in Neukölln, for example, the graduating teenage boys were asked to respond to the following query: "How Can the Legislator Contribute to the Preservation and Improvement of the 'German Race'?" And at the Friesen-Schule the young men receiving their *Abitur* were required to address the question: "Why Are the Jews a Danger for the German *Volk*"?[57]

As has been indicated, the list of topics available to help us discern the nature of biology exit exams is limited; the actual written and graded exams at our disposal are even more so. What has been analyzed, however, suggests that traditional biology topics along with those dealing exclusively with genetics were more common than those that were overtly political, at least at the outset of the Nazi era. There is some indication in this tiny sample that college-preparatory biology teachers offered a larger percentage of themes dealing with population policy, racial hygiene, and ethnology after the first few years of the Third Reich than they did at the beginning. However, as the above list of themes for biology exit exams from other Berlin schools indicate, the actual exams analyzed may be somewhat more skewed to the non-overtly political side of the spectrum than was normally the case. As will become clear from the following analysis, even those topics that were not overtly political were useful to the regime. Yet we can say with certainty that traditional topics were offered as late as 1938 and ideological themes were proposed as early as 1934.[58]

Consider, first, the topic offered for the 1934 biology exit exam at the Leibniz-Oberrealschule—the same institution that asked its graduating students to write on eugenics two years earlier. All seven male pupils were asked to demonstrate their knowledge of Mendel's experiment discussed several months earlier in the biology class of Dr. B. The title of their written exercise: "The Mendelian Laws of Genetics: Mono- and Dihybrid Crossing." Although the exams varied sharply in the quality of detail they provided and in their command of both technical terms and the German language, they all followed the same basic structure.

From an examination of several exit exams, it is obvious that Dr. B. had taught his students not only a brief history of the rediscovery of Mendel's laws, but the most up-to-date knowledge of genetics and cytology. Through handwritten illustrations of various crosses, the boys explained the principles of blending inheritance with reference to the experiment carried out by Correns on the flower *Mirabilis jalapa* (four-o'clock marvel of Peru). Through the use of Punnet squares, they were also able to give a theoretical justification for the expected ratios for Mendel's work with the edible pea with one and two characteristics. Some of the students deftly linked Mendel's work to contemporary findings dealing with chromosomes and microscopic cell division. But as one or two remarks by the boys suggest, this genetic information was especially important since it demonstrated the power of heredity and helped them better understand what was truly important: its role in humans. In that sense Dr. B's biology exit exams—although scientifically objective—were also serving a political function.[59]

The same can be said of the tests written by graduating young women that year at the nearby Viktoria-Luise-Schule. They were also required by their biology teacher, Dr. O., to write on a genetics topic: "The Meaning of Drosophila Experiments for Heredity." Here, too, the exams demonstrated state-of-art knowledge of genetics, especially the role of T. H. Morgan and the importance of the fruit fly as an experimental animal to verify Mendel's laws.[60] It is noteworthy, however, that two years later, in 1936, the teacher responsible for composing the topics for the biology exit exams at the Staatliche Augusta-Schule in Berlin, another higher secondary school for girls in the natural sciences, made no attempt to hide the political motivation behind his "pure genetics" question. He asked his female students to use the crossing of two races of Drosophila to demonstrate the "scientific" foundation of one of the Nuremberg Laws' regulations: "one-quarter Jews" may marry pure "Aryans" without any deleterious genetic results.[61]

By 1938, Rust's reforms for the secondary schools came into effect. Many of the schools that had stressed the natural sciences no longer did so. As such, far fewer biological exit exams have been preserved in the last year they were officially tested. Among them, however, some were decidedly ideological.

In the Moltke-Schule one of the topics that the male graduates could address was "The Birth Decline in Germany: The Facts and Their Meaning for the Present and Future of the German *Volk*, Its Causes, and the First Measures Taken by the Regime to Eliminate the Menacing Danger." H.S. selected this important national issue. Although the section of his exam dealing with the causes of Germany's population problem could have been written in the Weimar Republic, H.S.'s discussion of the means through which the government was tackling the problem could not belie the essay's political orientation. Since the "Nazi seizure of power,"

H.S. asserted, "there have been numerous measures instituted to raise the birth rate which have not failed in their effectiveness." "First, there must be a mental turn-around on the part of the *Volk*," H.S. reasoned, "such that it has faith in its government." Since the National Socialist takeover, "the entire *Volk* has trustworthily stood behind Adolf Hitler." The Nazi government, H.S. continued, passed a vast array of laws designed to alleviate the problem. In addition to legislation, economic and social institutions such as "Mother and Child" and the National Socialist Welfare Association made it easier to care for large families from poorer backgrounds. Moreover, "land settlement plans will depopulate the large cities"; the urbanites will, once again, rekindle their "natural relationship to their native soil." H.S. then reiterated his original point: "But could any of these developments have occurred if the most important one—faith in our leadership—had not returned?" "We want to believe that every German recognizes the danger that we are still in—despite the improvement in the birth rate—and realize that it is one of his tasks to protect the existence of our *Volk*," the teenager concluded. In his comments, H.S.'s biology instructor stressed the efforts made by the author to carefully delineate the regime's policies against birth rate decline. The teacher, Dr. B., obviously appreciated his intellectual diligence and gave H.S. a "2" (good) for his exam.

C.T. graduated from the Moltke-Schule that same year having written (from his teacher's standpoint) an exemplary biology exit exam on "The Racial Structure of the German *Volk*, the Danger of Racial Mixtures and the Measures Taken by the National Socialist State to Protect German Blood." This was clearly a topic on ethnology—one that could hardly be more in line with the ideals of Nazi pedagogical diehards. All the specifics mentioned in the official school textbooks on the subject of ethnology—most of them originating from Günther's work—could be found in C.T.'s exam: the six European races, the percentage of "Nordic blood" in the German population, the mathematics of determining a skull index, and "Nordics and Westphalians" as the races with the most desirable spiritual attributes. "The Nordic race is considered the efficiency race," C.T. wrote. Since it was forged in difficult climatic conditions, "everything that could not withstand [the race's] hard fate was destroyed by nature." As such, its mental qualities include "courage, rationality, a strong will and determination and also pride and a cool reserve." The Nordic individual, C.T. continued, is very efficient and always prepared. . . . He doesn't serve the national community by being a part of it, but by fighting and working for it. . . . The Westphalian person is someone who is steadfast; he is in many respects similar to the Nordic individual."

C.T. then turned to the question of "Nordification." Here he followed Günther, Lenz, and many Nazi biology pedagogues in equating the term with "the efficiency principle." As one would expect, racial mixing with Jews was the ultimate evil. The Jews, C.T. made clear, must be viewed as a "mental race," a view articulated by Lenz more than a decade earlier. "The Jew is not at all a nomad," C.T. explained. Rather, the Jew is "a parasite." To protect the national community from the danger of racial mixing that plagued Germany in the past, the Nazi government passed the Nuremberg Laws. C.T. went on to describe them in detail and ascertain who was permitted to marry whom. Dr. B., the *Studienrat,* felt that the exam displayed a "maturity in its judgment" and demonstrated the "capability of the author." He made but two negative comments in the text. Dr. B. took the boy to task for stating that the Near Eastern race (one of the two out of which the Jews were allegedly comprised) was more or less the same in its character traits as the Dinaric, one of the six European races. Clearly such an error of judgment could not be overlooked, even in an exam that received the highest possible grade, a "1" (very good). C.T.'s biology teacher also criticized him for using a word of foreign origin when a good German one was available.[62]

That several secondary school biology instructors somewhat fulfilled the role that Nazi pedagogue Benze articulated—to serve as experts for racial science in other subjects—appears to be confirmed by the exit exam on "The Biological Foundations of *Völkisch* Racial Care" with which this chapter began. As was noted, the two students writing on this topic were not satisfying their requirement for biology, but for German. Dr. F., a biology instructor at the Bismarck-Gymnasium, served as the "expert" on the topic.

As evidenced in the outline of the topic written on the first page of the exam, the two boys were clearly versed in the "big names" in the history of evolution and genetics. In particular, P.K. knew the meaning of Weismann's work, even if he could not spell his name correctly. For him, the German embryologist's studies proved for all time that the inheritance of acquired characteristics—"Lamarckism"—did not hold true. "Communism," P.K. continued, is based on this "faulty theory." As such, it supports "human equality."

The students had also mastered the basics of racial hygiene, as well as the standard line on ethnology. Their essays were passionate and appear to have been ideologically committed. Applying the principles of eugenics to "Aryans" was the task at hand. "Ethnology," P.K. maintained, is extremely important "in that it teaches us about the different characteristics of races." Although knowing racial characteristics might make

32 The first page of graduating high school student P.K.'s exit exam from the Bismarck Gymnasium in Berlin. The topic that this student chose as a qualification for the German language (not biology) in 1934 was "the biological foundations of *völkisch* racial care." The illustration contains the introduction to the topic "*völkisch* racial care as the practical application of the science of biology" as well as an outline of the main part of his essay. The conclusion stresses the "importance of *völkisch* racial care." We see that the teacher corrected P.K.'s spelling mistakes. Illustration courtesy of the Archiv und Gutachterstelle für Deutsches Schul- und Studienwesen/Landesarchiv, Berlin.

Körperzelle   Verteilung

Reifungsteilung
od: Reduktionsteilung.

Keim zellen mit halber
Chromosomenzahl.

Ab: 2.

Befruchtung   Neue Körperzelle.

33 A page in the middle of P.K.'s exam. On a prior page, the student explains the operation of Mendel's Second and Third Laws, using the garden pea as an example. Depicted in the student's illustration here is the microscopic confirmation of this process through an examination of mitosis and fertilization of the parasitic roundworm (which the student confuses with the tapeworm). P.K., however, makes a much more important mistake by saying that Darwin built upon Mendel's laws. As seen in the illustration, the biology teacher responsible for grading the content part of the exam states "wrong: Darwin did not know Mendel's laws." Illustration courtesy of the Archiv und Gutachterstelle für Deutsches Schul- und Studienwesen/Landesarchiv, Berlin.

34 The final page of P.K.'s exam. The student explains that the practical side to the science of biology is racial care; it alone can once again provide the *Volk* with inner (i.e., genetic) health. However, "the individual must recognize the importance of racial hygiene and support it in deed and with sacrifice. Regarding the question of the preservation of the race the following statement holds true: 'the good of the community comes before that of the individual.'" Interestingly, the biology teacher responsible for grading the exam's content thought it should have received a grade of "good" (equivalent to a B). The German language instructor, however, found that there were so many spelling and grammatical errors that he was only willing to rate the exam as "satisfactory" (equivalent to a C). P.K.'s final grade was "satisfactory"—perhaps demonstrating that racial knowledge was not always more important than a good command of the German language. Illustration courtesy of the Archiv und Gutachterstelle für Deutsches Schul- und Studienwesen/Landesarchiv, Berlin.

it easier to place people in what, for them, might be the biologically best occupations ("we could more easily place a person belonging to a leader race in a leadership position," if certain of that person's race), caution was necessary. Where the races are mixed, as is the case among the contemporary European races, a person "could have the physical traits of one race and the mental traits of another." As such, P.K.—like his "national comrade" C.T.—followed the official, pragmatic line that "race," at least for "Aryans," could not be determined by physical appearance. That having been said, it was clear for F.G. that humans "lost their healthy racial instinct" through "an artificial environment." This, however, was changing under National Socialism. P.K. closed his exit exam with the following reminder: "The individual must recognize the importance of racial hygiene and support it in deed and with sacrifice. With regard to the preservation of the race, the statement 'common good before private good' must stand."[63] At least in the case of these two German language exit exams and in the one biology essay preceding it, committed Nazi pedagogues had little reason to complain, either about what was being done for *völkisch* education in general or about secondary school biology instructors' support of such education, in particular.

What can we conclude from this seemingly contradictory material on human hereditary education in the secondary schools during the Third Reich? As was the case with Germany's biomedical researchers, secondary school biology instructors made the Faustian bargain with the Nazi regime. There was a symbiotic relationship between human heredity, in the form of racial science instruction, and National Socialist politics. Each reinforced the other. Without the importance placed on "heredity" and "race" by Nazi ideology and Hitler himself, it seems doubtful whether as much genetics (in its various guises) would have been taught in the German schools during the 1930s and 1940s. We have seen that biology educators lacked the requisite number of hours to teach eugenics and genetics effectively in every type of secondary school prior to the Third Reich, and these were precisely the biology subfields *Studienräte* believed would improve the status of their discipline. In other words, biology teachers had an interest even during the Weimar years in elevating the status of biology as a school subject by instructing their charges in socially and politically relevant material. Genetics and eugenics were such fields. We have seen how anxious many PhD biologists (both in the college-preparatory schools as well as in the universities and KWS research institutes) were to take up the pedagogical cross and support racial science after the "Nazi seizure of power." The sheer number of articles, handbooks, school texts, and extracurricular seminars written or supported by secondary school biology instructors testify to this commitment. As we have observed, German biology teachers had a sufficiently "usable past" to work with. In other words, there was a measure of intellectual continuity between the Weimar and Nazi regimes in the area of biology instruction.

Nazi politicians and pedagogues wished to ensure that no boy or girl be allowed to leave school without at least the rudiments of racial science education. According to Education Minister Rust, knowledge of this material was "an absolute prerequisite" for the "renewal of our *Volk*."[64] Although racial education was supposed to be communicated in all school subjects, it was naturally expected that biology instructors carry the lion's share of the burden. The National Socialist state needed their cooperation to construct and maintain the "racial state." Biomedical researchers such as Fischer, von Verschuer, and Rüdin had other roles to play in the overall structure of the Third Reich. As we have seen, they produced, legitimized, and, to a certain extent, popularized—both nationally and internationally—the human genetics knowledge required to make Germany the only fascist state where "race" became the criterion for citizenship and where human genetics fostered genocide. These individuals were far too important to spend time directly ensuring that the masses could make a "scientific" distinction between "national comrades" and "racial aliens." If not exclusively, this task generally fell to school biology instructors. The Hitler Youth might have preached the dangers of racial mixtures to its young cohorts, but only those trained in biology could really explain the "science" behind it. The higher secondary schools, in particular, also had a qualification function. Those who would later go on to study racial science at the university—certainly an interest of the Nazi regime—needed a proper grounding in the college-preparatory institutions by biology *Studienräte*.

As we have seen, human heredity school instruction functioned to promote racial hatred and reinforce the distinction between biological "insiders" and "outsiders" under the Nazi regime; its tendency to underscore politically motivated social and racial typecasting was radicalized during the war years. By that time, the Reich's biological enemies were simply deemed genetically "subhuman." Such instruction certainly helped loosen the traditional bonds of concern for the weakest members of society: the handicapped. Racial science instruction during the Third Reich dangerously narrowed the ethical universe of the youth; indeed, it helped to make morality dependent upon "heredity." In addition, racial science instruction, at least as it was taught, not only reinforced the separation between "Aryans" and "racial aliens." It also induced social conformity and efficiency—behavior necessary for the well-functioning of the Nazi regime.

Yet given the differences between the racial science ideal and racial science practice in the schools—and ambiguities did exit at virtually every turn—it is difficult to assess precisely how mutually reinforcing the relationship between school instructors of human heredity and the Nazi State was in the day to day classroom arena. As was suggested, it was probably not quite as perfectly symbiotic as was the relationship between biomedical professionals and the regime. The degree of the secondary school biology teachers' culpability in legitimizing a racist regime deserves to be questioned, in particular in comparison to that of KWS human geneticists. This raises difficult issues about the nature of complicity of various professional groups in a dictatorial regime like the National Socialist state. Can and should we expect more moral backbone from the producers of human genetic knowledge than from those who merely disseminate it in the schools? Was a school teacher far more a mere "cog in the machine" than a leading scientist during the Third Reich and therefore to be assessed differently? Does an educator have an obligation to not merely to encourage the young to think critically but to help forge its national identity? Such are the queries with which both author and readers have to grapple.

After having explored the international debut of human heredity prior to 1933 as well as having examined four case studies of the symbiotic relationship between human genetics and the Nazi state, it is now time to address how the international human genetics community faced the Nazi state's attempt to appropriate its discipline for its abject political purposes. We will begin where we left off in chapter 1.

SIX

The International Human Genetics Community Faces Nazi Germany

We have just examined our four case studies of the symbiotic relationship between human heredity and politics during the Third Reich. All were the result of the Faustian bargain made between German human geneticists and officials of the Nazi Party. Our first two sites—the Dahlem and Munich Kaiser Wilhelm Institutes—focused on the production of human genetic knowledge under the swastika. The second two cases, the use of professional scientific meetings and genetic and racial biology education in the secondary schools, were concerned with how such knowledge was disseminated under the conditions that existed during the Nazi period. If we recall the trajectory that human genetics and eugenics took on the international stage until 1933, it is fair to say that every racial hygiene policy measure resulting from the symbiotic relationship between human heredity and politics under National Socialism was either totally unique to Germany during the Third Reich or far surpassed, in its thoroughness, anything that existed elsewhere.

As an example, take the case of sterilization. Sterilization, as a form of negative eugenics, was much discussed prior to the Nazi seizure of power—not only in Germany but among many human geneticists worldwide. America took the lead in this procedure; indeed by 1933 somewhere between nine and twenty thousand mandatory sterilizations were legally performed in the United States, the majority in the state of

California.[1] By 1941—at a time when Germany had gone from sterilizing to euthanizing the "unfit"—the United States had robbed some thirty-six thousand individuals of their right to procreate.[2] In Nazi Germany, as we have seen, some twelve times the number of people was forced to undergo this procedure—most in the mere five years between 1934 and 1939. We will recall the role played by members of the KWIA in the secret and illegal forced sterilization of the so-called Rhineland Bastards—individuals rendered infertile not due to any alleged genetic defect but merely owing to "race." There was nothing equivalent to this practice in other countries where eugenics existed. In addition, although pro-natal policies had been championed as a form of positive eugenics by numerous human geneticists in many nations, no country went as far as Germany under the swastika. Provided that they were healthy and politically reliable "Aryans," German women were given financial incentives to bear children; at the same time they were forced out of professional employment. Genetically "valuable" women were given the Honor Cross for German Motherhood—a bronze one for four children, a silver one for six, and a gold one for eight. Abortion for "valuable" women was a serious crime for both patient and doctor. So-called racially valueless women, on the other hand, were discouraged from having progeny; in extreme cases, they were forced to abort their fetuses.[3] Again, this was a measure unique to Germany under the swastika.

More generally, National Socialist Germany could boast the dubious honor of being the world's only "racial state." Indeed Hitler's Reich was unique even among fascist countries in that "race" was the official cornerstone of politics; "race," not origin of birth or naturalization, became the criterion for citizenship during the Third Reich. The infamous 1935 Nuremberg Laws that deprived German Jews of their citizenship and forbade any sexual relationships between so-called Aryans and individuals of alien races (i.e., Jews, blacks, Roma, and Sinti) or the reams of anti-Jewish legislation were not adopted as a racial hygiene measure in any country not occupied or annexed by the Nazis.

If we turn to some of the outright biomedical crimes that were discussed in the context of our four cases, such as involvement with the "euthanasia" program—either directly or indirectly (by profiting from it scientifically or rhetorically legitimizing it)—or the individuals connected to medical experiments conducted at Auschwitz, there was no country that followed in Germany's footsteps. Human experimentation conducted in the United States during wartime—most notably the Tuskegee syphilis study—while transgressing all ethical boundaries accepted today, did ultimately have state-sanctioned checks placed upon it. There was never the degree of wanton disregard for human life that existed in Nazi Germany. Such practices were also never seen as eugenic measures. Although wartime Japan undertook bestial medical experimentation, like human experimentation in the United States, it, too, was not done under the banner of eugenics. The purpose of Japanese human experimentation, like the notorious high-altitude medical experiments carried out on prisoners of concentration camps like Dachau, was primarily to serve the interest of soldiers and the war machine.[4] While these heinous activities are, needless to say, in no way less reprehensible than what occurred with the aid of Mengele at Auschwitz, they were not designed as an instrument to further separate out "valuable" from "valueless" populations. Indeed, had the war lasted longer, the studies initiated by von Verschuer and Magnussen would have probably been used to scientifically legitimize the extermination of more racially undesirable individuals.

We have outlined the measures that differentiate German human genetics from that of other countries. The question that begs to be raised, however, is why these policies occurred in Germany under Nazism and not elsewhere. An attempt to address this critical query will be made in the conclusion. First, however, we need to return where we left off at the end of chapter 1. How did the international human genetics community react to the policies and practices undertaken in the name of eugenics during the Third Reich? What occurred at international meetings, especially at those held by IFEO? Did the activities and research of German human geneticists impact the direction of eugenics movements elsewhere? Were all countries equally affected? Developments between 1933 and 1945 outside the Reich are certainly an important component in assessing the uniqueness, or lack thereof, of human heredity under the swastika.

The Challenge of Reform Eugenics

Years before the Third Reich, numerous human geneticists, especially in the United States and Great Britain, began to distance themselves from the kind of eugenics that Davenport and other mainline adherents were propagating. Although the exact motives varied from individual to individual, in general, these so-called reform eugenicists felt that the accumulated insights into human genetics did not support most of the political and policy claims of the mainliners. Moreover, reformists believed that most of their mainliner brethren were blinded in their pronouncements by racial and class prejudice; their proclamations, the reformers felt, would serve to promote social inequality. Indeed many reformers, especially those in Britain, argued that any meaningful eugenics platform could only be implemented after a thoroughgoing economic reform of society. This did not, of course, mean that some of the reform eugenicists did not adhere to certain presuppositions held by the mainliners. On the whole, however, they were less sanguine in their views about the efficacy of eugenics to quickly transform the heredity stock of nations. They were more interested in advancing human geneticists as a prelude to legislating eugenics.

In the United States, Frederick Henry Osborn (1889–1981), secretary-treasurer of the American Eugenics Society and an expert in anthropology and population, was skeptical of any wholesale claims about the genetic inferiority of certain races and classes.[5] Herbert Jennings (1868–1947), a former student of Davenport, gained a position as a geneticist specializing in the heredity of microorganisms at the Johns Hopkins University. Although he believed that eugenics was theoretically useful, he did not share the optimism of the mainliners regarding the rapidity of eradicating defective genes. As Jennings argued in 1930, a scientific breakthrough that would allow the recognition of so-called normal carriers of deleterious genes was necessary before the gene pool could ever be cleansed of bad hereditary material—assuming, of course, that these millions of individual carriers could be persuaded to refrain from procreation. Moreover, the Hopkins professor maintained, until the environment was improved, it was difficult, if not impossible to know whether nature or nurture produced an undesirable behavior or disease. "Bad living conditions often produce the same kind of results that bad genes do," Jennings remarked. "Persons may become idle and worthless, insane or criminal or tuberculous [sic]—either through bad genes or bad living conditions, or through a combination of both."[6]

Certainly the most famous, and most political, of the American reform eugenicists was the world-renowned geneticist H. J. Muller. A committed socialist and supporter of the Soviet Union, Muller was even more radical in his attack against mainline eugenics than his other American colleagues. At the Third International Eugenics Conference in 1932 he declared open warfare on the mainliners in a talk entitled "The Dominance of Economics over Eugenics." Its flagrant Marxist position was anathema to Davenport, who chaired the Congress, and he did all he could to limit the time Muller had to deliver his paper. Insisting, however, that he be given a hearing, Muller began by offering a trenchant critique of social inequality under capitalism. Under the present economic system, Muller contended, financial, not eugenic, considerations govern human reproduction, and all hope of a better genetic future must remain "an ideal dream."[7] To legitimize capitalism, mainline eugenicists served as apologists, arguing—against scientific evidence—that race and class distinctions are the result of heredity. The capitalist system, Muller argued, "leads to a false appraisal of the genetic worth of individuals, and of vast groups, which results in entirely mistaken conceptions of eugenic needs." Only with a radical overhaul of the capitalist system was hereditary improvement possible. At that point, Muller emphasized, "the new eugenics will then come into its own and our science will not stand as a mockery." He ended his speech with a rallying cry: The opportunities for eugenics under socialism are "unlimited and inspiring. It is up to us, if we want eugenics that functions, to work for it in the only way now practical, by first turning our hand to help throw over the incubus of the old, outworn society." Muller's audience at the Congress was shocked; people on the political left who heard of his talk cheered. America was in the depths of the Great Depression.[8]

Many of Muller's British reform eugenics colleagues were also on the political left. They include the evolutionary biologist Julian Huxley (1887–1975), the population geneticist J. B. S. Haldane (1892–1964), and the experimental zoologist and medical statistician Lancelot Hogben (1895–1975). The only conservative in the group of reformers was the brilliant statistician and geneticist, R. A. Fisher (1890–1962). The first three men knew each other personally; whereas they would all later stand against the mainliners, only Hogben was initially against everything that Davenport and his supporters stood for.[9] He especially attacked the zeal in which mainliners attempted to apply their alleged scientific knowledge on the political stage. "In English law there is a wholesome provision which forbids public evidence until the case is closed. In science there is no penalty for contempt of court. It is a pity that there is not. The discussion of the genetical foundations of racial and occupational classes in human society calls for discipline, for restraint and for detachment. Nothing could make the exercise of these virtues more difficult than to force the issue into the political arena in the present state of knowledge."[10] As private citizens, Hogben maintained, those who would offer eugenic remedies are entitled to their opinions. However, when scientists speak in their public capacity, they should be "primarily concerned with sterilizing the instruments of research before undertaking surgical operations upon the body politic."[11] By the mid-thirties, however, developments in genetics and their left-wing politics eroded any lingering sympathies on the part of Huxley and Haldane for targeting the poor as intrinsically genetically unfit, something they originally had in common with their mainline colleagues. Their socialist politics and critique of mainliners did not mean, however, that they opposed eugenics per se. Indeed, their Marxist outlook rendered them more critical of aspects of liberalism related to eugenics than their nonsocialist colleagues. As Hogben remarked, "the belief in the sacred right of every individual to be a parent is a grossly individualistic doctrine surviving from the days when we accepted the right of parents to decide whether their children should be washed or schooled."[12] But more knowledge was needed before one knew which of these individuals should be denied the outdated "sacred right" of parenthood.

Fisher was different from other anti-mainliners. He was certainly scientifically sophisticated, as his groundbreaking treatise that would help launch population genetics, *The Genetical Theory of Natural Selection,* would demonstrate. Yet Fisher's numerous scientific accomplishments did not prevent him from appearing politically and temperamentally more like his German colleague von Verschuer than his fellow British reform eugenicists. Although his background in population genetics and statistics did not permit him to accept the biologically simplistic doctrines of mainliners, like von Verschuer, Fisher was both an excellent scientist and a man who was not able to shake off his class and racial prejudices. Like von Verschuer, he was also a devout Protestant. Fisher argued for state subsidies for so-called fit families—a policy that was implemented in Nazi Germany.[13]

Britain and the United States were not the only countries where human geneticists attacked the Davenport school. In Norway, Otto Mohr (1886–1967), a physician and cytologist and later one of the founders of medical genetics, had been critical of his mainline countryman Mjøen as least as early as 1915. He was a close personal friend of the famous American fruit fly geneticist T. H. Morgan, a man who also parted company with mainline eugenics early. Although Mohr was not originally opposed to eugenics, he became more and more critical of the conservative and racist implications of it as currently practiced. Moreover, like many of his British colleagues, he found most of what was said on the subject to be unscientific. He stressed the importance of environment in many so-called genetic diseases and remained on the opposite side of the divided Norwegian eugenics movement until Mjøen's death in 1939. Like the British reform eugenicists, Mohr was interested in radical social reform, including the practice of birth control for all women.[14]

The Swedish physician and human geneticist Gunnar Dahlberg (1893–

1956) also despised almost everything that the mainliners stood for. He was to the mainline Swedish eugenicist Lundborg what Mohr was to his countryman Mjøen. An outspoken man on the political left, Dahlberg had close contacts with the British group around Haldane. As we will see, he especially abhorred Nordic racism; Dahlberg would become an outspoken human geneticist against everything Nazi. Indeed, German human geneticists during the Third Reich would attempt to block his determination to succeed Lundborg at the Swedish Institute for Race Biology—the institute that had originally served as a model for Fischer's KWIA. Dahlberg demonstrated that he was a man who took his politics seriously when, as president of the University of Uppsala in neutral Sweden, he openly condemned the German invasion of Norway in 1940.[15]

The rise of the Third Reich with its notorious racial policies only intensified the suspicion, if not open hostility, that reform eugenicists and geneticists felt about the misuse of their science. In 1935, for example, L. C. Dunn, a distinguished American geneticist with no direct ties to eugenics, wrote John C. Merriman, president of the Carnegie Institute of Washington, the philanthropy that supported the ERO. Although the letter was primarily designed to critique the unscientific work undertaken at Davenport's Institute at Cold Spring Harbor, he used his close proximity to Germany (he was in Norway at the time) to attack Nazi racial policy while simultaneously warning of the danger of supporting mainline eugenics. Dunn reported that under the Nazis, there was a complete reversal of the relationship between scientific research and politics. "The incomplete knowledge of today," the geneticist argued, "much of it based on a theory of the state which has been influenced by the racial, class and religious prejudices of the group in power, has been embalmed in law." Whereas there might be some progress in reducing the number of elements deemed undesirable by the state, "the cost appears to be tremendous," Dunn continued. "The genealogical record offices have become powerful agencies of the state, and medical judgments when possible seem to be subservient to political purposes." And as a liberal scientist, what was most disturbing to Dunn was that the "the solution of the whole eugenic problem by fiat eliminates any rational solution by free competition of ideas and evidence."[16] The entire basis of his science appeared threatened in Germany.

Even earlier, in 1933, Muller wrote to the Rockefeller Foundation requesting that the philanthropy endorse a statement that made it clear that continued funding of a German institute would cease if its researchers were dismissed on grounds "other than their scientific work."

A research guest at the Kaiser Wilhelm Institute for Brain Research just outside Berlin at the time, Muller noticed that its world-renowned liberal codirectors, Oskar and Cécile Vogt, were in political danger. Muller naturally hoped the statement he requested from the American philanthropy would protect Jewish as well as politically suspect scientists.[17]

Huxley and Haldane were also direct in their response to the National Socialist brand of eugenics. Writing not only after the draconian German Sterilization Law but also after the infamous Nuremberg Laws, the two men attacked the underlying so-called scientific assumptions of both. In the preface to his 1936 coauthored book *We Europeans: A Survey of "Racial" Problems,* Huxley reminded his readers that "in a scientific age, prejudice and passion seek to clothe themselves in a garb of scientific respectability." For the Marxist biologist, nowhere was this more evident than with regard to "race." "A vast pseudo-science of 'racial biology,'" he lamented, "has been erected which serves to justify political ambitions, economic ends, social grudges, and class prejudices." In the body of his work, Huxley especially attacked the "'Nordic theory' which speedily became very popular in Germany." The "Nordic theory," he added, "has had a very great effect . . . in serving as the basis for the 'Aryan' doctrines upon which the Nazi regime is now being conducted." Huxley also added that such "pseudo-science" was used to change immigration laws in the United States. Here he was speaking of the role that mainline eugenicist Laughlin and the ERO played in legitimizing the Johnson Immigration Act of 1924—a policy that greatly limited the number of Eastern and Southern Europeans into the United States and made it even more difficult for Jews fleeing the Holocaust to embank on American shores. That such Nordic racism was pseudo-science was demonstrated to Huxley by claims that "Jesus Christ and Dante were turned into 'good Teutons.'"[18]

Writing two years later, Haldane, in his treatise *Heredity and Politics,* indirectly attacked the German eugenicist Rüdin in targeting schizophrenia as one of the diseases for which one could be forcibly sterilized. As the British Marxist biometrician argued, in order for the Germans to prevent the birth of one schizophrenic child, sixteen schizophrenics had to be sterilized, which would result in a loss of ten healthy children. "It is highly questionable," Haldane added, "whether the bad effects of this policy do not outweigh the good effects, at least from the point of view of Hitler, who wishes the German population to increase."[19] Blasting the notion that Jews could be distinguished from the non-Jewish population of Europe on the basis of physical and psychological traits, Haldane presented his readers with the absurd statement of the Nazi German Minister of Agriculture, Richard Walther Darré: "The Semites reject everything connected with the pig. The Nordic peoples, on the other hand give the pig the highest possible honour. . . . The Semites and the pig are faunal and thus physiological opposites." Sadly, Haldane continued, "evidence for this profound physiological discovery has not yet been published."[20]

Dahlberg, the close friend and colleague of the Haldane group, launched his own scathing attack on National Socialist racial policy. In a book originally written in Swedish designed as "an examination of the biological credentials of the Nazi creed," it was translated into English by Hogben and published in 1942 under the title *Race, Reason and Rubbish*.[21] At a time when the Nazis were transporting Jews all over Europe to be exterminated, the Swede Dahlberg wrote a remarkably accurate account of the development of anti-Semitism. Recognizing that it not only had its roots in religion, he saw clearly the importance of this form of traditional anti-Semitism for its newer, allegedly scientific garb. After a discussion of all the Nazi stereotypes of the physical and alleged mental differences between Jews and "Aryans," Dahlberg went on to demonstrate that Jews are often very much like the people around whom they live. For the geneticist, no rational person could believe that "the average divergence of external appearance"[22] between Jews and other Europeans is of fundamental importance. Moreover, considering the number of Jews who were Nobel Prize winners relative to their percentage of the population of various states, there can be no grounds for believing them to be inferior. From the standpoint of heredity, the Swede maintained, there "is no scientifically acceptable proof that Jews on average are worse than other folk. So far as inherited traits are concerned, there is absolutely no reason for maintaining that Jews represent a special type." After a discussion of why the minority status of Jews in European countries had always made it "possible to hate them," Dahlberg made a rational plea for treating all people as individuals rather as part of an alleged racial group.[23] Unfortunately, for the Jews in Germany and much of Europe, this plea fell on deaf ears.

Just prior to and during America's entry into the war, individual voices were raised against Nazi eugenics. In 1939, the future vice president under Roosevelt, Henry Wallace (1888–1965) spoke out against Nazi racism masquerading as science. The *New York Times* dubbed knowledgeable Americans who began to lash out against claims of Aryan supremacy "the minute men of science." Indeed, the *New York Times* science editor, Waldemar

Kaempffert, wrote that if geneticists do not "curb the 100-percenters [those who believed only in '100 percent American' blood] who see no good in any one who is not white, Protestant and 'Nordic,' we might as well abandon [eugenics] altogether as a guide in improving the social quality of humanity."[24]

But these comments came from disciplinary outsiders. In 1940, the reform eugenicist Osborn wrote his influential *Preface to Eugenics*. It was designed to offer a state of the art assessment of eugenics from the reform perspective. What is interesting is its ambivalent nature. On the one hand he offers conclusions that appear to critique eugenics in the Third Reich. Osborn mentions that there is no scientific evidence to support the claim that any nation is racially superior to another. Elsewhere he specifically states that "eugenics and democracy are significantly interrelated." Osborn goes on to argue that "the safeguards of democracy are essential to the development of a sound and humble eugenics."[25] On the other hand, he quotes evidence from Nazi racial scientists like Rüdin on schizophrenia and mentions the German population program, including sterilization, in neutral terms. Although he states that there is discrimination in that marriage loans are only given to Aryan families, the mode of discussion and his claim that "there is no evidence that racial discrimination complicated the sterilization program" in Nazi Germany, hardly qualifies his work as an outright attack against what the National Socialists were doing in the name of eugenics.[26]

This same ambivalence on Osborn's part can be seen in his willingness, in contrast to his British reform colleagues, to work together with American mainliners. In a letter written to Laughlin, the superintendent of the ERO and perhaps the most important mainliner in the United States, Osborn expressed the hope that the two would meet soon. After all, both were committed to eugenics. Osborn reminded Laughlin that "you felt that we had a very deep bond in our community of interest."[27] Secretary-treasurer of the American Eugenics Society at the time, Osborn served on the board of directors with Popenoe, another mainline eugenicist and coauthor, along with Gosney, of *Sterilization for Human Betterment: A Summary of the Results of 6,000 Operations in California*. As we have noted, this 1929 book served as both a model for the draconian Nazi Sterilization Law and simultaneously legitimized it.[28] Even von Verschuer, one of the leading human geneticists in Germany who made the Faustian bargain with the Nazi regime, was welcomed as a foreign member of the Society in 1940.[29] The lines, it would seem, between reform and mainline eugenics in the United States, was not as hard and fast as it was elsewhere—with the notable exception of the socialist geneticist Muller.

The Mainline American-German Connection

The rise of reform eugenics did not mean that the more conservative mainline colleagues in all countries that boasted a eugenics movement were silenced. This was especially true in the United States. Although Laughlin's ERO was criticized by reform eugenicists and even members of the philanthropy that funded his institute, the superintendent was able to secure money from the wealthy former lieutenant colonel and textile magnate Wickliffe Preston Draper—a man interested in the kind of mainline eugenics advocated by Laughlin and the Germans. In 1937, he was on the board of directors of the Pioneer Fund, an organization that still exists today. If one looks at the Pioneer Fund's web site today, one finds that its goal is "to advance the scientific study of heredity and human difference." From recent scholarship, however, we know that it was (and still is) involved in supporting all sorts of racist and right-wing projects. According to the Fund's original stated purpose, it was designed to "improve the character of the American people by encouraging the procreation of descendents of 'white persons who settled in the original thirteen colonies prior to the adoption of the constitution from related stocks' and to provide aid in conducting research on 'race betterment with special reference to people in the United States'" In fact, Laughlin served as its first president from 1937 to 1941.[30] Both before and during these years Laughlin used his position at the ERO and the Pioneer Fund to attack reform eugenicists, continue his work on "race" in America, and examine the birth rate among so-called superior groups—in the case of a 1937 study, the family life of American army aviators.[31]

As we have seen, German racial hygienists maintained close ties with Davenport, Laughlin's boss, since the first decade of the twentieth century. Nor did these ties loosen after WWI, as American mainliners did all they could to help bring back German eugenicists into the international movement. Germans and American mainliners worked closely together in the IEFO throughout the interwar years. By the rise of the Third Reich, American mainline eugenicists, who had served as a model for many of their German colleagues, were proud, if not envious, of their fellow racial hygienists in the new Reich. As one American mainliner put it, "the Germans are beating us at our own game."[32] Mainline Americans such as the openly racist Madison Grant (1865–1937), Clarence G. Campbell (1862–1938), Charles M. Goethe (1875–1966), Popenoe, Gosney, and, of course, Laughlin supported the Nazi racial policy, especially its sterilization program. That some were worried what this might mean for America

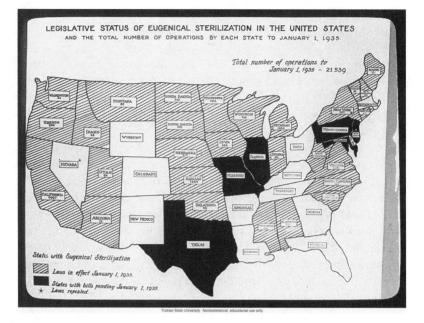

35 Map of the United States in 1935 showing the legislative status of mandatory sterilization. Photo courtesy of the Prickler Memorial Library, Truman State University.

can be seen in a letter by Goethe where he describes, among other things, the German eugenics laws. "However much one abhors dictatorship," Goethe remarked, "one is also impressed that Germany, by sterilization and by stimulating birthrates among the eugenically high-powered, is gaining an advantage over us as to future leadership."[33] In another note, he was even more explicit that what was at stake was national power. "The Germans are forging far ahead of us in this matter of accumulated [eugenic] data. They say they already have 4 Nobel prizes to 1 of ours, population considered, and that if we do not adopt their methods, they will run away from us with world leaders."[34] Nazi racial hygienists and their government used these affirmations of their policies to legitimize racial programs under the swastika. An examination of one episode dealing with Laughlin's contacts and research efforts clearly demonstrates the American-German connection during the Third Reich.

In July 1935, Laughlin wrote two letters to Eugen Fischer, who would be serving as president of the upcoming International Congress for the Scientific Investigation of Population Problems in Berlin. In one note to the Dahlem director he enclosed a paper accompanied by charts and pictures that he would have read for him at the Congress, as he was not planning to attend. "Your American colleagues," he stated, "wish the forthcoming Congress the success which it so richly deserves, both in the advancement of eugenics as a pure science, and its application to the practical solution of population problems."[35] Laughlin's second letter recommended W. P. Draper to Fischer. The trustee of the Pioneer Fund was planning to be in Germany at the time of the Congress and was interested in attending its sessions. "Colonel Draper," Laughlin wrote his friend Fischer, "has long been one of the staunchest supporters of eugenical research and policy in the United States." Two weeks later, Laughlin wrote Draper that Clarence Campbell would be attending the Congress and suggested that the textile magnate be appointed an official delegate. He would be the American colleague who would read Laughlin's paper. Draper indeed contacted Campbell and attended this major event.[36] At the well-attended international Congress, Campbell openly praised Hitler prior to his talk. This was used by the Nazi press as evidence of world support by international researchers for their racial policies. Back in the United States he tried to win support for National Socialist eugenics and refuted anti-Nazi propaganda.[37] On August 29, 1935, the *New York Times* published a factual report of Campbell's Berlin speech entitled "U.S. Eugenist Hails Nazi Racial Policy." It was printed in a column alongside the article "Nazis Behead Another." The science editor of the paper was so surprised that Laughlin offered space in his journal, the *Eugenical News,* to the racist talk Campbell held in Berlin that he asked the ERO director whether he was forced to do so or whether he, Laughlin, endorsed the views of Campbell. One member of the American Eugenics Research Association was so outraged at Campbell's speech that he felt forced to resign. "I hate to believe that the Eugenics Research Association, dedicated to eugenics as a science, has been prostituted to further false propaganda," the letter's author lamented.[38]

Meanwhile, the *Eugenical News*—the journal of The American Eugenics Research Association at the time—had been keeping its mainline readership abreast of developments in German racial hygiene, especially its Sterilization Law, since 1933.[39] In one report on sterilization under National Socialism, the German author praised his American colleagues. In everything dealing with Germany's efforts toward "improving the nation's health by applying biological laws," Dr. C. Thomalia argued, it has "learned from the United States."[40] The *Journal of Heredity,* certainly more renowned than *Eugenical News* as a scientific organ and boasting an editorial board comprised primarily of reform eugenicists, also printed information of the German sterilization project. In a 1935 article entitled "A Year of German Sterilization," the author Robert Cooke, the managing editor of the journal, expressed his view that "these glimpses at various phases of the most extensive experiment in sterilization for human betterment suggest a commendable conservatism in administration."[41] Whether he would have argued that some three years later is an open question. Cooke's article was not the only one to praise the German eugenics project. In a 1937 discussion of "Eugenics in Other Lands: A Survey of Recent Developments," Hilda von Hellmer Wullen, from New York, commented that Germany housed the "largest laboratory of eugenic experimentation in existence." Although she did not wish to "pass judgment" on the controversial platform of race purification, there was no doubt in her mind that "a concerted effort" was being made to apply the "Galtonian technique of altering the inborn quality of people through agencies under social control." Regarding the compulsory nature of the German Sterilization Law, von Hellmer Wullen agreed with the Nazi regime that such a measure can "be of little real value to humanity" were it merely voluntary. The German nation, she concluded, "has recognized before it is too late, (and before any other nation has taken any significant step in that direction), the biological importance of improving its racial stock by bringing into action all possible means at its disposal which may contribute to this important purpose."[42] In 1936, the year when "Baur-Fischer-Lenz" went into its fourth, revised edition, L. H. Snyder from Ohio State University could continue to express admiration that Lenz's treatment of the Jews and their mental traits were "still discussed with same dispassionate fairness as before."[43]

Both the *Journal of Heredity* and *Eugenical News* specifically carried articles dealing with "race" in Germany. Eugen Fischer's controversial talk in the Harnack-Haus on February 1, 1933, was reported in the first journal. *Eugenical News* printed a piece that argued that sixty thousand Jews were expelled from Germany. It was quickly refuted, in the same journal, by none other than Alfred Ploetz. The cofounder of German eugenics took exception to the word "expelled" and argued that most of the German Jews left of their own accord. "They went," Ploetz argued, "frightened by the Jewish reports of horror, because they feared, and unjustly, a pogrom." The old racial hygienist continued. "All the reports of horror (with the few exceptions of the few minor cruelties) were deliberately invented," he affirmed, "in order to hurt the new Government . . . in spite of the incessant Jewish propaganda of untruths"[44] Ploetz, who had attempted to hide his *völkisch,* anti-Semitic outlook prior to 1933, now took off his gloves.

To be fair, however, *Eugenical News* published an article by Franz Kallmann, a student of Ernst Rüdin who was forced to leave his post and country for racial reasons. Writing from his new position at the New York State Psychiatric Institute, Kallmann still argued for the need to limit the number of schizophrenics. As Rüdin's former student stated, "from a eugenic point of view, it is particularly disastrous that these patients not only continue to crowd mental hospitals all over the world, but also afford to society as a whole, an unceasing source of maladjusted cranks, asocial eccentrics and the lowest types of criminals offenders." Speaking to those in his newly adopted country, "even the faithful believer in the predominance of individual liberty will admit that mankind would be much happier without those . . . from the schizophrenic genotype."[45]

Both journals also reviewed important German eugenic texts such as those von Verschuer. *Eugenical News* gave a very favorable assessment of the German human geneticist's 1934 work *Erbpathologie* (*Hereditary Pathology*). "The word 'nation' no longer means a number of citizens living within certain boundaries," the journal proclaimed, "but a biological entity. This point of view also changes the obligation of the physician, whose service to the nation consists in caring for and safeguarding this population as a whole." The rave book review ended with the acclaim, "Dr. von Verschuer has successfully bridged the gap between medical practice and theoretic scientific research." It is fairly certain that Laughlin wrote the review for the journal.[46]

Less than two years after the Berlin Congress that Draper and Campbell attended, Laughlin wrote to the former asking support for an "applied eugenic" project that he was entrusted with in Connecticut. A survey was to be initiated, with Laughlin's help, to investigate the racial quality of the residents of the state. Although Connecticut had given $2,000 toward the survey, that was not nearly enough. Laughlin hoped that Draper would recognize the need of the state to conserve "its own foundational racial stocks and superior family strains" and donate money to the cause. "Any support the Pioneer Fund might give to this particular project should yield good returns in the field of applied American eugenics," Laughlin assured Draper. It was couched in a rhetoric that a textile magnate could easily understand. Laughlin then informed Draper that earlier money provided by the Fund to popularize the notorious Nazi eugenic film *Erbkrank* was put to good use. The film, which had anti-Semitic overtones and used trick photography to make mentally handicapped individuals look monstrous, was not only used in American schools, but was shown to a group of child welfare workers in Connecticut.[47] Needless to say, it was not viewed as in any way flawed by *Eugenical News*. In an article written by Laughlin, the film allegedly "expounds the economic, moral and biological costs of human handicap and inadequacy." The film, he added, depicts "no racial propaganda of any sort."[48] Laughlin's earlier private version of the write-up, however, excluded the last remark.[49]

Laughlin's letter to Draper, one might add, occurred the year after Laughlin was nominated for, and accepted, an honorary degree from the University of Heidelberg on its 550th anniversary, ostensibly for his important research. He was nominated by the Heidelberg Medical Faculty, in particular, by its dean, the future "euthanasia" physician, Carl Schneider.[50] The conferring of these honorary degrees was politically motivated. Indeed, the University believed that foreign dignitaries like Laughlin would be supported by the Foreign Office and Goebbels's infamous Reich Ministry for Popular Enlightenment and Propaganda to make Germany appear more respectable and for "foreign propaganda purposes."[51] The United States had more honorary degree recipients and more universities nominated for an award than any other country. Perhaps those in charge of the event did not wish to alter the fact that America sent more students to the University of Heidelberg than any other nation.[52] Owing to lack of time, Laughlin did not travel to Heidelberg to receive his degree personally. Instead, it was handed to him at the German General Consulate in New York. The ERO superintendent told the responsible individual at the consulate that "there was no other foreign university from which he would have rather received such an honor as from the University of Heidelberg."[53] He was congratulated on his award by von Verschuer—a man who certainly could run circles around Laughlin scientifically, but nonetheless shared with his American colleague a common racist and class-based eugenics philosophy. In his same letter to Laughlin, von Verschuer thanked the superintendent of the ERO for sending him eugenic literature for his new Institute in Frankfurt. "It is a special pleasure, von Verschuer added, that you take such a friendly interest in my work."[54]

We know that Hitler apparently was interested in the texts that certain mainline eugenicists had written. He asked both Leon Whitney, an expert on sterilization, and Grant for a copy of their works. The latter received a note from the Führer stating that the Passing of a Great Race, Grant's racist treatise, was Hitler's Bible.[55] When another book by Grant, The Conquest of a Continent, was translated into German in 1937, Grant asked Dahlem Director Fischer to "write a few words for the Foreword." As expected, he praised it and its author. Henry Fairfield Osborn (1857–1935), uncle of the reform eugenicist Frederick Henry Osborn, wrote its introduction. It was twice as long as Fischer's foreword.[56] Although Osborn argued that no race could be viewed as superior or inferior, this was something that

36 Eugen Fischer reading a copy of the American publication *Journal of Heredity,* ca. 1935. Photo courtesy of the Archiv der Max-Planck-Gesellschaft, Berlin Dahlem.

German racial hygienists such as Fischer and von Verschuer also stressed. Such terms as "superior" or "inferior," the German eugenicists continually emphasized, were not scientific. But by stating that the information Grant provided in his book "will take the wind out of the sails of enemies of racial theories who argue that racial worth is based on fairy tales," Osborn was obviously aiding Nazi policies. Indeed, at least for the professional German human geneticists who made the deal with the regime, Osborn's introduction was a welcome confirmation of their own views. His decision to travel to Frankfurt to receive an honorary degree from its university in 1934 also legitimized Nazi racial hygiene policy.[57]

Unfortunately we are in the dark regarding what average Americans thought either about Nazi eugenic practices or those of their own country. In July 1933, a Mrs. George Webb of Providence, Rhode Island, wrote Chancellor Hitler praising him for his initiative in establishing Germany's Sterilization Law.[58] It seems unlikely, however, that this letter was in any way typical of American sentiment with regard to National Socialist eugenics and racial policy. What is clear from the mountains of material that Laughlin collected on racial science and the "Jewish question" in Germany—much of it published in reputable papers like the *New York Times*—Americans could have been well informed about Nazi racial policies, had they wanted to. Included in what might be called Laughlin's anti-Jewish campaign archive are reports of anti-Semitic and racial issues infecting election campaigns in the United States.[59]

Among the numerous informative articles about Nazi racial practices in Laughlin's collection, there was an interesting letter written on November 24, 1933, to the editor of the *New York Herald Tribune*. During the prior month, the paper ran an article on the new Germany entitled "The Roots of Hitlerism." The piece obviously offended a well-placed German woman, and she titled her reply "In Praise of Hitlerism." As a person old enough to remember the Empire, the aftermath of the Great War, and the so-called chaos of the Weimar Republic, the author resented Hitler being viewed as a warmonger in the *New York Herald Tribune*. She defended the *Führer's* withdrawal from the League of Nations in 1933 as a necessary declaration that Germany would no longer be treated as a pariah state. In her "Praise of Hitlerism," she mentioned that the Third Reich redeemed "thousands from destitution and demoralization," giving them productive work in the form of "voluntary work camps" where, allegedly, they were "never so happy." The author also mentioned the good intentions of the Reich's Winter Relief Action. People all over the nation, she claimed, voluntarily gave up their Sunday dinner and ate a simple stew so that money could be raised for the poor during the winter months. With regard to the "Jewish question," the author admitted and deeply regretted "that the Jews have reason to feel bitter about the national revival." At the same time, she added, "they have brought the trouble on themselves by their misuse . . . of the influential position they enjoyed in Germany. They dominated in a manner quite out of proportion to their numbers."[60]

Although the markings on the article by Laughlin (he put lines alongside of the author's discussion of the "Jewish question" and the paragraph where she cited the "straightforward offers of peace and reconciliation made by Adolf Hitler")[61] tell us much about the superintendent and speak volumes about the way most Germans felt at the beginning of the Third Reich, the response becomes even more significant when we learn who penned the piece. The author was Eleonore von Trott zu Solz, a great granddaughter of John Jay (1745–1829), one of the so-called founding fathers of our country, chief justice of the United States, and coauthor of the *Federalist Papers*. The son of Eleonore von Trott zu Solz, Adam von Trott zu Solz, was apparently always opposed to Hitler. He was hanged as part of the July 20, 1944, military plot to assassinate the *Führer*. Eleonore became part of the Christian resistance to Hitler and his regime—exactly when is unclear. More interesting still, the Trott family lived close to the von Verschuers near the small town of Bebra in the northeastern

Hessen countryside. They were the two noble families who oversaw the area for centuries and frequently intermarried. Immediately after the war, Eleonore pleaded with von Verschuer to apologize for his actions during the Third Reich—something the human geneticist refused to do. His reluctance to admit blame at the time would impact the relationship between the two families to this very day.[62]

We know that the collection of documents containing Eleonore von Trott zu Solz's reply to the *New York Herald Tribune* was not merely of scholarly interest to Laughlin. In a 1934 letter addressed to "The Director of the Department of Eugenics of the Carnegie Institution of Washington D.C."—in other words, to Davenport—a Jewish general insurance broker complained about a recent report on immigration submitted by Laughlin to the Chamber of Commerce of the State of New York. In it Laughlin argued against any "exceptional admission for Jews who are refugees from persecution in Germany." He also recommended that there should not be "admission of any immigrant unless he has a definite country to which he may be deported, if the occasion demands." The author of the letter wondered whom Laughlin meant, if not the Jews, because "the Jew is the only one that has no country." The insurance broker reminded the addressee that if the ancestors of Laughlin had not been admitted to the United States through the "good grace of our liberal country . . . they would still roam barefooted over the Irish land and. . . . Dr. Harry H. Laughlin would not be a research man." The author insisted upon a complete explanation of the actions of Davenport's deputy, clearly an anti-Semite.[63]

The American mainline–German connection during the Nazi years appears to be unique within the international scientific community of human geneticists and eugenicists. Although, as we will see, there were close ties at this time between mainline Scandinavian and German racial hygiene advocates, it pales compared with those between conservative American adherents of the gospel of Galton and their colleagues who made the Faustian bargain with the National Socialist state. That being said, the connection was not symbiotic. It is clear that the Nazi officials and the German human geneticists who worked for them were ecstatic to receive American mainline praise. Most of this praise, however, was for the German Sterilization Law and its attempt to induce "fitter families" to have more children, not for later Nazi racial policy in general.

The American mainliners, for their part, did not profit from having Nazi recognition for their work. As the sometime reform eugenicist Osborn put it, it was a shame that the American public opposed "the excellent sterilization program in Germany because of its Nazi origin."[64] Indeed,

National Socialist support, especially after war broke out in Europe, helped to draw the curtain on mainline eugenics in the United States. The ERO, the organization of American mainline eugenics, was closed at the end of 1939.[65] Nazi racial policy was not the only factor in the demise of American mainline eugenics, but it was an important one. The decline of mainline American eugenics, of course, did not mean the end of eugenics in the United States per se. Nor did it signal the end of eugenic practices such as mandatory sterilization.

The German Usurpation of the IFEO and the Dissolution of the International Human Heredity and Eugenics Movement

We have already become acquainted with numerous examples of the astute politics of professional talk on the international stage during the Third Reich. German human geneticists hosted and attended professional conferences in their fields and used such occasions not only to meet colleagues they wished to influence but also to legitimize Nazi racial policy abroad. As was mentioned at the outset of chapter 1, the international reputation of German biomedical scientists was a critical intellectual resource for the Nazi regime, and the individual human geneticists as well as the National Socialist state knew how to exact the maximum advantage from this fact both before and during the war. However, there was a deliberate omission of any discussion of one international organization during the Third Reich: the IFEO. Certainly the most important umbrella organization for the international eugenics movement in the 1920s, the IFEO became increasingly strong and assertive in the years just prior to 1933. The rise of reform eugenics beginning in the early 1930s did not eliminate the mainliners who found their organizational home in the IFEO. It now remains to examine its fortunes during the National Socialism era. Owing to its virtual usurpation by the Germans with the help of their foreign supporters, the attempt on the part of the Nazi government to control organizations in the field of biomedicine outside its own country, and the negative foreign impact of National Socialist racial policies in general, the worldwide scientific community of human geneticists was deeply divided. By the outbreak of the war, it would be impossible to speak of an international movement in human heredity and eugenics. The occupation policies of the Nazis only served to exacerbate the situation. Indeed, the IFEO never convened during WWII. Moreover, Germany exploited its quick victories in Western Europe to bring all scientific organizations within its sphere of influence "in line."

The first meeting of the IFEO during the National Socialist era was held from July 18 to July 21, 1934, in Zurich. Indeed, Nazi officials were happy about the choice of location, as it seemed to ensure that there would be a strong pro-regime German presence among the delegates.[66] Rüdin, as we will recall, had been appointed president of the Organization in 1932 after Fischer had informed Davenport that he was too busy to accept the position. From the new president's point of view the conference would serve as an international referendum on Nazi racial policy, in general, and the Law for the Prevention of Genetically Diseased Offspring, in particular. Rüdin helped write the commentary to Germany's Sterilization Law and was certainly the most vocal advocate for it among German human geneticists.

Before the conference was held, there were signs that not all IFEO members were happy with the German Sterilization Law or the choice of Rüdin as president. There were Dutch and especially French colleagues who were opposed to the former.[67] A critique of the German practice was penned by Ludovic Naudeau and published in the paper *Le Phare* (*The Lighthouse*) in 1933. Naudeau was particularly concerned that "under Nazi terror" the newly established Hereditary Health Courts might decide to sterilize Germany's political adversaries. But he was also clearly against negative eugenic measures as they existed in the United States as well as Germany per se.[68] During that same year, Georges Schreiber, the vice president of the French Eugenic Society and French delegate to the IFEO, wrote a caustic article in the *Revue Anthropologique* (*Anthropological Review*) in which he questioned Rüdin's ability to serve as president at the upcoming meeting owing to his close ties to the Nazi regime and the role he played in the Nazi Sterilization Law. "The racial policy on the other side of the Rhine," Schreiber exclaimed, "is characterized by the Hitlerites as purely eugenic." In truth, he continued, "representatives of most other countries view it as dysgenic since an apparently talented portion of the [German] population is harassed and the book burning [of Jewish and politically left-wing authors] appears as evidence of the [German] desire for degeneration." Although President Rüdin is a "valued and esteemed researcher" who perhaps might find some Nazi racial measures "unfortunate," will he really be able to express his opinion openly at the conference?" Schreiber queried.[69] Several French eugenicists appeared to see clearly that the German Sterilization Law was part and parcel of a larger racial cleansing project that included anti-Semitism.

After receiving news from the pro-Nazi Norwegian delegate Mjøen[70] that the Dutch and French might make trouble at the conference, Rüdin ensured that a large and trustworthy delegation including von Verschuer,

Ploetz, and several others who had openly made the Faustian bargain with the regime were invited. In addition, prominent Nazis from party and state such as Walter Gross, Falk Ruttke, and Karl Astel joined them. On the other hand, anyone who was considered politically unreliable, like the Dresden eugenicist Rainer Fetscher, was prohibited from attending.

As was usual at such conferences, the president gave the opening speech. Rüdin made it clear that only "energetic measures to improve humanity" could prevent the "degeneration of cultured peoples." The "care of worthy hereditary material" and the "freedom" from the unfit could not, the Munich director emphasized, be achieved with "savvy philanthropic talk," but only with "goal-oriented action and iron consequence."[71] Somewhat later during his own talk, Nazi lawyer Falk Ruttke, the other commentator of the German Sterilization Law, told his audience from twelve countries how this would be accomplished in his nation. Our "entire legislation," Ruttke boasted, "is interspersed with racial hygiene perspectives." This did not merely include mandatory sterilization, but castration for habitual criminals, marriage loans for healthy, Aryan national comrades, and a Reich Hereditary Farm Law that protects the interests and property of healthy non-Jewish farmers. He assured the delegates that everything that was genetically nefarious for the German people would be "violently wiped out."[72]

As expected, there were critical voices heard at the conference. The renowned Dutch human geneticist Dr. G. P. Frets gave a talk entitled "The Development of Eugenics in the World" where he condemned the German Sterilization Law for its mandatory nature. He also emphasized the need for positive eugenic measures and the increase of population. It also appears that the delegates from Austria, the Netherlands, Poland, and Czechoslovakia were not favorably inclined toward Germany's practice of "collective sterilization." Schreiber, speaking for the French delegation, claimed that it was a "serious attack against individual freedom." Moreover, Schreiber added, one lacked exact knowledge of the genetic transmission of many of the diseases for which one could be legally sterilized in Germany. He bemoaned the fact that whereas "in France I am considered a 'progressive' eugenicist, the German delegation views me as 'antiquated.'" He noted that the president seemed very agitated whenever Schreiber spoke his mind. That having been said, he remarked that Rüdin, contrary to his earlier fears, did not react like a Nazi apologist, but rather with the "broad vision" of a man of "universal high culture."[73]

This image was of course just what Rüdin and the Nazi officials were after. German human genetic experts were supposed to appear scientific and urbane. This kind of demeanor rather than the shrill voice of a Nazi propagandist was far more effective in enhancing the symbiotic relationship between human heredity and Nazi politics. We have already observed this in chapter 4. The critique notwithstanding, things went well for the Germans in Zurich. Whereas the French and some of the Dutch colleagues had taken issue with the scope and mandatory nature of the Sterilization Law—and according to Rüdin sixteen thousand procedures were carried out since the proclamation of the Law on July 14, 1933[74]—the Americans, British, and Scandinavians in attendance appeared to have applauded Germany's success in this eugenic arena. And these were the very people most important to the German delegation. Moreover, the delegates, "all differences of opinion notwithstanding," could agree to two general resolutions: First, that many highly civilized countries were concerned about the threat of a "new large war." Such a war would be genetically anathema to "western culture." It goes without saying that such a declaration—prompted by Ploetz—was music to Hitler's ears. At the time the *Führer* was playing the peace card in Europe and the world. We already noted how the World Population Policy Conference in Berlin one year later was used to promote the same image of the Third Reich. Second, racial hygiene research and practice was a life or death matter for all states and could not be avoided. Those countries that had not yet undertaken eugenic research and measures should learn from those that have already done so.[75]

These resolutions were used as intellectual resources by the regime. For example, the *Völkischer Beobachter* carried a report entitled "Racial Hygienists against War Hysteria: International Recognition of German Racial Hygienic Legislation."[76] Other Nazi publications, like *Neues Volk* (*New Nation*), spoke about the "breakthrough of a new spirit" that was observable even outside Germany.[77] In an interview with a party publication, Rüdin reiterated that "it has been made clear to foreign countries, that German racial hygienists wish lasting peace in the world, as war destroys the fittest." Moreover, Rüdin added, attacking the problem of the unfit indirectly by eliminating defective germ cells through sterilization is ethically humane. "The destruction of life not worth living in the crude form in which it occurs in the plant and animal kingdom," Rüdin continued, "is not something we human beings can do."[78] As we have seen, the "euthanasia project" with which Rüdin was involved, certainly calls into question his above-mentioned declaration. It goes without saying that the continued positive feedback of the German Sterilization Law in Laughlin's *Eugenical News* only served to confirm the propaganda coup for the Nazi government as well as the German racial hygienists.

Although the World Population Policy Conference held in the Reich capital in 1935 was, as we have noted, by and large a great success for the Nazi eugenic cause, the negative fallout following the enactment of the Nuremberg Laws later that year, as well as the critical voices both in Zurich and in Berlin, caused Nazi racial policy makers to consider what could be done to further coordinate and control future international meetings, especially the IFEO. That such action was needed reminds us that although most left-wing human geneticists had long since shunned the mainline IFEO, even the moderate mainliners could not be completely trusted, especially as the American and British did not seem to be all that interested in the politics of the organization.[79] And there was always the possibility, according to the IFEO's statutes, that new eugenic-related institutes or organizations could join up as members.

As president, Rüdin did all he could do to fill the IFEO with sympathetic German institutions. For example, von Verschuer's Frankfurt Institute eagerly joined in 1935. During the following year, two further German institutes, that of psychiatrist and Rüdin admirer Kurt Pohlisch in Bonn as well as that of Ernst Rodenwaldt at the University of Heidelberg, added their names to the list.[80] Other German institutes would follow.

Rüdin had good reason to take this action. One of his ideological archenemies, the left-wing Swedish human geneticist Dahlberg, had been appointed head of the prestigious Institute for Racial Biology in Uppsala against the wishes of Swedish mainliners like Lundborg. Indeed, before his appointment to the professorship by the Swedish king, Dahlberg, who was part of the Haldane group in outlook and friendship, wrote a long letter to His Majesty. He complained that Rüdin, who was named as one of the experts who would advise the king regarding the best candidate for the post, could not, given his position in Germany, "give an objective verdict." This was especially true in this particular case, Dahlberg continued, as the Swedish geneticist stood "in marked opposition towards the views at present prevailing in German race biology, e.g. having publically spoken against anti-Semitism." The king accepted Dahlberg's argument, and Rüdin was replaced. Once appointed, Dahlberg completely transformed the Institute—the one that had initially served as the model for Fischer's future Dahlem KWIA—into a modern research center for human and medical genetics. He then applied for membership in the IFEO, perhaps if only to help preserve its neutrality. His membership was effectively stalled for five years; by that time he was admitted, the IFEO would never meet again.[81]

The progressive Norwegian medical geneticist Otto Mohr, the arch-enemy of pro-Nazi Mjøen, faced similar problems. Owing to his social democratic political views as well as his criticism of National Socialism and its Norwegian advocate, Mjøen, the aging racist mainliner made it clear to Rüdin that Mohr, with his "communist wife and friends" would be a terrible political risk and someone who would "wreck havoc" in the IFEO. But there were additional factors that could be held against him. In his popular book on genetics published in the 1920s, Mohr stated that he was skeptical of the efficacy of eugenics for the time being and empha-sized the need for environmental reforms. Moreover, he was known to be critical of the National Socialist truth that race crossing was harmful. "We have enough to do," Mjøen reminded his friend Rüdin, "to fight the destructive forces outside of our organization." If we allow such forces into the IFEO, "we are lost."[82]

If the German delegation at the IFEO conference in Zurich made the most of its presence to push through regime-friendly resolutions, its role at the 1936 conference in Scheveningen in the Netherlands demonstrates its intention of controlling, indeed usurping, the entire organization for the political purposes of the Nazi state. Some Nazi officials were so con-cerned about facing ideological opposition to National Socialist anti-Jewish policies in Holland that they considered holding the conference in Germany in order to have more control. This had worked only a year earlier in Berlin. For whatever reason, the Reich Ministry of Education spoke out against this plan, but demanded that a "numerical and quali-tative exceptional" representation of German human geneticists be sent into battle. In addition, members of the Nazi government should not accompany them, as "the German point of view" in discussions is much better defended from the mouths of the scientists themselves. This again points to the role of the reputation of these scientists as intellectual re-sources for the Nazi state.[83]

Rüdin was even more careful selecting the German delegation for the 1936 conference than he was two years earlier. Although he could judge the scientific credentials of those who should attend, he left their po-litical qualifications up to the National Socialist Professors' Association. Hermann Muckermann, the former Jesuit director of the KWIA, although desired by the Dutch delegation, was refused permission to form part of the German delegation by Rüdin. Once it was clear that Rüdin would, once again, be delegation leader, the Munich director wrote to the Ger-man Congress Center (DKZ) and stressed the importance of a large Ger-man presence. The upcoming international IFEO meeting, Rüdin assured

Dr. Knapp of the DKZ, "is from the perspective of the [international] judgment and status of our racial hygiene legislation of the utmost importance for our country." As such, he continued, he kindly requested that everything possible to secure an "appropriate" number of German delegates be undertaken. Moreover, Rüdin added, it would be appropriate to find hard currency for the female spouses of the German delegates who planned to attend. Although such money for spouses was recently denied for another conference, the upcoming IFEO meeting was in another league. This one, the Munich director assured Knapp, was about enhancing the "prestige of German racial hygienic legislation, measures and ideas." As the conference in the Netherlands will also be a social event with "foreigners of both genders," the German woman "with her exact knowledge" of the situation in her homeland would have the opportunity to make "propaganda for a nazified and eugenically inclined Germany."

It is unclear whether Rüdin was ultimately successful in securing propaganda money for the German spouses, although the DKZ rejected his pleas. The Munich director then turned to his Jena friend and colleague Karl Astel, a man whose well-placed connections within the NSDAP might help him secure the needed hard currency from party sources.[84] Whatever happened with the spouses, when the German delegates were finally selected, the fifteen individuals who passed both the scientific and political litmus test dominated, in terms of sheer number, the other participants from America and Europe.[85] Indeed, the United States had only three representatives. France and Great Britain had two and three, respectively; Norway and Sweden sent only one each. Only the Dutch hosts, with five delegates, boasted more than all other countries besides Germany.[86]

There was another reason that Rüdin was particularly concerned about the upcoming IFEO meeting in the Netherlands: a new president of the Organization needed to be elected. The Munich director had served his four-year term—most of it during the National Socialist era. He had used his position to legitimize Nazi racial policy within the IFEO. Unfortunately, it was not acceptable for more than one president to come from any one country. Now the question remained: which individual could be elected who was at least friendly to the Nazi regime and not German? As the Dutch Frets was not acceptable to the Germans and none of the overtly pro-Nazi Scandinavians—even Fischer's stated choice of the pro-Nazi Mjøen—were willing to serve for reasons of health or age, there were a limited number of tolerable potential candidates. Interestingly, even most of those individuals who were not completely anti-Nazi were unwilling to stand for election. As it turns out, a student of the Swede Lundborg, Torsten Sjögren became the Germans' choice. He had been the person Rüdin hoped, with the help of Davenport, to push as successor to Lundborg at the Swedish Institute for Race Biology instead of the antiracist Dahlberg. Sjögren's subsequent election, however, could not have been better for Rüdin, the German delegation, and the Nazi state. Under his watch, the IFEO became an even more docile instrument for Nazi racial policy propaganda.[87] The Nazi government could be doubly satisfied, as Fischer was named vice-president.

Not long after his election, Sjögren wrote the former IFEO president asking for advice. He used the opportunity to assure Rüdin that his "admiration for the genetic and racial hygienic research and results" in Germany "goes without saying." Indeed, the Swede viewed its work in this area as "a brilliant model." He reported that he would do everything in his power to see to it that the IFEO met in Berlin in 1938. The new president thanked Rüdin for his invitation to come to his Munich Institute to study.[88] During his stay for a multiweek course at the GRIP, Sjögren agreed with his host that the *Archiv für Rassen-und Gesellschaftsbiologie,* the main German eugenics journal, should also become the official organ of the IFEO.

Unfortunately for the German delegation at the Scheveningen meeting of the IFEO, there were still enough Dutch eugenicists to defeat the motion to hold the next IFEO conference in Germany. Instead, it would be hosted in Poland, Hungary, or Estonia. Sjögren and the Germans, however, did not give in easily. President Sjögren attempted to convince the organization that instead of meeting as the IFEO, the body should be part of a Racial Hygiene World Congress held in Berlin. Here, however, he met with the opposition of Cora Hodson, longtime British secretary of the IFEO. Hodson, although partial to the Nazi eugenics project, did not wish to split the Organization by forcing a meeting in Berlin. Perhaps there was still some sense of British fair play on her part. In the last analysis, however, Sjögren and the Germans succeeded in ousting Hodson from her position. Mjøen's wife took over the IFEO's secretarial functions.

This move, of course, eliminated any obstacle from holding a conference in Berlin. For political reasons, however, Nazi officials wanted it to be held in Vienna. After Austria was "brought home to the Reich" in 1938, Nazi bigwigs felt that using Vienna as a venue would be good advertizing for racial hygiene in the *Ostmark* (the former provinces of Austria). It was no longer to be a mere IFEO meeting. Rather, it was billed as the Fourth International Congress for Racial Hygiene (Eugenics)—in other words, as a continuation of the pre–National Socialist tradition of such international meetings that began in London in 1912. Reich Minister of the Interior Frick, the same person who presided over the 1935 World Population Policy Conference in Berlin, would be the honorary host. Rüdin, Fischer, Lenz, and German eugenicists, including the world-renowned geneticist von Wettstein, were on the steering committee. Rüdin, who was to be acting president at the Vienna meeting, invited most of the leading eugenicists in the world. However, on a list prepared with the names of these individuals, Rüdin put a question mark next to those people who were seen as critical of National Socialist racial policy; he also placed the words "Jew" or "Jewish-Bolshevik" alongside Germany's "racial enemies." Scheduled to be held in August, 1940, the congress never took place. Both the outbreak of Germany's "racial war" and the beginning of regime's "euthanasia project" made such a conference impossible. Reich Minister for Education, Bernhard Rust, announced that the Congress would unfortunately have to be postponed until after the war.[89] There would be no further international eugenics meetings during the Third Reich.

Rüdin also tried to exert influence on international congresses other than the IFEO. In August 1937, the Seventh International Congress of Genetics was scheduled to be held in Moscow. A year or so prior to the planned Congress, Soviet geneticists, including the head of the Maxim Gorky Institute for Medical Genetics and Congress program organizer, Levit, together with antiracist American and British colleagues such as Walter Landauer, Clarence Little, and Julian Huxley, proposed a special general section entitled "Human Genetics and Race Theories." Although the details of the background to this proposal need not concern us, suffice it to say that a German communist geneticist, Julius Schaxel (1887–1943), formerly a professor at Jena before immigrating to the Soviet Union after he was dismissed from his university chair in 1933, was the leading person behind the decision. Schaxel, who was also kicked out of the German Genetics Society in 1935 owing to his "crude insults against German science and its representatives," wanted to organize "a responsible forum of serious researchers" to discuss the National Socialist racial theory from a professional genetic point of view.[90] At any rate the plan for a session on "Human Genetics and Race Theories" at the Moscow meeting was announced in the relevant journals, including the prestigious *Journal of Heredity*. According to a circular put out by the International Committee of Genetics, abstracts for all sessions of the meeting were to be sent to the Organization Committee in Moscow by February 15, 1937.[91]

The plan to hold such a special session obviously designed to attack the German delegation and its government racial policies—policies that claimed to be based on the latest science—was more than a mere thorn in the side of the regime. Indeed, the dreaded Gestapo warned that the German human geneticists would be used for hostile purposes and strongly urged that the German delegation not attend. The German Racial Policy Office, the party, the Ministry of Education, the Foreign Office, and Deputy *Führer* Rudolf Hess's Office all agreed with the Gestapo's position. Virtually all state and party offices suggested that Germany and foreign "friends of Germany and its racial policy" boycott the Moscow meeting as it allegedly would be used for the "political aims of Bolshevik propaganda." Only in the event that a respectable boycott proved impossible to organize would a small group of "especially well-selected and well-prepared German researchers" be sent to the Soviet Union.[92]

As it turns out, the Nazi regime did not need to make a final decision on the matter. The Soviet government decided at first to postpone the Moscow conference for an unspecified time. Somewhat later, it announced that it would not host the Seventh Genetics Conference at all. The Germans were probably not the only ones who were happy about this decision; Davenport had signed a statement in which he indicated that he was personally opposed to holding a conference in the Soviet Union. Probably the mainline Swedish geneticists contacted by Rüdin in 1936 about the matter were also relieved by the turn of events.[93]

The reasons for the Soviet Union's decision not to proceed as planned are complex; they cannot, however, be separated from the Great Terror unleashed by Stalin that was demoralizing and destroying the USSR at the time. More immediately relevant to the proposed conference, however, was the general Soviet attack on the part of well-placed and aggressive Lysenkoists against Mendelian genetics and eugenics as "fascist science." The caustic attacks continued despite all efforts by Levit and his Moscow Institute to counter Nazi racism through first-rate research. Although, as we have seen, Levit first appeared successful in his pursuit of medical genetics (as well as eugenics questions under cover of medical genetics), his luck changed in late 1936. As part of the large series of purges taking place, Communist Party members who were accused of "Menschevizing idealism" and had spent time abroad, as had Levit, were prime targets of Stalin's wrath. In November 1936, the party official in charge of science in Moscow denounced Levit as an "abettor of Nazi doctrines." At a session of the Lenin All-Union Academy of Agricultural Sciences, matters came to a head. H. J. Muller, a friend of the Soviet Union, had set up a genetics laboratory first in Leningrad and then in Moscow between 1933 and 1936. The American geneticist did not help his Soviet colleagues when, at the Lenin-Union academy, he openly attacked the Lysenkoists.

His reform eugenic treatise, *Out of the Night,* was completed in the Soviet Union. Contrary to Muller's hopes, the work greatly displeased Stalin, and the Soviet leader ordered an attack against it. Muller was lucky to escape the country with his life.

His colleagues, however, did not share Muller's good fortune. Agol, who, like Levit, had studied with Muller in Texas, was arrested as an "enemy of the people" and shot. Levit was subsequently removed from his position at the Maxim Gorky Institute in Moscow in 1937. Several months later, in early 1938, he was apprehended by the Soviet secret service and was accused of being an American spy. It appears that he was killed sometime that year.[94] The elimination of Levit marked the definitive end of eugenics in the Soviet Union.

During the time when the Conference was officially "postponed" in the Soviet Union until a later date, the international genetics community was concerned that something unpleasant was taking place. A series of letters between Norwegian reform eugenicist Mohr, the Croatian American geneticist Milislav Demerec (1895–1966), and A. I. Muralov and N. I. Vavilov, members of the Soviet Program Committee, documents the attempt to find out if, and when, the Conference would be held and whether there was any truth to a "very alarming statement" published in December 1936 in the *New York Times.* The respected New York City newspaper reported that there was "a serious schism" among Soviet geneticists and that some were even arrested. Although Mohr and other left-wing practitioners would have liked to have waited for a definitive answer from the Soviet government, it was clear that other geneticists were no longer anxious to go to Moscow. As Mohr reported to Muralov and Vavilov with some regret, his colleagues wondered whether it would be better to host the meetings in a country "that was not in the forefront of political attention."[95]

Ultimately, the Congress was held in Edinburgh from August 22 to 31, 1939. Fears of the outbreak of war hung over the meetings like a dark and threatening cloud. The renowned plant geneticist Vavilov, who was supposed to preside over the Congress, did not attend. Nor did any of his Soviet colleagues deliver papers. In 1940, Vavilov was accused of "wrecking" Soviet agriculture. He died in a Soviet prison camp in 1943. Although a *New York Times* article predicted that the "democracies [will] have big delegations" and that "some Germans will stay away," the thirty-two politically handpicked scientists from the Reich who traveled to Scotland were a formidable presence. In truth, the German scholars probably spent more time listening to the radio than discussing the newest developments in genetics. They were sitting on packed luggage, so to speak, ready to leave at a moment's notice. The German Foreign Office in fact did order the Reich delegation to leave following the signing of the infamous German-Soviet Non-Aggression Pact on August 23. Soon after, most other continental European participants made their way home, fearing an outbreak of hostilities. During the last few days, the Congress was largely reduced to an exchange between British and American geneticists.[96]

That having been said, the meeting served as a venue for a very eventful undertaking. Watson Davis, director of the American Science Service, asked world leaders who planned to attend the Congress in Edinburgh the following query: "How could the world's population be improved most effectively genetically?" The response came to be known as the *Geneticists' Manifesto*. Based in large part on Muller's 1935 treatise, *Out of the Night*, the *Manifesto* was not only an outspoken plea for a reform eugenics based on the socialist transformation of society but also a sharp critique against racism among mainline eugenicists in general and Nazi racial policies in particular. As a reflection of Muller's Marxist perspective, he viewed all racism, the Nazi variety included, as a by-product of capitalism. "The removal of race prejudice and the unscientific doctrine that good or bad genes are the monopoly of a particular people or of persons with features of a given kind will not be possible," the *Manifesto* stated, ". . . before the conditions which make for war and economic exploitation have been eliminated." Only after a new socialist economic order and a more profound understanding of genetics would it be possible for a eugenics that would be worthy of the name—one that would raise the mental and physical genetic level of all individuals. The *Manifesto* was signed by twenty-three of the most influential reform eugenicists, including the Marxist British group around Huxley, Theodosius Dobzhansky, Joseph Needham, Dahlberg, and of course Muller. It was later published in the influential British journal *Nature*. Interestingly, the text also appeared in the mainline journal *Eugenical News*.[97]

Although a clear eugenics alternative was presented at the Congress in Edinburgh, the reformers were not internationally unified. As was mentioned, individuals such as Muller, Dahlberg, Mohr, and the British Marxists shunned the IFEO. They communicated among themselves, but they never formed a separate international eugenics society to propagate their ideas. Marginalized during the National Socialist era in the German-dominated IFEO, in 1935 the Latin eugenicists formed their own organization, La Fédération Internationale des Sociétés d'Eugénique Latine (the International Federation of Latin Eugenic Societies). It held its first congress in Paris in 1937. The papers at this meeting reflected their pro-natalist, positive eugenic program. The Fédération included Central and South American eugenic practitioners, a group within the Latin eugenicists.[98]

Hence, by the outbreak of war, there were certainly alternatives to the Nazi variety of eugenics, but as far as the continuation of Nazi racial policy was concerned, it mattered little, if at all. Needless to say, after Germany began its "racial war" with an attack on Poland on September 1, 1939, any international eugenic cooperation with the Reich on the part of Western Europe was impossible. We have seen that hard-core American racists like Laughlin did not immediately break off contact with Germany. But with the American entry into the war, even the few American mainliners who were still active had no further professional relations with the enemy. With revelations of Nazi atrocities, eugenics, at least its mainline variety, was discredited in the United States. As we have mentioned, however, eugenic practices such as sterilization continued.

Within Germany, the war unleashed a twofold development. Eugenics, even as understood by the German human geneticists who made the Faustian bargain with the Nazi state, did not initially include the outright killing of individuals deemed genetically or racially defective. In 1939, however, the transition was made in Germany from the sterilization of the "unfit" to their physical liquidation as part of the "euthanasia" project. Although there were no lack of willing medical professionals who anxiously sought to enhance their careers by serving as "euthanasia doctors" or hoped to profit scientifically from those murdered, the leap to the systematic killing of the "unfit" did not originate from the human geneticists themselves. Nor were these researchers directly responsible for the decision to undertake the "Final Solution," however much their rhetoric and service to the state aided this historically unique and ethically unfathomable mission. Research has outlined the ideological, professional, and methodological connections between the destruction of so-called useless eaters and the mass killing of "racial undesirables" in Germany's slave and extermination camps after 1941.[99] We have seen how von Verschuer, Magnussen, Nachtsheim, and Rüdin were involved with these murderous procedures. Even German school biology teachers made their (albeit modest) contributions to it.

The other development during the war years involved the attempt by the Nazi regime to exert direct control over the professional institutions in Germany and occupied and neutral countries. This naturally included biomedical organizations. We have seen that even prior to the war the Nazi government took a very active interest in the composition of the German delegation at international professional conferences and how a deal emerged between the German human geneticists who attended such meetings and the National Socialist state. There were other instances, however, where matters were more one-sided. For example, in 1935 a Bureau of Human Heredity was established in London. Its alleged purpose was to serve as a clearinghouse for new research on human genetics. The Chairman of the Bureau was the British geneticist R. Ruggles Gates; Hodson, at the time secretary of the IFEO, was elected honorary secretary of the Bureau. At the beginning of 1937, the Reichsausschuß für Volksgesundheitsdienst (the German Commission for Public Health Service), an organization within the German Ministry of the Interior, wrote a letter to Rüdin questioning why Germany should be involved in the Bureau. The problem was that Haldane, an outspoken critic of Nazi Germany, was on the executive committee. As Dr. Lemme of the Reichsausschuß bluntly put it, given Haldane's role in the organization Germany's "collaboration with the Bureau of Heredity is not acceptable." Lemme demanded a response from the Munich director; when Rüdin did not immediately answer, he wrote him another letter. In Rüdin's reply—one directed not only to Lemme but also to another Nazi medical bureaucrat as well as to the prestigious medical journal the *Münchener Medizinische Wochenschrift* (*Munich Medical Weekly*)—Rüdin listed all the reasons why such collaboration was not desirable. He was probably honest in many of his own reservations he offered Lemme, but the point is that he was more or less forced to comply with the wishes of the state.[100]

During the war, the German government was anxious to exert the *Führer* principle on organizations to an ever greater extent than before. In early 1940, an important bureaucrat in the Reich Ministry of the Interior and administrator of the child "euthanasia" program, Herbert Linden (1899–1945), wrote a letter to Rüdin regarding the further coordination of the German Society for Racial Hygiene. Linden queried whether separate local sections in various cities really needed to exist, as was the case in the newly annexed Austria. It was necessary, "especially in this field," Linden continued, "to have the closest possible cooperation with the Party." Linden made it known that his Ministry would insist that Deputy *Führer* Rudolf Hess would have a say in any change in the chairmanship of the Society. He reminded the Munich director that "all forms of training in the field of Genetic and Racial Care were under the bailiwick of the Racial Policy Office of the NSDAP." Moreover, in 1935 the German Society for Racial Hygiene was informed that when it held public meetings it needed the approval of the party. Linden reprimanded Rüdin for an earlier letter in which he claimed that the Society was merely responsible to the Reich Ministry of the Interior, i.e., a state ministry and not the Nazi Party.[101] If it was not totally transparent to German human geneticists in the first years of the Nazi regime, it was certainly crystal clear now which of the two parties to the Faustian bargain had the upper hand.

Certainly the most grandiose Nazi government plan for international biomedical institutions was its attempt to create a "new order" for all scientific life in Europe. Preparations for this new scientific order were discussed at meetings hosted by the Reich Ministry of Education in late 1940 and early 1941. The first meeting was held on November 12, 1940, and was hosted by SS member and head of the German Research Council, Dr. Rudolf Mentzel. In addition to several party members and bureaucrats, there were representatives from the professional disciplines present. Eugen Fischer spoke for most of the anthropological and biomedical sciences; von Wettstein served for biology.[102]

The Herculean task that the ministry set for itself was the question of whether the seat of all professional organizations in occupied countries could or should be transferred to Germany, or how, otherwise, it could dominate all the scientific disciplines in the present and future Greater Germany.[103] Speaking for his field, von Wettstein argued that the correct plan for biology would be to "bring the European [organizations] in order under German leadership." "After the War," the plant geneticist continued, "the struggle will have to be fought out with America anyway." At that point, scientific prowess will decide whether the right of leadership will lie with Europe or the United States.[104]

Fischer, for his part, informed those present at the meeting that Germany already had the "main influence" over the IFEO.[105] His view of the French Institut international d'anthropologie (International Institute of Anthropology) in Paris was anything but supportive. According to Fischer, it was Germanophobic going back to the days of the League of Nations in the 1920s. The Reich never joined the organization. Apparently in a letter to the Ministry of Education, the Dahlem director stated that he was opposed to the Institut both because many of its French members were Lamarckian and because its director was opposed to the National Socialists. Fischer suggested that one send his former KWIA assistant, SS member Horst Geyer, to help the occupation forces in Paris collect information on the anti-German past of the Institut. The Dahlem director's goal was that "this notorious Institut disappear from the Earth," once and for all. From a memorandum dated 1941, the Ministry of Education had obviously taken up Fischer's request to investigate the French organization; the Dahlem director's original damning report on the Institut was appended.[106]

Discussing the question of whether the new order in science would proceed in an "evolutionary" or "revolutionary," manner, the general consensus of those at the meeting was that in "exploiting the given situation, all means should be employed to create a fundamentally new organization of scientific cooperation." This might entail working with existing foreign organizations, "depending on the particular conditions." In the scientific struggle with the United States, Germany's present situation is "organizationally and financially better than ever before." The participants suggested the creation of a reliable German instrument to secure its position within the international organization of science.[107] This was necessary, as a later report documented, since the new scientific order was not merely an "independent problem" but nothing less than a facet of Germany's foreign policy in the field of science together with its cultural and political impact. Moreover, "following the spirit of the European revolution," international congresses on the continent, as they functioned in the past, would cease to exist.[108]

Summarizing the role that German human geneticists played on the international stage from the first decade of the twentieth century, when their fields first made their debut, until the midwar years, when at least Fischer took part in a gargantuan plan for a "new order" of European science under German hegemony, is no easy matter. If, however, we limit our scope to their participation within the international eugenics movement we find more unity among them than among practitioners of other nations dedicated to the Galtonian gospel.

Although the banner of German racial hygiene was embraced by individuals of all political persuasions, those active in the international movement in terms of attending and hosting conferences were, by and large, politically conservative mainliners. This was true both before and after the "Great War." The aftermath of WWI served not merely to temporarily polarize the international eugenics movement, but permanently alienated German racial hygienists and precipitated their move to the political right. Moreover, many of the active German racial hygienists were first and foremost human geneticists who secured their training through medicine—a profession that enjoyed an inordinate amount of social prestige and viewed serving the *Volksgemeinschaft* as its unique calling.[109] Although physicians also dominated French eugenics, these individuals thought of themselves first as medical doctors and then, if at all, as geneticists of a Lamarckian persuasion.

During the mid-1920s, Great Britain, the United States, and the Scandinavian and the Benelux countries either had active mainline and reform eugenicists who attended the same conferences or the reformers among them played an active role at other international professional meetings. Outside Germany, reformers frequently critiqued their fellow mainline countrymen openly. Whereas there was a so-called Berlin and Munich contingent of the German Society for Racial Hygiene at this time—the former allegedly representing a moderate eugenics that did not emphasize the hereditary superiority of "Aryans," while the Munich group stressed class and race as criteria of eugenic fitness—we know that forming any hard and fast dichotomy out of these differences is problematic. This is because two of the most active members of the Berlin group in the late 1920s and early 1930s, Fischer and von Verschuer, were secret Nordic enthusiasts. Their silence on the Nordic issue was merely political. Moreover, most of the moderates were not active on the international circuit.

Latin eugenicists, as we have mentioned, were by and large Lamarckian, and while they attended eugenic-related international conferences with Mendelian practitioners until the mid-1930s, they were not taken seriously by the latter. During the Third Reich the Lamarckians were heavily attacked by the German delegation. Soviet human geneticists, while Mendelians, never attended a eugenics conference, as their interest in the field had to be hidden. By the time it came to hosting a genetics meeting where the topic of eugenics and Nazi racism could be discussed, namely, at the proposed Seventh International Genetics Conference in 1937, the Soviet government refused to hold the gathering. Most Soviet colleagues were already under attack by the Lysenkoists or worse.

Without wishing to overemphasize the degree of unity among the German human geneticists who played an important role in the international arena, it would not be an exaggeration to state that most of those eugenicists who were not kicked out of their posts owing to race or political ideology and hence no longer had access to international meetings during the Third Reich had already offered their support to the Nazi state. As we mentioned, most were German conservatives who had enough in common with National Socialism that they were hopeful about the coming of the new order. As it became clear that they would profit professionally from a state that made race and heredity the cornerstones of its *Weltanschauung,* the German racial hygienists had no reason to resent representing their government at international conferences and within the international eugenics movement. Those like Fischer, von Verschuer, and Rüdin believed in the power of genetics; even if they occasionally had quibbles with individual facets of Nazi racial policy, the official Nazi ideology's emphasis on the hereditary nature of physical and mental characteristics was in accord with their own science. Their national conservative or *völkisch* outlook made it easy for them to accept and propagate the regime's anti-Semitic policies at home and abroad.

37 Photograph taken by the famous Jewish Russian-American photographer Roman Vishniac, featuring his daughter Mara in front of a Berlin store window, 1933. Mara poses in front of a device for measuring the difference in size between Aryan and non-Aryan skulls. The photo demonstrates that the so-called science of racial research was not just hidden behind the walls of the academy. Copyright Mara Vishniac Kohn; courtesy of the International Center of Photography.

What is clear is that in the international arena, as well as at home, German human geneticists during the Third Reich could capitalize on their mainline eugenic tradition and entered into a symbiotic relationship with the National Socialist state. German human geneticists as well as their government served as mutually beneficial resources. The international *renommée* of the German scientists was perhaps their most valuable resource for the Nazi state. The outcome of this symbiosis was that eugenics was discredited, and there was never a new international movement. For decades the term "eugenics" was by and large taboo. The word still has a largely negative connotation today.

The Road Not Taken Elsewhere: Was There Something Unique about Human Heredity during the Third Reich?

There were in the memory of mankind Genghis Khans and Eugen Fischers, but never before had a Genghis Khan joined hands with an Eugen Fischer. For this reason, the blow was deadly efficient.[1] MAX WEINREICH, AUTHOR OF *HITLER'S PROFESSORS*

Over the course of this book, we have examined four different venues designed to illuminate the relationship between human heredity and politics under National Socialism. Two of these case studies, we will recall, focused on the production of human genetics knowledge in the most important research sites for the investigation of human heredity and eugenics in Germany, the Dahlem KWIA and the Munich GRIP. The other examples highlighted the dissemination of this knowledge at professional meetings and in German secondary school biology classrooms. It is hoped that the symbiosis between human genetics and Nazi politics is clear for all to see. In addition, I have attempted to lay bare exactly how human genetics researchers and National Socialist officials in question served as mutually beneficial resources. Hence, the four case studies answered one of the queries raised in the introduction: What induced so many trained German human geneticists to make a deal with Nazi bureaucrats in the first place, and what did the senior partner to this Faustian bargain, the Nazi state and its functionaries, gain from these scientists? Moreover, biomedical research scientists working at the KWS both before and during the Third Reich were placed within the context of the international human genetics and eugenics movement prior to and after 1933. Only by examining the fortunes of human heredity and eugenics in Germany on the international stage during a period extending almost a half century are we in the position to offer an answer to the two most central questions posed in the introduction: First, what, if anything, was uniquely "Nazi" about human heredity under the swastika? Second, how effectively does this peculiarity explain why some important representatives of the human genetics community embarked on a path that led to the moral abyss, for both themselves and their science?

A German *Sonderweg* in Biology?

The question of whether there was something unique about human heredity under National Socialism is part of a much broader query that has preoccupied German historians since the end of last world war: Was there something special about Germany and its history that accounts for the rise of Nazism and its accompanying brutalities at home and abroad? Several answers have been offered during the past sixty years.[2] Among the best known, if now controversial, explanations is the *Sonderweg* ("special path") thesis. Articulated in the 1970s and 1980s by members of the Bielefeld school of social history, most notably Hans-Ulrich Wehler and Jürgen Kocka, the *Sonderweg* thesis argues that Germany's path of development differed in numerous important respects—including the lack of a politically mature middle class and the absence of a strong liberal tradition—from other Western countries. These specific peculiarities set the stage for Germany's inevitable and tragic path toward Nazism. Critics of the *Sonderweg* thesis, however, claim it assumes some normative path of development toward an industrial, advanced capitalist state and is thus misguided. In the case of science—the issue under discussion here—opponents of a German *Sonderweg* at least imply that there is no necessarily "correct" way in which human heredity and politics impact each other.[3]

In an attempt to construct a position between advocates and critics of the *Sonderweg* thesis, the late German social historian Detlev Peukert argued that the German path of modernization—which ultimately, but not *inevitably*, let to Nazism—was one of many possible paths open to Western industrial-capitalist societies. Germany's trajectory in the first half of the past century represents an extreme example of social and political developments that occurred in part, and to varying degrees, in other countries.[4] It would appear that Peukert's view is relevant to the case of human genetics and eugenics in Germany during the Third Reich.

Applying Peukert's position to the subject of this book, an analysis of the international human heredity community in the first third of the twentieth century, suggests that there was nothing intrinsically peculiar or unique in the history of German human genetics and eugenics prior to 1933 that made the transgression of all moral boundaries in the biomedical sciences such as we have seen in the KWIA-Auschwitz connection or the GRIP-"euthanasia" relationship inevitable. We have come to learn that the intricately related fields of human genetics and eugenics represent socially constructed knowledge. As such, it should not surprise us that their practitioners in the thirty or so countries that boasted eugenics movements differed in terms of their allegiance to a specific mechanism of heredity, their methodology of choice (either positive or negative eugenics or both), their political persuasion and professional training, their willingness to legislate genetic fitness, and their openness for ideologies of Nordic supremacy. If one compares the countries that adopted the "gospel of Galton," one frequently finds mainliners and reforms within the same nation. Professional organizations also incorporated such divisions. Indeed even during the first several years of the Third Reich, conservative mainliners, liberal opponents of Nordic supremacy, and Latin eugenicists attended the same international conferences. As was mentioned in chapter 1, it appears that the only common denominator among eugenicists everywhere was their penchant to consciously or unconsciously view human beings as human resources whose numbers could be manipulated for some transindividual purpose. We have seen that the power of genetics to solve social problems in a state interventionist framework was accepted in capitalist, socialist, and non-Western countries alike.

The conclusion that there was no one special feature about racial hygiene in Germany not merely prior to 1933 but throughout the interwar years is further strengthened by a recent study in which the KWIA was compared to two non-German research institutes for human heredity: the American ERO and the Soviet Maxim Gorky Institute for Medical Genetics. Both similarities and differences abounded between all three institutes. There was "no single factor—be it the professional background of the scientists, the kind of research undertaken, the source of funding, the administrative style of the directorate or even the existence or nonexistence of a dictatorial regime—that somehow made the Fischer institute unique during the Weimar Republic or under the Third Reich."[5] Even the exploitation of vulnerable individuals or populations without rights for experimental purposes was not a unique feature of Nazism. During WWII the United States' notorious syphilis study on African Americans came close to pushing the boundary of ethically acceptable research, if indeed it did not transcend it. Wartime Japan engaged in numerous experiments on local inhabitants of occupied Manchuria for military purposes. It is estimated that in one Japanese unit alone over three thousand individuals died while serving as human guinea pigs during a ten-year period.[6]

Historical Contingency and the Nazi Symbiosis

That having been said, we somehow believe that there *was* something different about the path that human genetics research took under National Socialism. In no other country and at no other time were the ethical boundaries of traditional research practice so brutally violated as in Germany under the swastika. Although Nazi Germany may not have been the only regime that engaged in human experimentation during the war, it was the only one to do so, at least in part, in the name of eugenics (not all forms of human experimentation in the Third Reich were related to racial hygiene). In addition, much of the research conducted by certain KWIA personnel like von Verschuer, Magnussen, and Nachtsheim, as well as Deussen at the GRIP, was deliberately designed to further eliminate unwanted populations. This tragic trajectory cannot be explained by what might possibly be the only significant anomaly between German racial hygiene prior to 1933 and movements in other countries: the absence of a sustained and visible critique by geneticists and the larger population in the Reich of the very practice of eugenics. Why this was so and whether such public silence did not take place in other countries with eugenics movements has yet to become the subject of scholarly debate. What is clear is that the destruction of a viable civil society during the Third Reich rendered any such criticism, not only of eugenics per se but of the extreme racist form practiced to the forced exclusion of other varieties, all but impossible.

Drawing on the examples of our four case studies, Germany's role in the context of the history of the international human genetics and eugenics movement as well as the in-depth comparison of one of the Reich's most prestigious institutes, the KWIA, with other similar non-German research centers, we come to the following conclusion. If there was anything uniquely "Nazi" about human heredity and eugenics in the Third Reich, it pertains to the *particular way in which human genetics interfaced with National Socialist politics and how they served as resources for one other. The symbiosis that ensued between human heredity and the broad political context of Nazism served to radicalize them both.*[7] As we have seen, during the prewar years, the Faustian bargain between German human geneticists and officials of the National Socialist state led to greatly increased funding for politically cooperative and enthusiastic biomedical scientists; the knowledge subsequently produced helped legitimize the "racial state" at home. Even school children were instructed in governmental racial policy and learned that it was allegedly based on vanguard science. Abroad, the Nazi "racial state" was defended on the international stage by renowned German human geneticists who in turn profited from the political usefulness of their international *renommée.*

During Germany's "racial war," the symbiosis turned deadly. It changed the opportunities for biomedical research as well as the nature of the "research material." The special nature of this union created unexpected and highly coveted possibilities to engage in scientific investigations on populations without rights that would not have been possible under other political and ideological circumstances. The newly acquired "research material" could then provide the continued scientific legitimization and expertise necessary to further execute racial policy under the swastika. Even school biology textbooks employed during this period served to further dehumanize populations targeted for extermination and to legitimize the brutal two-front war against racial undesirables at home and abroad.

The specific historical conditions that set the stage for the unique relationship between human genetics and politics under the swastika was "the coming together of a dictatorial regime for which race and heredity served as an ideological cornerstone and a mature eugenics movement for which race and heredity functioned as its epistemological categories." This intersection was certainly exacerbated by the dire economic and social tensions plaguing the late years of the Weimar Republic after the outset of the Great Depression in Germany in 1930—conditions that were more extreme there than in other countries.[8] In addition, the still highly charged negative legacy of the "Great War" remained a part of the national memory. We have observed the impact of the former in the two institutional histories of the KWIA and the GRIP. The fallout from the latter was evident in German racial hygienists' shift to the political right and their attitude toward international professional meetings in the years immediately following WWI.

That these historically contingent factors help explain the uniqueness of the form that human genetics took in Germany may be gleaned by a final, brief comparison with the situations in the Soviet Union and the United States, countries that were political opposites and harbored fundamentally different projects for the science of human heredity.

As in Nazi Germany, human genetics in the Soviet Union was certainly pursued in the context of a dictatorial regime. Soviet practitioners, like their colleagues in Germany, openly collaborated with their government. Yet official Soviet ideology, Marxism-Leninism, opposed, rather than reinforced, a scientific paradigm focusing on race and nature. Perhaps most importantly, although Levit, as head of the Maxim Gorky Institute, was forced to "overlook" the crude brutalities of his regime—as were Fischer, von Verschuer, and Rüdin in Nazi Germany—the latter made a *direct contribution* to those brutalities through the various activities of their institutes. Levit as well as the German KWS directors served oppressive regimes to advance their own scientific interests, but only in the case of the latter did the science itself (in the political context of the Third Reich) function to legitimize and exacerbate the government's inhumanity.

In the United States, capitalist ideology might have been quite compatible with the mainline eugenic thought that was pursued in Nazi Germany. Indeed we have observed the intimate ties between mainline American eugenicist like Laughlin and conservative German racial hygienists both before and after 1933. But by the 1930s, the Unites States' racist brand of eugenics was represented by individuals who, unlike the German human geneticists, could boast no international scientific renown. In addition, the ERO's funding agency, the Carnegie Institute of Washington, forced it to close its doors in 1939. The work undertaken at the ERO had become a scientific and political embarrassment. Neither had the United States adopted the "gospel of Galton" as its official ideology nor was it ever a one-party state. Even when some of its citizens' liberties were threatened by the work of American human genetics research—as in the case of mandatory sterilization—the United States continued to retain a healthy civil society, even during the troubled years of the Great Depression. The existence of such a society with a plurality of ideologies served as a barrier to the adoption of the sort of radical state policy that was practiced in Nazi Germany.[9]

As has been demonstrated, both the intellectual content and the international context of human genetics research served as scientific capital for a regime desiring that its racial policies be based upon the most up-to-date science. Fischer, von Verschuer, and Rüdin's international reputations were no less important to Nazi officials than their numerous racial science investigations.[10] Once the deal was sealed with the German human geneticists, the relevant research institutes, national and international professional conferences, and even high school biology classrooms served as intellectual resources for the Nazi state.

When National Socialist politics became a resource for the science of human heredity itself, more changed than research budgets. Nazi institutions and programs related to the racial policy goals of the state altered research practice as well.[11] This was especially true after the beginning of Germany's "racial war." As we have seen in chapters 2 and 3, in the context of the Third Reich, the sudden availability of a great reservoir of potential human subjects for research purposes provided a window of opportunity for some biomedical scientists who could otherwise never avail themselves of such prospects in the absence of a world war unleashed by Nazi Germany. We will recall that the potential human subjects were people who had been incarcerated in concentration camps, extermination camps, or "euthanasia" hospitals and stripped of all rights and dignity because of Nazi racial policies.

A Constellation of Motivations

It is certainly not out of place at this juncture to speculate about the motivations that resulted in research transcending all ethical boundaries. In the introduction, it was suggested that unbridled research enthusiasm was frequently cited as the motivating force for the German human geneticists' actions. As this study has shown, this factor was surely at work. However, that there is no known case of such barbaric operations undertaken on individuals who were not first deemed "barbaric," "useless-eaters," "subhuman," or "parasites" through Nazi racial policy rhetoric should give us pause to think. It is highly unlikely that unbridled research enthusiasm was the only motivation.

With respect to the above-mentioned statement, we might well ask whether the dehumanization process that was part and parcel of the pre-1945 international eugenics movement itself—a program predicated upon increasing the stock of "more valuable" and reducing or eliminating "less valuable" "hereditary material"—was not also responsible for the ensuing tragedy. Naturally, we must also consider career opportunism. We know, for example, that such motivation played a large role among "euthanasia" physicians in carrying out their tasks. It also helps explain why such a large portion of the German medical profession joined the Nazi Party.[12] One may also assume that ideological support for all Nazi racial policies led German human geneticists to undertake such ethically compromising research. Although there is not a shred of doubt that there was a high level of ideological compatibility between the political outlook of our KWS directors and the Nazi worldview, there is no evidence to suggest that Fischer, von Verschuer, or Rüdin supported, let alone initiated, these killing projects. What is certain is that Rüdin (along with his student Deussen) and von Verschuer were willing to profit scientifically from a fait accompli. In other words, it appears that our German human geneticists found enough common ground with the most extreme Nazi racial practices to enable them to utilize the new "fortuitous" circumstances that were presented to them during the war for their own professional ends without having to support them openly.

It seems probable that a constellation of such factors served as motivating forces for those German human geneticists who transgressed all ethical boundaries in their scientific research. There are, however, two other considerations worth discussing at this point. We have mentioned on several occasions that the biomedical researchers under discussion would have never become accomplices or profiteers in mass murder in the absence of a Nazi-style "racial war." Other studies have demonstrated that large segments of the German population during the late war years were affected by the general process of "moral numbing." Just because our German human geneticists, and to a lesser degree high school biology teachers, contributed to the brutality of the regime through their science or their pedagogy does not mean they cannot also have been morally dehumanized themselves.[13]

If we consider the notorious speech that Fischer held in Paris in late 1941 where he all but denied that "Jewish Bolsheviks" were part of the human species, we can see this rhetorical brutalization at work. After the halt of Germany's victory on the Eastern Front and in the midst of a conflict there that became more barbaric with each passing month, Hitler's equation of Jews with Bolsheviks was certainly shared by German soldiers fighting in the Soviet Union.[14] We can imagine how German propaganda at home bombarded "national comrades" with pronouncements regarding the "Jewish Bolsheviks." It seems likely that the internalization of such ubiquitous rhetoric might have further radicalized the nationalistic Fischer—now a party member whose son was fighting in the Soviet Union[15]—who was certainly anti-Semitic to begin with.

Finally, we must take seriously the stepwise process of the radicalization of biomedical research and its application under the swastika as perhaps the most significant, if profane, of the factors at work.[16] Such consideration is necessary if we not only wish to answer the question of what caused some German human geneticists to cross the line of moral culpability during the war but also understand why these same individuals continued to work for the Nazis even after the true nature of the government was clear. As we have seen in the introduction, the German human geneticists at the focus of this study were not moral monsters in 1933. Nor could they or anyone else have projected the "end game" of the what started out as a coalition between the National Conservatives and Nazis that was dedicated to the hereditary well-being of the German *Volk*. This was a government that was a welcome change to the social and financial chaos of the late Weimar Republic, not only for our human geneticists, but for a large percentage of the non-Jewish German population. "Murderous science" was not yet in the cards.

If we once again take Fischer as an example, we will recall that in his first speech during the Third Reich, the Dahlem director articulated his views on race as he had done in the past. It probably did not occur to him that evening in the Harnack House that his scientific expertise on this question would not only be challenged but condemned by his new superiors. We have seen that it took him a while to recognize that he was no longer dealing with just any regime. He negotiated this new situation and placed his Institute in the service of the new order. He had also justified the existence of the renowned Dahlem research center during the Weimar Republic by promoting its service to the state. Genetic health and eugenics were also important to Weimar politicians.

We may well ask at what point did the compromises and accommodations made by Fischer with government officials during the early years of the Third Reich transcend the bounds of morality? When exactly did his government-sponsored research and teaching activities—tasks begun under a democracy—turn criminal? Were Fischer and other German human geneticists aware of when they crossed the ethical point of no return? More significantly, after making the first concessions, would it have been possible for them to extricate themselves from their professional activities in 1941, 1938, or even 1934?

In psychology, the theory of cognitive dissonance maintains that individuals cannot hold contradictory ideas and thus avoid actions or behaviors that are at odds with their beliefs. They resist reflecting upon or considering anything that would make them uncomfortable. In other words, it is necessary to do everything possible to affirm the view or action one has adopted and not allow cognitive contradictions to surface.[17] In this particular case, once our human geneticists identified themselves at some level with the Nazi regime they worked for, it became extremely difficult for them to call their position and actions into question. The longer they held on to this Nazi worldview and their place within it, the harder it became for these human geneticists to adopt a critical stance against it. If they entertained any belief that would undermine their duties as civil servants of the Third Reich dedicated to the genetic health of the *Volk*—for example, that it is un-Christian to work for a government that commits physical violence against "respectable" German Jews (recall that Fischer, like most German conservative nationalists, made a distinction among Jews at the beginning of the Third Reich)—that belief might well be changed to conform to their duties as racial hygienic experts under the swastika. The avoidance of cognitive dissonance is a way for individuals to lead what for them is a normal life. It is certainly not limited to negotiating one's way through the Third Reich.

Returning again to the stepwise nature of our German human geneticists' professional involvement in the Third Reich, it might be useful to turn our attention to a classic historical study. Forty-five years ago historian William Sheridan Allen pointed out in *The Nazi Seizure of Power* that one of the main problems for the population of the northwest German town of Northeim was that of perception; people saw one or another side of National Socialism, "but none saw it in its full hideousness." Earlier in his work, the author discussed why none of the Northeimers offered resistance to the new regime. The answer: what one act during the early months of the Third Reich could have turned Northeim from a democracy to a dictatorship? There was a series of semilegal actions that the Nazis implemented over a period of the first six months of the regime's existence—none of which singly facilitated a dictatorial revolution—but the sum total of these transformed the town from a republic to a dictatorship. During the mature dictatorship, Allen argued, it was the Northeimers' accommodation to the new realities of the Third Reich that was paramount for the state's functioning. But their accommodation was predicated on a regime whose officials did not demand enthusiasm from the entire population.[18]

We may ask: what one act of accommodation made by our German human geneticists turned them from respectable scientists to morally culpable researchers? Like the Northeimers, Fischer, von Verschuer, and Rüdin might well have known at some point during the Third Reich that things had changed radically for them. But did they realize when and why? Did they not merely accommodate themselves, step by step, to the new political realities and do what they always did: secure money and subjects for their research? And did not the officials of the regime frequently refuse to offer invitations to human geneticists for party meetings when enthusiasm, not professional accommodation, was demanded? Did

Nazi officials have to accept that the human geneticists could be pushed only so far? Yet, their step-by-step accommodations with the regime certainly worked to the full advantage of the National Socialist state. When Fischer held his talk at the Harnack House in February 1933, could he have imagined the nature of the speech he would give in occupied Paris nearly eight years later? There can be little doubt that his talk served the interest of the regime's extermination policy.

As historian of biology Garland Allen has argued, there is an important lesson for us to learn from the actions of these German human geneticists. How many small compromises and concessions are we willing to make even today, Allen queried, "as budgets tighten, funding sources become reoriented (and perhaps not in directions a scientist [or scholar] would have chosen), and institutional or job-related expectations are changed? How many steps does it take to cross that fatal line?" Continuing in the same vein, Allen noted that "we humans are remarkably adaptable in many respects; but the most dangerous adaptations are those we do not consciously examine or that we try to deny." German human geneticists were probably "no more culpable for their accommodations to the new requirements placed before them than other members of society." According to Allen, "those who went the farthest did so in a particular context and ended up crossing the boundary of moral culpability." Needless to say, this in no way excuses or legitimizes these egregious ethical transgressions. It does, however, render them more human—perhaps all too human. Indeed, unless we are careful in considering our choices, we too can wind up on a path we may not wish to travel and find ourselves at a moral dead end.[19]

Although the Faustian bargain we have examined is a product of historical contingency rather than inevitability, the sheer human suffering unleashed by this Nazi symbiosis should caution present and future researchers in human genetics to remain vigilant against wielding the "sword of [their] science" against human dignity. We threaten to inflict a terrible disservice to the countless victims of this deadly symbiosis if we do not.

Acknowledgments

The subject matter of this book has preoccupied me for quite a long time. It would not be an exaggeration to say that I was already thinking about the relationship between human genetics and Nazi politics when I decided to write my dissertation on the origins of the German eugenics movement during the Empire. Although I would have liked to jump into the "hotter topic" of biomedicine under National Socialism, I had a feeling more than twenty years ago that one could not adequately discuss the relationship between human heredity, German eugenics, and Nazi racial politics without understanding the early history of racial hygiene under the Empire and the Weimar Republic. Somewhat later, I recognized that the unholy trajectory that human heredity and racial hygiene would take in Germany needed to be imbedded in a larger history: that of the international eugenics movement. As is often the case in academia, other scholars beat me to the punch. Not long after I published my revised dissertation, *Race Hygiene and National Efficiency: The Eugenics of Wilhelm Schallmayer* (Berkeley and New York: University of California Press, 1987), several important books on the history of German biomedicine and eugenics in the first half of the twentieth century, including, of course, the Third Reich, were suddenly available. For years, I abandoned the hope of writing my own account, believing that it would be superfluous. Yet by the turn of the century, I realized that the available literature was, for one reason or another, not suitable to use in advanced upper-division undergraduate courses in the history of science, Nazi history, or Holocaust history. More important, perhaps, these books did not answer the query that has remained with me ever since the mid-1980s: what, if anything, was unique about human heredity and eugenics under the swastika? The present volume is the result of a pedagogical need and my lack of complete intellectual satisfaction with the existing literature. When, in 2002, one of my students got tired of my queries and criticisms of other readings on the topic, she said, "well, why don't you write a book on it yourself."

I still felt light-years away from such an undertaking until I received an offer, in spring 2001, to become part of the intellectual community of full-time and guest scholars working on the large, five-year Max Planck Society Presidential Committee established to investigate the Kaiser Wilhelm Society (KWS) during the National Socialist Era. Here is where my first thanks goes: Mark Walker persuaded me to apply for a fellowship for the summer of 2002 to work on a comparative institutional history of the Kaiser Wilhelm Institute for Anthropology, Human Heredity, and Eugenics (KWIA) and two similar non-German research centers. In doing so, he changed my academic trajectory, and for that I will always be grateful. I owe a debt of gratitude to the project manager at the time, Carola Sachse, for having the confidence in me to afford me a fellowship, as well as the two senior heads of the committee, Reinhard Rürup and Wolfgang Schieder. However, without the willingness of Garland Allen and Mark Adams—who agreed to put their vast knowledge of the American Eugenics Record Office and the Maxim Gorky Institute for Medical Genetics, respectively, to use for what would become a lengthy joint article—my long stint in Berlin would have never came to pass. An additional Max Planck Fellowship for the KWS project enabled me to spend the next summer working on aspects of human genetics and politics at the KWIA. Over the course of this time, I had the pleasure of intellectual exchanges— far too numerous to mention—with the full-time scholars on the project team, Carola Sachse, Susanne Heim, Helmut Maier, Benoît Massin, Florian Schmaltz, Bernd Gausemeier, and, in particular, Michael Schüring. Their knowledge of the available archival material on the KWS, and the KWIA specifically, made the task of negotiating the wealth of documents much easier. I was also enriched by having had the opportunity to discuss my work with other guest fellows such as Helga Satzinger, Achim Trunk, and Richard Beyler.

By the end of my first summer in Berlin, the initial ideas for this book emerged, but it took an extraordinary incident to persuade me to bring the current volume into better focus. In the spring semester 2003, I offered my upper-division course on Nazi Germany at Clarkson University. As Clarkson is a technological university with few history majors, even students in so-called advanced courses frequently have little background coming into the seminar. In this particular class, however, I found an exceptional student. A highly gifted mechanical engineering major, Thomas M. Berez, also happened to be a budding historian. He was so talented that when I read his first paper, I suggested that we do a book review together of Nicosia and Huener's *Medicine and Medical Ethics in Nazi Germany,* an edited collection of essays on biomedicine in the Third Reich that we were reading in class. It was subsequently published in *Isis* 94 (2003): 541–543.

Thomas's first foray into the world of scholarship only wetted his appetite for more. After spending a summer with me in Berlin learning more about history of science, he decided to consider seriously pursuing a career in this field, rather than in engineering. With such an unusual commitment from such an exceptionally talented and disciplined individual, I asked him if he would consider working on a book project with me—the topic of the present volume. I decided to apply for an NSF grant and list him as my junior partner in this research endeavor. It speaks volumes about the Science, Technology and Society Program within the NSF, and the broad vision of the then Program Officer, Keith Benson, as well as the anonymous reviewers and panelists, that Thomas and I received a generous grant. With this essential financial support (Proposal #0349845), we spent the summer and fall of 2004 collecting material from over fifteen archives in Germany and the United States. The NSF was also generous enough to provide us with a supplemental grant (Proposal #0614307) somewhat later. In addition, I was able to make progress on the book while I was the Daimler-Chrysler Fellow at the American Academy in Berlin in fall 2006. Clarkson University made a material contribution to this book by both providing me release time and underwriting part of the costs of the numerous illustrations included in this volume. It goes without saying that my colleagues in the Department of Humanities and Social Sciences have, during the time I have been working on this book, always shown a keen interest in this work and have offered their moral support. This is particularly true of Lewis Hinchman, now emeritus professor of political science.

Although I cannot remember the names of all the numerous archivists whose invaluable aid made the enormous data collection for this study possible (fortunately some of the material is now digitalized), a few people deserve special mention. First, I must thank the entire staff at the Archive of the Max Planck Society in Berlin-Dahelm, especially the head archivist, Dr. Marion Kazemi, the archival assistant, Bernd Hoffmann, and the media and photo specialist, Susanne Uebele. What they provided both Thomas and me in terms of assistance, copying privileges, and access to documents and photos goes beyond the call of duty. They also granted me permission to use numerous photos for this study. In addition, chapter 3 of this volume would have been impossible to write without the help of the head of the Historical Archives of the Kaiser Wilhelm Institute for Psychiatry in Munich, Dr. Matthias Weber. Not only did he provide us with access to all relevant documents, his own expertise on Ernst Rüdin enabled him of offer important insights into the former research center and its director. I thank him for his permission to publish many illustrations from the former KWI for Psychiatry as well as his willingness to comment on a draft version of chapter 3. In this regard, I am also grateful for the critical comments of Volker Roelcke, another eminent scholar of the history of German psychiatry during the Third Reich.

For allowing me access to a largely unknown, but extraordinarily rich, archival source in the history of German secondary school education during the twentieth century, I am beholden to Christine Bernebée-Say of the Archiv und Gutachterstelle für Deutsches Schul- und Studienwesen in Berlin. It was she who directed me to biology exit exams (and those in other subjects) that form an important section of chapter 5. These documents are presently housed in the Landesarchiv Berlin under the direction of Dr. Volker Viergutz. I thank both Ms. Bernebée-Say and Dr. Viergutz for permission to publish an example of these rare exit exams. I am also indebted to Dr. Christian Ritzi of the Bibliothek für Bildungsgeschichtliche Forschung in Berlin for helping me locate numerous school biology textbooks and handbooks for the Weimar and Nazi periods. Dr. Jutta Schmidt and Mechtild Fischer (the latter from the University Archive Duisburg-Essen) granted me permission to publish several examples of school wall hangings from the Third Reich. Many of the documents regarding professional conferences and talks in chapter 4 were only secured through the diligence of Karl-Heinz Roth and his staff at the Stiftung für Sozialgeschichte des 20. Jahrhunderts in Bremen.

A large number of illustrations that appear in this study can be viewed on the Image Archive on the American Eugenics Movement (http://www .eugenicsarchive.org/eugenics/). I would like to thank the American Philosophical Society Library, the International Center for Photography, and, more particularly, Elaine M. Doak, head of Special Collections at the Truman State Library Archive, as well as Clare Clark and Susan Lauter of the Dolan DNA Learning Center of Cold Spring Harbor, for granting me permission to use individual photos as well as aiding me in their acquisition. I am also grateful to the Lilly Library, Indiana University, for allowing me to publish rare pictures from the H. J. Muller collection.

My special thanks goes to the two anonymous reviewers of the manuscript. Their careful reading and valuable suggestions for revision were indispensable for the final product. This study would have been weaker without the care, guidance, and attention to details of numerous individuals at the University of Chicago Press, especially the acquisitions editor, Karen Merikangas Darling. I am responsible for any remaining deficiencies.

Suitable funding, rich source material, and dedicated staff members of a press are necessary, but not sufficient, conditions for the completion of a project of this scope. Scholars, like other people, need moral and intellectual sustenance to help keep them going when the going gets tough. Given the emotionally difficult and ethically sensitive nature of the topic of this volume, it was easy to lose courage. In those instances, I frequently thought of Helmut Freiherr von Verschuer, the son of one of the book's major protagonists, Otmar Freiherr von Verschuer. He has not only been willing to provide me with any material concerning his father still in his possession (most of the personal and professional papers are now in the Archive of the Max Planck Society). More surprisingly, perhaps, he has welcomed me in his home on numerous occasions, opened the way for interviews with relatives, friends, and students of Otmar von Verschuer and, finally, a couple of years ago, against the warning of some of his acquaintances, supported my desire to undertake a biography of his controversial late father. I am happy to say that I have now also secured the requisite financial support for this new project, and, as I mentioned in the introduction, work on the volume is already underway.

Three other people—whom I am proud to call both friends and colleagues—have also provided mental and moral support. I am fortunate to be able to count two of the most knowledgeable individuals in the area of biomedicine during the Third Reich among them: Hans-Peter Kröner and Hans-Walter Schmuhl. Both are as generous as friends as they are first-class scholars. Both have given graciously of their time—on numerous occasions—in the interest of this study. This book would be far poorer without their innumerable contributions to my thoughts about the human genetics–politics symbiosis during the Third Reich. Garland Allen, certainly the most seasoned scholar of the history of American genetics and eugenics, has become an intellectual mentor of sorts. Always a giver of good advice, Gar's passion for scholarship combined with that of social justice (indeed the two are inseparable for him) is a credit to our profession, and he has constantly reminded me why this study needs to be done. In addition, I would like to thank Jonathan Harwood and Omer Bartov for their gracious support of my work, in general, and this project, in particular. They have helped me in innumerable ways over the years, and I am certain this book would have never been completed without their aid.

Finally, Thomas Berez—once my student, now my junior colleague as he nears completion of his PhD at Johns Hopkins—made this book what it is. In seeking his collaboration, I wanted assistance in writing a study that advanced undergraduates and laypeople could read. More importantly, I desired him to help me put into practice several messages that I have tried to get across in the class he first took with me. They include the following: history is about interpretation; historians have no monopoly on the truth and no window into the ultimate motivation of their protagonists; history is not black and white but comes in shades of gray; and, in the last analysis, the reader, not the author, should pronounce moral judgment, when relevant. Leaving aside the endless hours of photocopying sources and reading archival material—not mere grunt work but rather essential tasks of any scholar—Thomas proofread every chapter, made numerous editorial and some substantive corrections, and wrote a section of the introduction. The history of biomedicine in the Third Reich is, of course, my field of expertise, not his. I mapped out the structure of the book, supervised all the research and, of course, analyzed it. Yet although there is but one name listed as author of this book, it is in a peculiar way, a joint project. This was especially true of its inception. There are no words adequate to express what I owe Thomas. I will forever be grateful that I was able to experience the excitement of having a graduate-school-level undergraduate share this rare intellectual adventure.

Before ending, I must mention the person who has seen this work through at great personal sacrifice, but always with a loving heart. To my husband, Wilhelm Schauer, I will simply say thank you for everything. This study, whatever its flaws, exists because you never stopped believing in me.

Archival Sources

ADK Archiv des Diakonischen Werkes

ADS Archiv und Gutachterstelle für Deutsches Schul- und Studienwesen

APL Archives of the American Philosophical Library

BAB Bundesarchiv-Berlin-Lichterfelde

GSPK Geheimes Staatsarchiv Preußischer Kulturbesitz

HHA Hessisches Hauptstaatsarchiv, Wiesbaden

Laughlin Papers Harry H. Laughlin Papers, Prickler Memorial Library, Truman State University, Kirksville, MO

Lilly Library Indiana University, Bloomington, IN

MPG-Archiv Archiv des Max-Planck-Gesellschaft

MPIP-HA Archiv der Max-Planck-Instituts für Psychiatrie

PAAA Politisches Archiv des Auswärtigen Amts

RAC Rockefeller Foundation Archives

SfS Archiv der Stiftung für Sozialgeschichte des 20. Jahrhunderts, Bremen

STA-HH Staatsarchiv Hamburg

UAF Universitäts-Archiv der Albert-Ludwigs-Universität, Freiburg im Breisgau

UAHD Universitätsarchiv Heidelberg

UAM Universitätsarchiv Münster

UFAR Universitätsarchiv Frankfurt, Akten des Rektors

Von Verschuer Papers Private Papers, Helmut Freiherr von Verschuer, Nentershausen

Notes

INTRODUCTION

1. Werner Siedentrop, "Biologie," in *Höhere Schule—wozu? Sinn und Aufgabe,* ed. Rudolf Bohm (Leipzig: Quelle und Meyer, 1935), p. 80. Translations of quotations are my own, unless otherwise noted.
2. This tour through Staufen is based on the author's travels to the city taken in July 2004, mainly as a diversion from researching material for this book.
3. Some of these more famous treatments of Faustus are Johann Wolfgang von Goethe's two-part drama *Faust,* Charles Gounod's opera *Faust,* and Thomas Mann's novel *Doctor Faustus: The Life of the German Composer Adrian Leverkühn as Told by a Friend.*
4. This analysis of Marlowe's Faustus is taken from Christopher Marlowe, *Dr. Faustus,* ed. Roma Gill (New York: W. W. Norton and Company, 1990).
5. Daniel J. Kevles, *In the Name of Eugenics: Genetics and the Uses of Human Heredity* (New York: Knopf, 1985). The quote is found on p. ix.
6. Although the term "Faustian bargain" was mentioned by Robert Jay Lifton in 1986, it was not used specifically to describe the relationship of human geneticists to the Nazi state. He spoke more generally about medical doctors, primarily in respect of the end point of such a bargain: medical killing. Robert Jay Lifton, *The Nazi Doctors: Medical Killing and the Psychology of Genocide* (New York: Basic Books, 1986). Only more recently has the term been applied by the author and others to a larger group of biomedical professionals, and not only with regard to the "end game" of the "bargain."

Even a cursory search on the Internet demonstrates the ubiquity of this phrase now for describing the relationship between German professionals and the Nazi state. In the physical sciences, the phrase was made known by Richard Rhodes, *The Making of the Atomic Bomb* (New York: Simon & Schuster Paperbacks, 1986), p. 166. The most recent use of the phrase is perhaps Michael Neufeld's *Von Braun: Dreamer of Space, Engineer of War* (New York: Alfred Knopf, 2007), p. 5.

7. The most recent English translation of Benno Müller-Hill's classic work is *Murderous Science: The Elimination by Scientific Selection of Jews, Gypsies, and Others in Germany, 1933–1945,* 2nd ed. (Cold Spring Harbor, NY: Cold Spring Harbor Press, 1998).

8. For an idea of just how many scholarly books and articles are available on the general subject of biomedicine under the Nazis, see Carola Sachse and Benoît Massin, *Biowissenschaftliche Forschung an Kaiser-Wilhelm-Instituten und die Verbrechen des NS-Regimes* (Preprint no. 3 from the Research Program "History of the Kaiser Wilhelm Society in the National Socialist Era," 2000), and Sheila Faith Weiss, *Humangenetik und Politik als wechselseitige Ressourcen: Das Kaiser-Wilhelm-Institut für Anthropologie, menschliche Erblehre und Eugenik im "Dritten Reich"* (Preprint no. 17 from the Research Program "History of the Kaiser Wilhelm Society in the National Socialist Era," 2004), pp. 9–10. There are numerous additional works that have been published since these bibliographies and historiographic overviews were compiled.

9. The classic work of this genre is that of Alexander Mitscherlich and Fred Mielke, eds., *Doctors of Infamy: The Story of the Nazi Medical Crimes,* trans. Heinz Norden (New York: Henry Schuman, 1949). Naturally, Müller-Hill's study would fall into this category.

10. Although the works of Ernst Klee have been of service to the historical community, they are not, in and of themselves, histories. See Ernst Klee, *"Euthanasie" im NS-Staat: Die "Vernichtung lebensunwerten Lebens"* (Frankfurt am Main: S. Fischer, 1983), *Was sie taten—was sie wurden: Ärzte, Juristen und andere Beteiligte am Kranken- oder Judenmord* (Frankfurt am Main: S. Fischer, 1998), *Auschwitz, die NS-Medizin und ihre Opfer* (Frankfurt am Main: S. Fischer, 2001), *Deutsche Medizin im Dritten Reich: Karrieren vor und nach 1945* (Frankfurt am Main: S. Fischer, 2001). Lifton's *Nazi Doctors* does offer a theoretical framework to explain the deeds of these medical professionals. I do not, however, find its psychoanalytic framework satisfactory.

11. The term "witches' Sabbath" is taken from Mitscherlich and Mielke, *Doctors of Infamy,* pp. x–xi.

12. Edwin Black, *War against the Weak: Eugenics and America's Campaign to Create a Master Race* (New York: Four Walls Eight Windows, 2003); http://ukpmc .ac.uk/articlerender.cgi?artid=525446.

13. Michael Burleigh, *Death and Deliverance: "Euthanasia" in Germany c. 1900–1945* (Cambridge: Cambridge University Press, 1994); Henry Friedlander,

The Origins of Nazi Genocide: From Euthanasia to the Final Solution (Chapel Hill: University of North Carolina Press, 1995); Michael Kater, *Doctors under Hitler* (Chapel Hill: University of North Carolina Press, 1989).

14. Niels C. Lösch, *Rasse als Konstrukt: Leben und Werk Eugen Fischers* (Frankfurt am Main: Peter Lang, 1997); Matthias M. Weber, *Ernst Rüdin: Eine kritische Biographie* (Berlin: Springer Verlag, 1993); Ulf Schmidt, *Karl Brandt: The Nazi Doctor. Medicine and Power in the Third Reich* (London: Hambledon Continuum, 2007). The term "racial state" is taken from Michael Burleigh and Wolfgang Wippermann's *The Racial State: Germany, 1933–1945* (Cambridge: Cambridge University Press, 1991).

15. The following authors have written numerous works that have demonstrated that medically trained professionals (assuming that they were not racial or political enemies of the Nazi state) and their medical specialties did not suffer under National Socialism. Owing to lack of space, I will list their earliest or most significant work dealing with this question. Gerhard Baader and Ulrich Schultz, eds., *Medizin und Nationalsozialismus: Tabuisierte Vergangenheit—Ungebrochene Tradition?* Dokumentation des Gesundheitstages Berlin 1980 (West Berlin: Verlagsgesellschaft Gesundheit mbH, 1980); Karl Heinz Roth, *Erfassung zur Vernichtung: Von der Sozialhygiene zum "Gesetz über Sterbehilfe"* (Berlin: Verlagsgesellschaft Gesundheit, 1984); Götz Aly et al., eds., *Cleansing the Fatherland: Nazi Medicine and Racial Hygiene* (Baltimore: Johns Hopkins Press, 1994); Robert Proctor, *Racial Hygiene: Medicine under the Nazis* (Cambridge and London: Harvard University Press, 1988); Hans-Walter Schmuhl, *Rassenhygiene, Nationalsozialismus, Euthanasie: Von der Verhütung zur Vernichtung "lebensunwerten Lebens," 1890–1945* (Göttingen: Vandenhoeck & Ruprecht, 1987); Peter Weingart et al., *Rasse, Blut und Gene: Geschichte der Eugenik und Rassenhygiene in Deutschland* (Frankfurt am Main: Suhrkamp, 1988); Paul J. Weindling, *Health, Race and German Politics between National Unification and Nazism, 1871–1945* (Cambridge: Cambridge University Press, 1989).

16. For information on the Max Planck Society–sponsored research program, see http://www.mpiwg-berlin.mpg.de/KWG/engl.htm.

The following links to the Max Planck Society Web site list all the publications related to the project between the years 2002 and 2005:

http://www.magazindt.mpg.de/forschungsergebnisse/
wissVeroeffentlichungen/archivListenJahrbuch/2002/
praesidentenkommission/A-Z/index.html;

http://www.imprs.mpg.de/forschungsergebnisse/
wissVeroeffentlichungen/archivListenJahrbuch/2003/
praesidentenkommission/A-Z/index.html;

http://www.juniorgroups.mpg.de/forschungsergebnisse/
wissVeroeffentlichungen/archivListenJahrbuch/2004/
praesidentenkommission/A-Z/;

http://www.filme.mpg.de/forschungsergebnisse/wissVeroeffentlichungen/
archivListenJahrbuch/2005/praesidentenkommission/A-Z/.

17. Kristie Macrakis, *Surviving the Swastika* (New York: Oxford University Press, 1993).

18. Hans-Peter Kröner, *Von der Rassenhygiene zur Humangenetik: Das Kaiser-Wilhelm-Institut für Anthropologie, menschliche Erblehre und Eugenik nach dem Kriege* (Stuttgart: Gustav Fischer, 1997).

19. The monographs relevant to the names listed include Bernd Gausemeier, *Natürliche Ordnungen, und politische Allianzen: Biologische und biochemische Forschung an Kaiser-Wilhelm-Instituten, 1933–1945* (Göttingen: Wallstein, 2005); Susanne Heim, *Kalorien, Kautschuk, Karrieren: Pflanzenzüchtung und landwirtschaftliche Forschung in Kaiser-Wilhelm–Instituten, 1933–1945* (Göttingen: Wallstein, 2003); Wolfgang Schieder und Achim Trunk, eds., *Adolf Butenandt und die Kaiser-Wilhelm-Gesellschaft: Wissenschaft, Industrie und Politik im "Dritten Reich"* (Göttingen: Wallstein Verlag, 2004); Helga Satzinger, *Krankheiten als Rasse: Politische und wissenschaftliche Dimensionen eines Forschungsprogramms am Kaiser-Wilhelm-Institut für Hirnforschung (1919–1939)*, in *Rassenforschung an Kaiser-Wilhelm-Instituten vor und nach 1933*, ed. Hans-Walter Schmuhl (Göttingen: Wallstein, 2003). Those wishing a short introduction to this new historiography can see Sheila Faith Weiss's "Essay Review: Racial Science and Genetics at the Kaiser Wilhelm Society," *Journal of the History of Biology* 38 (2005): 367–79.

20. These state-of-the art publications are collected in an edited volume by Susanne Heim, Carola Sachse, and Mark Walker entitled *The Kaiser Wilhelm Society under National Socialism* (New York: Cambridge University Press, 2009). In addition, two fine monographs are also accessible to the English-speaking public: in particular, Schmuhl's exhaustive and meticulously researched institutional study of the KWIA and Heim's fascinating account of plant breeding research, economic autarchy, and "Eastern policy" under the swastika. Susanne Heim, *Plant Breeding, and Agrarian Research in Kaiser Wilhelm Institutes, 1933–1945: Calories, Caoutchouc, and Careers*, Boston Studies in the Philosophy of Science vol. 260 (Heidelberg: Springer, 2008); Hans-Walter Schmuhl, *The Kaiser Wilhelm Institute for Anthropology, Human Heredity and Eugenics, 1927–1945: Crossing Boundaries*, Boston Studies in the Philosophy of Science vol. 259 (Heidelberg: Springer, 2008). The translation of the title of this latter book is unfortunate, as it does not really convey the originality of the German title. The English book should have been titled "Boundary Crossings: The Kaiser Wilhelm Institute for Anthropology, Human Heredity and Eugenics." The study talks about how the Institute crossed three boundaries during its existence.

21. There is one important exception to the lack of readable works designed for the nonspecialist on the history of human genetics. This is the fine study by Diane B. Paul, *Controlling Human Heredity, 1865 to the Present* (Atlantic

Highlands, NJ: Humanities Press, 1995). Paul's book is focused, however, on Anglo-American eugenics. Francis R. Nicosia and Jonathan Huener, eds., *Medicine and Medical Ethics in Nazi Germany: Origins, Practices, Legacies* (New York: Berghahn Books, 2002), is an excellent and highly readable anthology. As its title suggests, however, it is about Nazi medicine and medical ethics, not specifically about human heredity in Germany under the swastika.

22. Stefan Kühl's useful book, *Die Internationale der Rassisten: Aufstieg und Niedergang der internationalen Bewegung für Eugenik und Rassenhygiene im 20. Jahrhundert* (Frankfurt am Main: Campus, 1997), presents an overview of the international eugenics movement; its focus is not specifically on the way in which German racial hygiene was similar to, or differed from, other national traditions.

23. Omer Bartov, "Ordinary Monsters: Perpetrator Motivations and Monocausal Explanations," in *Germany's War and the Holocaust: Disputed Histories,* ed. Omer Bartov (Ithaca, NY: Cornell University Press, 2003), p. 136.

24. Mitchell Ash, "Wissenschaft und Politik als Ressourcen für einander," in *Wissenschaften und Wissenschaftspolitik: Bestandsaufnahmen zu Formationen, Brüchen und Kontinuitäten im Deutschland des 20. Jahrhunderts,* ed. Rüdiger vom Bruch and Brigitte Kaderas (Stuttgart: Steiner, 2002), pp. 32–49.

25. *Humangenetik und Politik als wechselseitige Ressourcen: Das Kaiser-Wilhelm-Institut für Anthropologie, menschliche Erblehre und Eugenik, 1927–1945* (Ergebnisse 17, Vorabdrucke aus dem Forschungsprogramm "Geschichte der Kaiser-Wilhelm-Gesellschaft im Nationalsozialismus" der Präsidentenkommision der Max-Planck-Gesellschaft, Berlin, 2004).

26. See, for example, Heim, Sachse, and Walker, eds., *The Kaiser Wilhelm Institute under National Socialism,* p. 2.

27. Eric A. Johnson and Karl-Heinz Reuband, *What We Knew: Terror, Mass Murder, and Everyday Life in Nazi Germany: An Oral History* (Cambridge, MA: Basic Books, 2005), p. xv.

28. Richard J. Evans, *The Coming of the Third Reich* (New York: Penguin Press, 2004). The quote is on p. xx.

29. See James Watson's afterword to Müller-Hill's *Murderous Science,* p. 197.

30. Throughout the book, the terms "human heredity" and "human genetics" will be used interchangeably, although in the very early years of the twentieth century, using them as synonyms may appear somewhat ahistorical.

31. When discussing eugenics in Germany, I will use the terms "eugenics" and "racial hygiene" interchangeably, recognizing, of course, the difference in political meaning that using one or the other word in the Weimar Republic had for German human geneticists. In the Third Reich, racial hygiene was almost universally used, as the double meaning of the term (genetic improvement of the human race vs. genetic improvement of a particular

anthropological race) was in the interest of those researchers remaining at their posts as well as the Nazi regime.

1. For the thesis that interwar eugenics was spurred on worldwide by international political opportunities such as the League of Nations, see Deborah Barrett and Charles Kurzman, "Globalizing Social Movement Theory: The Case of Eugenics," *Theory and Society* 33 (2004): 487–527.
2. Peter J. Bowler, *The Mendelian Revolution: The Emergence of Hereditarian Concepts in Modern Science and Society* (London: Athlone Press, 1989), pp. 93–109; Diane B. Paul, *Controlling Human Heredity, 1865 to the Present* (Atlantic Highlands, NJ: Humanities Press, 1995), p. 46.
3. Bowler, *The Mendelian Revolution*, pp. 106–7.
4. Ibid., p. 110.
5. The Galton quote is taken from Susan Bachrach, "Introduction," in *Deadly Medicine: Creating the Master Race*, ed. Susan Bachrach and Dieter Kuntz (Chapel Hill: University of North Carolina Press, 2004), p. 3.
6. Galton quote taken from Daniel J. Kevles, *In the Name of Eugenics: Genetics and the Uses of Human Heredity* (New York: Knopf, 1985), p. 3.
7. Sheila Faith Weiss, "The Race Hygiene Movement in Germany," *The Wellborn Science: Eugenics in Germany, France, Brazil, and Russia*, ed. Mark B. Adams (New York: Oxford University Press, 1990), p. 11. The article is only concerned with Germany, but these two contexts are relevant for Britain and the United States as well.
8. Ibid., p. 12.
9. Quoted from Sheila Faith Weiss, *Race Hygiene and National Efficiency: The Eugenics of Wilhelm Schallmayer* (Berkeley: University of California Press, 1987), p. 29.
10. Ibid.
11. Weiss, "The Race Hygiene Movement," pp. 13–14. The quote is taken from p. 14.
12. Pauline M. H. Mazumdar, *Eugenics, Human Genetics and Human Failings* (London and New York: Routledge, 1992), pp. 1–6.
13. Paul, *Controlling Human Heredity*, p. 18; Davenport quote taken from Garland E. Allen, "The Eugenics Record Office at Cold Spring Harbor, 1910.1940," *Osiris*, 2nd ser., 2 (1986): 225.; Mark B. Adams, Garland E. Allen, and Sheila Faith Weiss, "Human Heredity and Politics: A Comparative Institutional Study of the Eugenics Record Office at Cold Spring Harbor (United States), the Kaiser Wilhelm Institute for Anthropology, Human Heredity and Eugenics (Germany), and the Maxim Gorky Medical Genetics Institute (USSR)," *Osiris*, 2nd ser., 20 (2005): 236.
14. Weiss, *Race Hygiene and National Efficiency*, pp. 6–26; Sheila Faith Weiss, "German Eugenics, 1890–1933," in *Deadly Medicine*, pp. 17–18.

15. Stefan Kühl, *Die Internationale der Rassisten: Aufstieg und Niedergang der internationalen Bewegung für Eugenik und Rassenhygiene im 20. Jahrhundert* (Frankfurt am Main: Campus, 1997), p. 24.

16. Hans-Walter Schmuhl, *Grenzüberschreitungen: Das Kaiser-Wilhelm-Institut für Anthropology, menschliche Erblehre und Eugenik, 1927–1945* (Göttingen: Wallstein, 2005), p. 20.

17. Adams, Allen, and Weiss, "Human Heredity and Politics," pp. 236–37.

18. Paul, *Controlling Human Heredity*, p. 18.

19. Angus McLaren, *Our Own Master Race: Eugenics in Canada, 1885–1945* (Toronto: McClelland and Stewart, 1990), pp. 9–10.

20. Gunnar Broberg, "Scandinavia: An Introduction," *Eugenics and the Welfare State: Sterilization Policy in Denmark, Sweden, Norway, and Finland,* ed. Gunnar Broberg and Niels Roll-Hansen (East Lansing: Michigan State University Press, 1996), pp. 3–4.

21. Nils Roll-Hansen, "Norwegian Eugenics: Sterilization as Social Reform," *Eugenics and the Welfare State,* pp. 155–57.

22. See the numerous letters between Mjøen and Davenport in the Charles B. Davenport papers, American Philosophical Library (APL), B:D27, Series 1. The term "reform eugenics" was coined by Kevles in his standard work, *In the Name of Eugenics.* He used it to distinguish the eugenic efforts of left-wing human geneticists from those who accepted various racial and class prejudices. Kevles termed the latter "mainline eugenicists." Kevles's distinction between these two groups as well as his terminology has been accepted by the majority of historians of eugenics including this author.

23. Gunnar Broberg and Mattias Tydén, "Eugenics in Sweden: Efficient Care," *Eugenics and the Welfare State,* pp. 77–90. Lundborg quote taken from p. 84.

24. The apt term "protean concept" to describe eugenics was coined by Paul, *Controlling Human Heredity*, p. 3.

25. William H. Schneider, "The Eugenics Movement in France, 1890–1940," *The Wellborn Science,* pp. 70–73; the quote is on p. 72.

26. Nancy Leys Stepan, "Eugenics in Brazil," *The Wellborn Science,* pp. 110–15; quote on p. 119.

27. The story of socialist eugenics in Germany is fully explored in Michael Schwartz, *Sozialistische Eugenik: Eugenische Sozialtechnologien in Debatten und Politik der deutschen Sozialdemokratie, 1890–1933* (Berlin: Dietz Verlag, 1995).

28. For a discussion of eugenics in the Soviet Union, see Mark B. Adam, "Eugenics in Russia, 1900–1940," *The Wellborn Science,* pp. 163–216.

29. For a discussion of left-wing British eugenicists, see Diane B. Paul, "Eugenics and the Left," *Journal of the History of Ideas* 45 (1984): 567–90.

30. Richard Cleminson, *Anarchism, Science and Sex: Eugenics in Eastern Spain, 1900–1937* (New York: Peter Lang, 2000; Maria Bucur, *Eugenics and Modernization in Interwar Romania* (Pittsburg: University of Pittsburg Press, 2002).

31. Yuehtsen Chung and Yuehtsen Juliette Chung, *Struggle for National Survival: Eugenics in Sino-Japanese Contexts, 1896–1945* (London: Taylor and Francis, 2002); Chloe Campbell, *Race and Empire: Eugenics in Colonial Kenya* (Manchester: Manchester University Press, 2007).

32. Eugen Fischer to Charles Davenport, December 12, 1908, Charles B. Davenport papers, APL, B:D27, Series 1.

33. Akademischer Senat der Universität Freiburg an das Grossherzogliche Unterrichtsministerium, February 20, 1909, Universitäts-Archiv der Albert-Ludwigs-Universität, Freiburg i Br. (UAF), B24/794; Sheila Faith Weiss, "Human Genetics and Politics as Mutually Beneficial Resources: The Case of the Kaiser Wilhelm Institute for Anthropology, Human Heredity and Eugenics during the Third Reich," *Journal of the History of Biology* 39 (2006): 48–49.

34. Eugen Fischer to Charles Davenport, December 12, 1908, Charles B. Davenport papers, APL, B:D27, Fischer Folder; Weiss, "Human Genetics and Politics as Mutually Beneficially Resources," p. 48n17.

35. Charles Davenport to Ernst Rüdin, June 3, 1912; Ernst Rüdin to Charles Davenport, November 23, 1912, Charles B. Davenport papers, APL, B:D27, Rüdin Folder.

36. For a discussion of the early life and career of Ernst Rüdin, see Matthias M. Weber, *Ernst Rüdin: Eine kritische Biographie* (Berlin: Springer Verlag, 1993), pp. 17–77. On Rüdin's views concerning Aryan ideology, see especially pp. 60–61.

37. Leonard Darwin, "Introduction," *Problems in Eugenics: Papers Communicated to the First International Eugenics Congress Held at the University of London, July 24th to 30th, 1912* (London: Eugenics Education Society, 1912).

38. Adolphe Pinard, "La Puériculture avant la Procreation," *Problems in Eugenics,* pp. 458–59,

39. Kühl, *Die Internationale der Rassisten,* pp. 26–33.

40. Agnes Bluhm, "Rassenhygiene und Ärztliche Geburtshilfe," *Problems in Eugenics,* pp. 379–87.

41. Kühl, *Die Internationale der Rassisten,* p. 28.

42. Ibid., pp. 32–37.

43. Rudolf Hagemann, "Deutsche Gesellschaft für Vererbungswissenschaft 1921–1938/1945, www.gfgenetik.de/deutsch/geschichte/hagemann.pdf.

44. Kühl, *Internationale der Rassisten,* pp. 40–47. All quotes taken from Kühl.

45. Weiss, *Race Hygiene and National Efficiency,* p. 142.

46. Kaiserliches Konsulat Chicago to the Auswärtiges Amt, March 17, 1901, Bundesarchiv-Berlin-Lichterfelde BAB R 86-2371/18.

47. Paul J. Weindling, *Health, Race and German Politics between National Unification and Nazism, 1871–1945* (Cambridge: Cambridge University Press, 1989), p. 281.

48. "Einladung zu einer Versammlung für die Bevölkerungsfrage mit dem Gegenstande: 'Der Neuaufbau des Familienlebens nach dem Krieg' 7., 8. und

9. November in Darmstadt" BAB R 83-2371/68; Weiss, *Race Hygiene and National Efficiency*, pp. 140–41.

49. Berliner Gesellschaft für Rassenhygiene, June 15, 1917, BAB R 83-2371/77.

50. Weindling, *Health, Race and German Politics*, p. 284.

51. Weiss, "German Eugenics, 1890–1933," p. 22.

52. Barrett and Kurzman, "Globalizing Social Movement Theory," p. 506. Quote taken from this page.

53. Ibid., p. 507; Osborn quote taken from this page.

54. Elof Axel Carlson, *The Unfit: A History of a Bad Idea* (Cold Spring Harbor, NY: Cold Spring Harbor Press, 2001), pp. 269–71; Kühl, *Die Internationale der Rassisten*, pp. 53–59 and 71. Quotes taken from Kühl, pp. 55–56.

55. Kühl, *Die Internationale der Rassisten*, pp. 59–61.

56. See, for example, the letter written from Laughlin to Lundborg thanking him for his latest book, *The Racial Characteristics of the Swedish Nation*. Laughlin to Lundborg, March 24, 1927, D-2-2:17, Harry H. Laughlin Papers (Laughlin Papers), Prickler Memorial Library, Truman State University, Kirksville, MO.

57. Kühl, *Die Internationale der Rassisten*, p. 61.

58. Erwin Baur to Colleague, November 24, 1920, Charles B. Davenport papers, APL, B:D27, Baur Folder.

59. Charles Davenport to Erwin Baur, March 30, 1923; Erwin Baur to Charles Davenport, May 1, 1923, Charles B. Davenport papers, APL, B:D27, Baur Folder.

60. Charles Davenport to Fritz Lenz, July 13, 1923, Charles B. Davenport papers, APL, B:D27, Lenz Folder.

61. Fritz Lenz to Charles Davenport, August 8, 1923, Charles B. Davenport papers, APL, B:D27, Lenz Folder.

62. Sheila Faith Weiss, "The Race Hygiene Movement in Germany," *Osiris*, 2nd ser., 3 (1987): 216; Heiner Fangerau, *Etablierung eines Rassenhygienischen Standardwerkes 1921–1941: Der Baur-Fischer-Lenz im Spiegel der Zeitgenössischen Rezensionsliteratur* (Frankfurt am Main: Peter Lang, 2001); Charles Davenport to Lucien Howe, May 7, 1928, www.eugenicsarchiv.org/ID#256.

63. Kühl, *Die Internationale der Rassisten*, pp. 71, 73, and 76–77

64. Charles Davenport to Ernst Rüdin, December 5, 1929; Ernst Rüdin to Charles Davenport, December 12, 1929, Charles B. Davenport papers, APL, B:D27, Rüdin Folder.

65. Kühl, *Die Internationale der Rassisten*, pp. 71–72.

66. Ibid., pp. 86–94.

67. Stefan Kühl, *The Nazi Connection: Eugenics, American Racism and German National Socialism* (New York: Oxford University Press, 1994), p. 13. The quote is taken from Kühl, p. 3.

68. The quote is taken in full from Kevles, *In the Name of Eugenics*, p. 111. For a full discussion of the Buck vs. Bell case, see Paul A. Lombardo, *Three*

Generations, No Imbeciles: Eugenics, the Supreme Court and Buck vs. Bell (Baltimore: The Johns Hopkins University Press, 2008).

69. Rüdin, for example, possessed a copy of the Supreme Court ruling in his files. "A Judgment of the Supreme Court of the USA," Archiv der Max-Planck-Gesellschaft (MPG-Archiv), ZA 131 Nachlass Ernst Rüdin, Box 3.

70. Fritz Lenz to Charles Davenport, January 28, 1924; Charles Davenport to Fritz Lenz, November 9, 1927, Charles B. Davenport papers, APL, B:D27, Lenz Folder; Kühl, *The Nazi Connection,* p. 19.

71. Charles Davenport to Hermann Muckermann, October 6, 1928; Hermann Muckermann to Charles Davenport, December 18, 1928; Hans Nachtsheim to Charles Davenport, November 25, 1922; Charles Davenport to Hans Nachtsheim, March 3, 1926; Charles Davenport to Hans Nachtsheim, May 10, 1926; Günter Just to Davenport, May 10, 1927, Charles B. Davenport papers, APL, B:D27, Muckermann, Nachtsheim, and Just folders.

72. Eugen Fischer to Charles Davenport, October 24, 1932, "Fragebogen zur Untersuchung der Rassenkreuzung," Charles B. Davenport papers, APL, B: D27, Fischer Folder.

73. Weindling, *Health, Race and German Politics,* pp. 399–430.

74. Auswärtiges Amt an das Reichsministerium des Innern, December 25, 1925, BAB R 86-2371/252–53; *Annals of Eugenics: Journal for the Scientific Study of Racial Problems,* July 1925, R86-2371/309–10; Thüringisches Wirtschaftsministerium an das Reichsministerium des Innern, July 16, 1923, R86-2374/16–17; Oberregierungsrat Dr. Hesse an das Reichsministerium, October 15, 1923, R86-2374/24–27.

75. Quote taken from Weber, *Ernst Rüdin,* pp. 114–15.

76. Ibid., pp. 116–19.

77. Ibid, pp. 105–9.

78. Ibid., pp. 135, 141–46

79. Schmuhl, *Grenzüberschreitungen,* pp. 32–42. According to Hagemann in an article entitled "Deutsche Gesellschaft für Vererbungswissenschaft 1921–1938/1945," the founding year was 1920, not 1922; http://www.gfgenetik .de/deutsch/geschichte/hagemann.pdf.

80. Schmuhl, *Grenzüberschreitungen,* p. 42.

81. Weiss, "Human Genetics and Politics as Mutually Beneficial Resources," p. 50n21.

82. Hagemann, http://www.gfgenetik.de/deutsch/geschichte/hagemann.pdf.

83. MPG-Archiv, Abt. I, Rep. 3, Nr. 23, p. 357.

84. Fischer-Davenport, July 19, 1929, MPG-Archiv, Abt. I, Rep. 3, Nr. 23.

85. Adams, Allen, and Weiss, "Human Heredity and Politics," pp. 248–52.

86. Program of the sessions of The Third International Congress of Eugenics, Laughlin Papers, C-2-5:6.

87. Charles B. Davenport to Alfred Ploetz, June 3, 1932, Charles B. Davenport papers, APL, B:D27, Ploetz Folder.

88. Carlson, *The Unfit,* p.271; Kühl, *The Nazi Connection,* p. 22; www .eugenicsarchiv.org/ID#1069,1074.

89. Augustine Cline, "This Date in History: Eugenics Laws in America," http:// atheism.about.com/b/a/152211.htm; Jonathan Gottshall, "The Cutting Edge: Eugenics and Sterilization in California, 1909–1945,"http://www .gottshall.com/thesis/index.html.

90. Paul, *Controlling Human Heredity,* p. 83.

91. Kevles, *In the Name of Eugenics,* p. 114.

92. Schmuhl, *Grenzüberschreitungen,* p. 134.

93. Mary Fulbrook, *The Divided Nation: History of Germany, 1918–1990* (Oxford: Oxford University Press, 1992), p. 57; "Depression in the United States," http://encarta.msn.com/encyclopedia_761584403/Great_Depression_in_ the_United_States.html.

94. "Hermann Muckermann, "Illustrationen zu der Frage: Wohlfahrtspflege und Eugenik," *Eugenik* 2 (1931): 41–42.

95. "Geleitwort," *Eugenik* 1 (1930), n.p.

96. For a full discussion of religious leaders and eugenics in the United States, see Christine Rosen, *Preaching Eugenics: Religious Leaders and the American Eugenics Movement* (New York: Oxford University Press, 2004).

97. In this context we only need to think of Friedrich Naumann, the German Protestant who wished to offer a Christian alternative to Marxism and Social Democracy as a way of aiding the poor and solving the "social question." Martin Kitchen, *A History of Modern Germany, 1800–2000* (Oxford: Blackwell, 2006), p. 186.

98. For an in-depth discussion of the Innere Mission and its role during the Great Depression in Germany, see Sabine Schleiermacher, *Sozialethik im Spannungsfeld von Sozial- und Rassenhygiene: Der Mediziner Hans Harmsen im Centralausschuß für die Innere Mission* (Husum: Matthiesen Verlag, 1998), in particular, pp. 196–236.

99. For a discussion of the Gosney-Popenoe book, see Alexandra Minna Stern, *Eugenic Nation: Faults and Frontiers of Better Breeding in Modern America* (Berkeley: University of California Press, 2005), chap. 3; Harry Bruinius, *Better for All the World: The Secret History of Forced Sterilization and America's Quest for Racial Purity,* (New York: Alfred Knopf, 2006), pp. 272–73.

100. Paul, *Controlling Human Heredity,* p. 89.

101. Schmuhl, *Grenzüberschreitungen,* pp. 130–31.

102. Protocol in Treysa bei Kassel, May 18–20, 1931, Archiv des Diakonischen Werkes (ADK), CA/G 381, 56, 58–59, 82. First quote on pp. 58–59; the title of von Verschuer's talk is on p. 56.

103. Ibid., pp. 58–59.

104. Schleiermacher, *Sozialethik im Spannungsfeld,* p. 231.

105. Schmuhl, *Grenzüberschreitungen,* pp. 138–41.

106. Ibid., pp. 141–45.

107. Ibid., p. 143.

108. Kevles, *In the Name of Eugenics*, p. 168.

109. Schmuhl, *Grenzüberschreitungen*, pp. 145–48.

110. Alfred Ploetz to Eugen Fischer, March 9, 1930, MPG-Archiv, ZA 131, Nachlass Rüdin, Box 3. Ploetz obviously sent a copy of his letter to Fischer to Rüdin as well.

CHAPTER TWO

1. MPG-Archiv, Abt. I, Rep. 1A, Nr. 2399, p. 5; Abt. I, Rep. 1A, Nr. 2403, pp. 65a–69.

2. MPG-Archiv, Abt. I, Rep. 1A, Nr. 2404, pp. 11–12; the role of Gütt at the *Kuratorium* meeting has been duly noted by Kristie Macrakis, *Surviving the Swastika* (New York: Oxford University Press, 1993), p. 127.

3. In 1944, Otmar von Verschuer, the then director of the Dahlem Institute, suggested that the name "Eugen-Fischer-Institut" be added to its original in honor of his mentor and first Institute director, Fischer. Albert Vögler to Otmar von Verschuer, March 11, 1944, MPG-Archiv, Abt. I, Rep. 1A, 2400, p. 259.

4. For use of this term, see Hans-Walter Schmuhl, *Grenzüberschreitungen: Das Kaiser-Wilhelm-Institut für Anthropology, menschliche Erblehre und Eugenik, 1927–1945* (Göttingen: Wallstein, 2005), p. 314.

5. "Kaiser-Wilhelm-Institut für Anthropologie," *Deutsche Allgemeine Zeitung,* September 16, 1927, in MPG-Archiv, Abt. IX, Rep. 2.

6. Auszug aus dem Protokoll der Sitzung der KWG-Senats am June 19, 1926, MPG-Archiv, Abt. I, Rep. 1A, Nr. 2411, p. 2.

7. Peter Weingart et al., *Rasse, Blut und Gene: Geschichte der Eugenik und Rassenhygiene in Deutschland* (Frankfurt am Main: Suhrkamp, 1988), pp. 243–44; Niels C. Lösch, *Rasse als Konstrukt: Leben und Werk Eugen Fischers* (Frankfurt am Main: Peter Lang, 1997), p. 177. Quote taken from Lösch.

8. Paul J. Weindling, *Health, Race and German Politics between National Unification and Nazism, 1871–1945* (Cambridge: Cambridge University Press, 1989), pp. 433–34; Lösch, *Rasse als Konstrukt,* pp. 170 and 176–77. Quote cited from Lösch, p. 175.

9. Von Harnack to Preußischer Landtag, July 18, 1926, MPG-Archiv, Abt. I, Rep. 1A, Nr. 2411, p. 3.

10. The phrase "theoretical institute" is employed by Fischer. See *"Kaiser-Wilhelm-Institut für Anthropologie, menschliche Erblehre und Eugenik,"* MPG-Archiv, Abt. I, Rep. 3, Nr. 4, p. 5; Lösch, *Rasse als Konstrukt,* pp. 182–83.

11. Quoted from Schmuhl, *Grenzüberschreitungen,* p. 47.

12. Ibid., p. 44.

13. Eugen Fischer, "Aufgaben der Anthropologie, menschlichen Erblichkeitslehre und Eugenik," *Die Naturwissenschaften* 14 (1926): 749.

14. Ibid., pp. 753–54.

15. Benoît Massin, "Rasse und Vererbung als Beruf," in *Rassenforschung an Kaiser-Wilhelm-Instituten vor und nach 1933,* ed. Hans-Walter Schmuhl (Göttingen: Wallstein, 2003), p. 193.

16. Fischer, "Aufgaben," pp. 749–50. For a more detailed view of the relationship between Fischer and Boas, especially during the Nazi period, see Hans-Walter Schmuhl, "Feindbewegungen Das Kaiser-Wilhelm-Institut fur Anthropologie menschliche Erblehre und Eugenik und seine Auseinaudersetzung ein ander setzung met Franz Boas, 1927–1942," in *Kulturrelativismus und Antirassismus Der Anthropologe Franz Boas (1858–1942),* ed. Hans-Walter Schmuhl (Bielefeld: Transcript, 2009), pp. 187–209.

17. Ibid., 751–52.

18. Ibid., p. 754.

19. Ibid.

20. Ibid., pp. 754–55. Also quoted in Schmuhl, *Grenzüberschreitungen,* p. 46.

21. Schmuhl, *Grenzüberschreitungen,* p. 57.

22. "Kaiser-Wilhelm-Institut für Anthropologie," MPG-Archiv, Abt. IX, Rep. 2, 1928.

23. Quote cited from Schmuhl, *Grenzüberschreitungen,* p. 58.

24. For a discussion of the "Harnack principle," see Rudolf Vierhaus, "Bemerkungen zum sogenannten Harnack-Prinzip: Mythos und Realität," in *Die Kaiser-Wilhelm/Max-Planck-Gesellschaft und ihre Institute: Das Harnack-Prinzip,* ed. Bernhard vom Brocke and Hubert Laitko (Berlin and New York: Walter de Gruyter, 1996), pp. 129–38.

25. Eugen Fischer, "Das Kaiser-Wilhelm-Institut für Anthropologie, menschliche Erblehre und Eugenik," *Zeitschrift für Morphologie und Anthropologie* 27 (1928): 147–48; "Kaiser-Wilhelm-Institut für Anthropologie, menschliche Erblehre und Eugenik in Berlin Dahlem, " MPG-Archiv, Abt. I, Rep. 3, Nr. 4.

26. These points concerning the Dahlem Institute as a research center that studied human beings from a genetic perspective and the crossing of boundaries, in this case scientific ones, are discussed in Schmuhl, *Grenzüberschreitungen,* pp. 11–18; quote on p. 11.

27. Quote taken from F. Schmidt-Ott to E. Day, September 5, 1929, Rockefeller Foundation Archives (RAC) folder 187, box 20, series 717 S, RG 1.1. Quote taken from a letter from R. A. Lambert to A. Gregg, May 5, 1932, folder 63, box 10, series 717 A, RG 1.1, RAC. Information on Fischer's survey of Prussian civil servants is documented in a report dated May 5, 1931, BAB R86/2370. Also see Lösch, *Rasse als Konstrukt,* pp. 199–201; and Weindling, *Health, Race and German Politics,* pp. 466–69.

28. For a discussion of Muckermann's six hundred speeches, see MPG-Archiv, Abt. I, Rep. 1A, Nr. 832, p. 233. Von Verschuer studied with Fischer in Freiburg in 1921–22. MPG-Archiv, Abt. III, Rep. 86A, Nr. 1, Von Verschuer, *Erbe-Umwelt-Führung,* chap. 4, "Medizinstudium, 1919–1923," p. 11.

29. MPG-Archiv, Abt. III, Rep. 86A, Nr. 3-1, von Verschuer, *Erbe-Umwelt-Führung,* section titled "Mein wissenschaftlicher Weg," p. 2.

30. Von Verschuer, *Erbe-Umwelt-Führung*, chap. 3, "Kriegsjahre, 1914–1918," p. 43.

31. Quoted from Schmuhl, *Grenzüberschreitungen*, p. 72.

32. Ibid., p. 73.

33. MPG-Archiv, Abt. III, Rep. 86A, Nr. 3-1, *Erbe-Umwelt-Führung*, chap. 4, "Medizinstudium, 1919–1923," pp. 4–7, 11, 14.

34. Schmuhl, *Grenzüberschreitungen*, pp. 74–75.

35. Von Verschuer, *Erbe-Umwelt-Führung*, chap. 5, "Assistentenjahre in Tübingen, 1923–1927," p. 11.

36. Alan Gregg's officer's diary, December 13, 1929, RG 12.1, RAC.

37. MPG-Archiv, Abt. III, Rep. 86 A, Nr. 3-1, *Erbe-Umwelt-Führung*, section entitled "Mein wissenschaftlicher Weg," pp. 5–10. According to one newspaper report, von Verschuer had registered about eight hundred twins by 1931. "Das Geheimnis der Zwillinge," *Rheinische Zeitung*, November 8, 1931, in MPG-Archiv, Abt. IX, Rep. 1, Verschuer; Eugen Fischer, "Kaiser-Wilhelm-Institut für Anthropologie, menschliche Erblehre und Eugenik," in *25 Jahre Kaiser-Wilhelm-Gesellschaft zur Förderung der Wissenschaft*, ed. Max Planck (Berlin: Julius Springer, 1936), Bd. 2, p. 349.

38. Schmuhl, *Grenzüberschreitungen*, p. 108.

39. Weindling, *Health, Race, and German Politics*, pp. 436–37. For funding on tubercular twins, see MPG-Archiv, Abt. I, Rep. 1A, Nr. 2412, p. 157; and in the same folder, "Beihilfe zur Tuberkuloseforschung betr." (no number).

40. Schmuhl, *Grenzüberschreitungen*, pp. 228–29.

41. Ibid., pp. 130–31. Quote on p. 130. See also chap 1., pp. 59–62.

42. Lösch, *Rasse als Konstrukt*, pp. 216–20 and 339.

43. Ibid., pp. 231–53.

44. The plant geneticist Hans Stubbe of the KWI for Breeding Research, as well as two of his Institute colleagues, were also subject to political attacks. Susanne Heim, "'*Die reine Luft der wissenschaftlichen Forschung': Zum Selbstverständnis der Wissenschaftler der Kaiser-Wilhelm-Gesellschaft*" (Preprint no. 7 from the Research Program "History of the Kaiser Wilhelm Society in the National Socialist Era," 2002), pp. 12–13. Nor were Oskar and Cécile Vogt of the KWI for Brain Research spared political denunciations—allegedly for their pacifism and liberalism. Helga Satzinger, "Krankheiten als Rasse: Politische und wissenschaftliche Dimensionen eines internationalen Forschungsprogramms am Kaiser-Wilhelm-Institut für Hirnforschung (1919–1939)," in *Rassenforschung an Kaiser-Wilhelm-Instituten*, pp. 146–89.

45. The Racial Policy Office of the NSDAP, the Reich Ministry of the Interior, the Reich Ministry of Education, the Race and Settlement Main Office of the SS, the Reich Foreign Office, the Reich Ministry for Popular Enlightenment and Propaganda, the Reich Chancellery, and, finally, the Deputy Führer's Office were all involved with the "Fischer case"! BAB R1501 126243, 126244, 126245, 126245/1.

46. For the NSDAP's unwillingness to see Fischer as the Rector of the University of Berlin, see BAB R 4901 1255/433. Regarding Fischer's unwillingness to serve as president of the IFEO, see his letter to Charles Davenport from July 29, 1932, in MPG-Archiv, Abt. I, Rep. 3, Nr. 23. For the claim that Fischer worked well with Weimar politicians, see BAB R1501 126243/291, 289, 303.

47. MPG-Archiv, Abt. I, Rep. 1A, Nr. 808, pp. 93–96. The quotes are on p. 94. Müller-Hill was the first to mention this important lecture given by the KWIA director in Benno Müller-Hill, *Murderous Science: The Elimination by Scientific Selection of Jews, Gypsies, and Others in Germany, 1933–1945*, 2nd ed. (Cold Spring Harbor, NY: Cold Spring Harbor Press, 1998) p. 10. Macrakis has also noted the relevance of Fischer's controversial speech, *Surviving the Swastika*, p. 127.

48. BAB R1501 126243/289.

49. Schmuhl, *Grenzüberschreitungen*, p. 157. Muckermann was ready to work with the Nazis on eugenic policy even after the National Socialists came to power. He held a talk entitled "The Race in the *Völkisch* State" at the University of Munich on June 12, 1933, Nachlaß Ernst Rüdin (used with the permission of Edith Zerbin-Rüdin), MPG-Archiv, ZA 131 Rüdin, Box 2.

50. BAB R1501 126243/304; quote is taken from Schmuhl, *Grenzüberschreitungen*, p. 169.

51. *Der Begriff des völkischen Staates, biologisch betrachtet* (Berlin: Preussische Druckerei und Verlags-Aktiengesellschaft Berlin, 1933), p. 14. A similar discourse on the "Jewish question" is also adopted by von Verschuer in his textbook, *Leitfaden der Rassenhygiene* (Leipzig: Georg Thieme, 1941), p. 125. Whether the two men consciously negotiated this response is unknown, but not unlikely, as von Verschuer commented elsewhere to Fischer that, in the "Jewish question," he always wished to go "hand in hand" with his mentor. Von Verschuer to Fischer May 20, 1937, MPG-Archiv, Abt. II, Rep. 86A, Nr. 291-3.

52. BAB R1501 126245/218.

53. This assessment was given by Fischer in his unpublished autobiographical account, "50 Jahre im Dienste der menschlichen Erbforschung und Anthropologie," p. 107.

54. For von Verschuer's *völkisch* background, see MPG-Archiv, Abt. III, Rep. 86A, Nr. 3-1, *Erbe-Umwelt-Führung*, section titled "Mein wissenschaftlicher Weg"; and Otmar Freiherr von Verschuer, "Rassenhygiene," in *Deutsche Politik: Ein völkisches Handbuch*, ed. Kyffhäuser-Verband (Frankfurt am Main: Englert and Schlosser, 1925), *Allgemeiner Teil*, pp. 1–15; *Besonderer Teil*, pp. 1–16. Von Verschuer, "Vererbung, Auslese und Rassenhygiene," *Der deutsche Gedanke* 2 (1925): 744–51; on his anti-Semitism, see BAB R1501 126243/303.

55. Apparently, Streicher's notorious anti-Semitic hate publication was extremely profitable for its publisher. It made Streicher a millionaire. See http://www.spiegel.de/sptv/reportage/0,1518,101879,00.html.

56. MPG-Archiv, Abt. I, Rep. 1A, Nr. 2404, pp. 22–26.

57. MPG-Archiv, Abt. 1, Rep. 1A, Nr. 2404, p. 17; Abt. I, Rep. 3, Nr. 9 *Tätigkeits-bericht* Jg. 1932/33.

58. The quotations are taken from Schmuhl, *Grenzüberschreitungen,* pp. 178 and 182.

59. Regarding Fischer's complaints about the funding of his Institute during the final years of the Weimar Republic, see Eugen Fischer to the KWG-Generalverwaltung, May 1, 1931 and October 1, 1931, MPG-Archiv, Abt. I, Rep. 1A, Nr. 2405, pp. 66–67 and 75; the quote is on p. 67. Fischer's threat to close the Institute is found in MPG-Archiv, Abt. I, Rep. 3, *Tätigkeitsb-ericht* Jg. 1931/32, Nr. 8. The quote can also be found there. For evidence regarding the actual financial situation with regard to research projects at the Dahlem Institute, see Schmuhl, *Grenzüberschreitungen,* p. 68. Quote to the RF is taken from a letter from Fischer to RF, February 16, 1932, RAC, folder 63, box 10, series 717A, RG 1.1.

60. Schmuhl, *Grenzüberschreitungen,* p. 190.

61. For Fischer's questioning of the budget of the KWI for Brain Research, see MPG-Archiv, Abt. 1, Rep. 1A, Nr. 2406, pp. 145–47. The sums for the KWIA listed above are taken from "Vermögensübersicht und Einnahmen-und Ausgabenrechnung" reports for 1933, 1934, and 1937. MPG-Archiv, Abt. I, Rep. 1A, Nr. 2399, pp. 47, 45, and 153; for information on the final sum of 170,000 RM and Fischer's reference to Gütt, see MPG-Archiv, Abt. I, Rep. 1A, Nr. 2413, p. 18b; Abt. I, Rep. 1A, Nr. 2406, p. 149.

62. Schmuhl, *Grenzüberschreitungen,* pp. 192–93.

63. Ibid., pp. 195–96.

64. MPG-Archiv, Abt. I, Rep. 3, Nr. 10 *Tätigkeitsbericht* Jg. 1933/34, Nr. 11. *Tätigkeitsbericht* Jg. 1934/35.

65. Ibid.

66. Schmuhl, *Grenzüberschreitungen,* pp. 196–202.

67. For the Institute's departmental reshuffling upon von Verschuer's departure, see MPG-Archiv, Abt. I, Rep. 3, Nr. 12.

68. Quoted from Schmuhl, *Grenzüberschreitungen,* p. 205.

69. The quote is taken from ibid., p. 207.

70. For von Verschuer's success in Frankfurt, see Benoît Massin, "Mengele, die Zwillingsforschung und die 'Auschwitz-Dahlem-Connection,'" in *Die Verbindung nach Auschwitz: Biowissenschaften und Menschenversuche an Kaiser-Wilhelm-Instituten: Dokumentation eines Symposiums in Juni 2001,* ed. Carola Sachse (Göttingen: Wallenstein, 2003), pp. 204–7.

71. MPG-Archiv, Abt. I, Rep. 3, Nr. 12.

72. Massin, "Die fünf Methoden der Rassengenetik" (unpublished ms.)

73. Massin devotes a section of his unpublished manuscript to the political nature of this research. He argues that it is political in two ways: that it (1) served to legitimize the existence of the concept of race that was controversial, at least among people such as Franz Boas, and (2) was used

to support and advance Nazi racial policy. I have modified his two broad categories and turned them into three. Massin, "Die fünf Methoden der Rassengenetik."

74. Eugen Fischer and Gerhard Kittel, *Das antike Weltjudentum. Tatsachen, Texte, Bilder,* Schriften des Reichsinstituts für Geschichte des neuen Deutschlands, Forschungen zur Judenfrage, Bd. 7 (Hamburg: Hanseatische Verlagsanstalt Aktiengesellschaft, 1943). The two "racial portraits" taken of Jews in the ghetto of Litzmannstadt (Lodz) appear on pp. 117 and 121. MPG-Archiv, Abt. I , Rep. 1A, Nr. 2409, p. 80a.

75. For von Verschuer's correspondence with Hodson, see von Verschuer to C. B. S. Hodson, April 2, 1938, MPG-Archiv, Abt. III, Rep. 86A, Nr. 204. Massin provides several examples of how the study of what von Verschuer called "harmless" traits could be important for Nazi racial policy, particularly racial expert testimony. Massin, "Die fünf Methoden der Rassengenetik."

76. MPG-Archiv, Abt. I, Rep. 1A, Nr. 2399, p. 52.

77. Gisela Bock, *Zwangssterilisation und Nationalsozialismus: Studien zur Rassenpolitik und Frauenpolitik,* Schriften des Zentralinstituts für sozialwissenschaftliche Forschung der Freien Universität Berlin, 48 (Opladen: Westdeutscher Verlag, 1986), pp. 80–84.

78. Kaiser Wilhelm Institute for Anthropology, RAC folder 46, box 4, series 1.1, RG 6.1.

79. Schmuhl, *Grenzüberschreitungen,* pp. 285–87. Italics in the quotation (Schmuhl, pp. 286–87) in the original.

80. For a detailed discussion of this extralegal project and the KWIA's direct involvement with it, see Reiner Pommerin, *Sterilisierung der Rheinlandbastarde: Das Schicksal einer farbigen deutschen Minderheit, 1918–1937* (Düsseldorf: Droste, 1979). See also, Lösch, *Rasse als Konstrukt,* pp. 344–48; Hans-Peter Kröner, *Von der Rassenhygiene zur Humangenetik: Das Kaiser-Wilhelm-Institut für Anthropologie, menschliche Erblehre und Eugenik nach dem Kriege* (Stuttgart: Gustav Fischer, 1997), pp. 45–46; and Carola Sachse and Benoît Massin, *Biowissenschaftliche Forschung an Kaiser-Wilhelm-Instituten und die Verbrechen des NS-Regimes* (Preprint no. 3 from the Research Program "History of the Kaiser Wilhelm Society in the National Socialist Era," 2000), p. 21.

81. Schmuhl, *Grenzüberschreitungen,* pp. 296–99.

82. BAB R1501 126252/16; Gisela Bock, "Nazi Sterilization and Reproductive Politics," in *Deadly Medicine: Creating the Master Race,* ed Susan Bachrach and Dieter Kuntz (Chapel Hill: University of North Carolina Press), p. 69.

83. Lösch, *Rasse als Konstrukt,* p. 355.

84. Quote and analysis taken from Schmuhl, *Grenzüberschreitungen,* pp. 238–39.

85. For the Dahlem Institute's formulation of expert testimonials, see MPG-Archiv, Abt. I, Rep. 1A, Nr. 2399, pp. 80 and 86.

86. Lösch, *Rasse als Konstrukt,* p. 340; Sachse and Massin, *Biowissenschaftliche Forschung,* p. 22.

87. For KWIA scientists' use of testimonials as raw data, see MPG-Archiv, Abt. I, Rep. 3, Nr. 16, *Jahresbericht* 1938/39. Von Verschuer's quote can be found in Otmar Freiherr von Verschuer, *Leitfaden der Rassenhygiene* (Leipzig: Georg Thieme, 1941), p. 230.

88. Quotes cited from Schmuhl, *Grenzüberschreitungen*, pp. 307–9.

89. "Die körperlichen Rassenmerkmale des Judentums," *Völkischer Beobachter*, January 15, 1939, p. 5. I would like to thank Susanne Heim for making this article available to us.

90. Verschuer to Fischer, May 20, 1937, and May 11, 1937, MPG-Archiv, Abt. III, Rep. 86 A, pp. 291–93.

91. "Tätigkeitsbericht der Kaiser-Wilhelm-Gesellschaft zur Förderung der Wissenschaften," in *Die Naturwissenschaften* 6 (1934): 349. The report from von Verschuer is taken from Schmuhl, *Grenzüberschreitungen*, p. 265.

92. MPG-Archiv, Abt. I, Rep. 1A, Nr. 2399, p. 52; Abt. I, Rep. 1A, Nr. 2404, pp. 49–49a; Fischer's lecture (with pictures) is published as Eugen Fischer, "Menschliche Erb-und Rassenlehre als Grundlage einer Bevölkerungspolitik: Vortrag vor Gemeinschaftsleitern und Referendaren, gehalten im Gemeinschaftslager Hanns Kerrl am 20. Juni 1934," in *Leben in der Justiz: Vorträge und Erlebnisse aus der ersten Schulungswoche preußischer Gemeinschaftsleiter,* ed. Heinrich Richter (Berlin: Industrieverlag Spaeth & Linde, 1934), pp. 114–27.

93. MPG-Archiv, Abt. I, Rep. 1A, Nr. 2404, p. 49a. Fischer claims that twenty SS men were sent to him, but in a report from Walter Gross, a list of twenty-one for the first course is given. BAB R1501 126252/1/404. For a discussion of the "negative atmosphere" in the Institute, see Schmuhl, *Grenzüberschreitungen,* p. 267.

94. MPG-Archiv, Abt. I, Rep. 1A, Nr. 2406, "Aktenvermerk," May 25, 1934; BAB R1501 126252/1/404.

95. Lösch, *Rasse als Konstrukt*, pp. 356–61; Schmuhl, *Grenzüberschreitungen*, p. 268.

96. The term "the racial state" is taken from Michael Burleigh and Wolfgang Wippermann's book by the same title, *The Racial State: Germany, 1933–1945* (Cambridge: University of Cambridge Press, 1991).

97. Many of the press releases for these KWS-sponsored talks by Society members can be found in MPG-Archiv, Abt. I, Rep. 1A, Nr. 832, p. 41; and Nr. 808, p. 23.

98. Schmuhl, *Grenzüberschreitungen*, pp. 324–26.

99. Cited from ibid., pp. 326–27.

100. Ibid., pp. 318–20.

101. Kaiser Wilhelm Institute for Anthropology, RAC folder 46, box 4, series 1.1, RG 6.1.

102. Ute Deichmann, *Biologen unter Hitler: Vertreibung, Karrieren, Forschung* (Frankfurt am Main: Campus, 1992), pp. 132–33; Schmuhl, *Grenzüberschreitungen,* pp. 317–19.

103. Fischer to Verschuer, August 3, 1940, MPG-Archiv, Abt. III, Rep. 86 A, pp. 291–93.

104. Schmuhl, *Grenzüberschreitungen*, p. 323.

105. Niederschrift über die Sitzung des Kuratoriums des Kaiser-Wilhelm-Instituts für Anthropologie, menschliche Erblehre und Eugenik, Anlage 2, January 9, 1941, MPG-Archiv, Abt. I, Rep. 1A, 2400, p. 190

106. For a discussion of Magnussen, see Hans Hesse, *Augen aus Auschwitz: Ein Lehrstück über nationalsozialistischen Rassenwahn und medizinische Forschungen: Der Fall Dr. Karin Magnussen* (Essen: Klartext, 2001); Ernst Klee, *Deutsche Medizin im Dritten Reich: Karrieren vor und nach 1945* (Frankfurt am Main: Fischer 2001), pp. 348–71; The information that Magnussen officially started her career at the KWIA with a stipend from Fischer was taken from Schmuhl, *Grenzüberschreitungen*, p. 364.

107. Niederschrift über die Sitzung des Kuratoriums des Kaiser-Wilhelm-Instituts für Anthropologie, menschliche Erblehre und Eugenik, Anlage 2, January 9, 1941, MPG-Archiv, Abt. I, Rep. 1A, Nr. 2400, p. 193.

108. Quote taken from Schmuhl, *Grenzüberschreitungen*, p. 334.

109. "Aktenvermerk," September 19, 1939, BAB NS 21 352.

110. Quote taken from Schmuhl, *Grenzüberschreitungen*, pp. 346–47.

111. Ibid., pp. 344–50.

112. A discussion of Fischer's successes is given in Lösch, *Rasse als Konstrukt*, pp. 274–77. For Fischer's party membership, see Hessisches Hauptstaatsarchiv, Wiesbaden (HHA), Abt. 520/Ro 1298/47 Spruchkammerakte Fischer. Von Verschuer's postwar argument about the KWIA directorship can be seen in HHA, Abt. 520/F FZ5261 Spruchkammerakte von Verschuer. Although the future director of the KWIA became a party member in July 1940, he apparently did not receive his membership card until 1941. For this information and the exact date of Fischer's party membership, see Schmuhl, *Grenzüberschreitungen*, p. 401.

113. For Fischer's letter to von Verschuer about the KWIA directorship, see Fischer to von Verschuer, September 10, 1934, MPG-Archiv, Abt. III, Rep. 86A, Nr. 291-1.

114. The assessment that von Verschuer's scientific reputation was exceedingly high at the time he was made KWIA director is based on recently discovered letters of reference and testimonials of both German and non-German geneticists and other biomedical professionals in connection with von Verschuer's postwar attempt to receive a professorship in the Medical Faculty of the University of Frankfurt. I am indebted to Florian Schmaltz who first made me aware of this *Dekanatsarchiv* as a treasure drove of heretofore unexplored archival material.

115. Von Verschuer delivered a lecture entitled "Twin Research from the Time of Francis Galton to the Present-Day," on June 8, 1939. It was published in *Proceedings of the Royal Society of London* 128 (1939): 62–81.

116. Fischer to von Verschuer, September 18, 1942, MPG-Archiv, Abt. III, Rep. 86A, Nr. 291-4.

117. A brief note about Mengele: Although he was viewed as a "guest researcher" at the KWIA during his brief stay in Dahlem before going to Auschwitz, Mengele was never an official member of the Fischer Institute with a contract from the KWS. He did, however, undertake expert racial testimonies at the KWIA. Mengele was allegedly von Verschuer's favorite student from his Frankfurt days, and the new KWIA director would have liked to have given Mengele an assistantship in Dahlem. Other "duties" prevented this from happening. Sachse and Massin, *Biowissenschaftliche Forschung*, pp. 24–26. The von Verschuer–Mengele research connection was first described by Miklos Nyiszli in his book *Auschwitz: A Doctor's Eyewitness Account of Mengele's Infamous Death Camp*, trans. Tibère Kremer and Richard Seaver (New York: Seaver Books, 1986); it is also discussed in Müller-Hill, *Murderous Science*.

118. This new context for von Verschuer's work with Mengele has been provided by Schmuhl, *Grenzüberschreitungen*, p. 511–22.

119. For the most up-to-date and detailed account of this project, see Schmuhl, *Grenzüberschreitungen*, pp. 502–10.

120. Magnussen's iris structure project is discussed in Lösch, *Rasse als Konstrukt*, p. 410; Massin, "Mengele, die Zwillingsforschung und die Auschwitz-Dahlem Connection,'" in *Die Verbindung nach Auschwitz*, pp. 240–52; and Hans Hesse, *Augen aus Auschwitz*, p. 58. The research on this subject is synthesized in Schmuhl, *Grenzüberschreitungen*, pp. 482–502.

121. Klee, *Deutsche Medizin im Dritten Reich*, p. 370.

122. Sachse and Massin, *Biowissenschaftliche Forschung*, pp. 36–37; Diane Paul, "Genetics under the Swastika," *Dimensions: A Journal of Holocaust Studies* 10 (1996): 26.

123. Schmuhl, *Grenzüberschreitungen*, pp. 480–82.

124. The term "biobanks" was coined by Alexander von Schwerin and used in his multiauthored article, Gerhard Baader et al., "Pathways to Human Experimentation, 1933–1945: Germany, Japan and the United States," *Osiris* 20 (2005): 211.

125. Von Verschuer to Bautzmann, July 2, 1942; von Verschuer to Stadtmüller, November 15, 1942, MPG-Archiv, Abt. III, Rep. 86 B, Nr. 36. I wish to thank Alexander von Schwerin for making these documents available to me.

126. Schmuhl, *Grenzüberschreitungen*, pp. 337–38.

127. Quoted from Schmuhl, *Grenzüberschreitungen*, p. 397.

128. "Tätigkeitsbericht 1943/44," MPG-Archiv, Abt. I, Rep. 3.

129. Massin, "Mengele, Zwillingsforschung, und die 'Auschwitz-Dahlem Connection,'" in *Die Verbindung nach Auschwitz*, pp. 210–17.

130. Von Verschuer to C. B. S. Hodson, April 2, 1938, MPG-Archiv, Abt. III, Rep. 86A, Nr. 369.

131. "Tätigkeitsbericht 1943/44," MPG-Archiv, Abt. I, Rep. 3.
132. For a discussion of biomedical-military network and its usefulness to Nachtsheim during the war, see Alexander von Schwerin's impressive study, *Experimentalisierung des Menschen: Der Genetiker Hans Nachtsheim und die vergleichende Erbpathologie, 1920–1945* (Göttingen: Wallstein, 2004).
133. Almost all the scholars included in Sachse's excellent anthology *Die Verbindung nach Auschwitz* suggest unbridled research as an explanation for these investigators' unethical deeds. This revealing partial quote from Mengele was taken from Schmuhl, *Grenzüberschreitungen,* p. 477.
134. Helmut von Verschuer, Otmar von Verschuer's son, reports a conversation that his mother had with Mengele when the latter was at the von Verschuer home in Berlin for a visit. When she asked whether his duties were difficult, she claimed that Mengele refused to discuss his work saying only "It's dreadful. I cannot talk about it." Müller-Hill, *Murderous Science,* p. 128.
135. The Fischer quote is taken from ibid., p. 66. Von Harnack's statement is given in "Kaiser-Wilhelm-Institut für Anthropologie," *Deutsche Allgemeine Zeitung,* September 16, 1927 in MPG-Archiv, Abt. IX, Rep. 2.
136. Schmuhl, *Grenzüberschreitungen,* pp. 448–49. This transition will also be revealed in chapter 4 when we examine Fischer's statement on Jews in occupied France. Fischer's more radical stance on the "Jewish question" is also supported by his willingness to include blatantly anti-Semitic remarks in his coauthored book with Kittel on Jews in the ancient world in 1943—at a time when the Holocaust was at its height.

CHAPTER THREE

1. Therese Rüdin to Anita Ploetz, March 20, 1933. I thank Matthias Weber for making a copy of this letter available to me. It is also quoted in Matthias M. Weber, *Ernst Rüdin: Eine kritische Biographie* (Berlin: Springer Verlag, 1993), pp. 179–80.
2. Benno Müller-Hill, *Murderous Science: The Elimination by Scientific Selection of Jews, Gypsies, and Others in Germany, 1933–1945,* 2nd ed. (Cold Spring Harbor, NY: Cold Spring Harbor Press, 1998), p. 131.
3. Gregg to Pearce, November 24, 1925, RAC folder 54, box 9, series 717A, RG 1.1; Glum to Kraepelin, December 18, 1924, MPG-Archiv, Abt. I, Rep. 1A, Nr. 2429, pp. 129–30; Walther Spielmeyer, "Die Eröffnung der Neubaues des Deutschen Forschungsanstalt für Psychiatrie (Kaiser-Wilhelm-Institut) in München," *Zeitschrift für die gesamte Neurologie und Psychiatrie* 117 (1928): 180–81; Weber, "Psychiatric Research and Science Policy in Germany: The History of the Deutsche Forschungsanstalt für Psychiatrie (German Institute for Psychiatric Research) in Munich from 1917 to 1945," *History of Psychiatry* 11 (2000): 246–47.

4. Forschungsanstalt für Psychiatrie Munich, RAC folder 58, box 10, series 717A, RG 1.1; quote on p. 1.

5. Deutsche Forschungsanstalt für Psychiatrie, Inaugural Ceremony, RAC folder 54, box 9, series 717A, RG 1.1; Harnack quote on p. 247; Krupp von Bohlen and Halbach quotes on pp. 247–48.

6. Felix Plaut, June 13, 1928, pp. 262 e–i, "Some of Kraeplen's Fundamental Conceptions concerning the 'Deutsche Forschungsanstalt für Psychiatrie,'" RAC folder, 54, box 9, series 717A, RG 1.1. Plaut's German speech is printed in Felix Plaut, "Einige Grundgedanken Kraepelins über die Deutsche Forschungsanstalt für Psychiatrie," *Zeitschrift für die gesamte Neurologie und Psychiatrie* 117 (1928): 191–94.

7. Weber, "Psychiatric Research and Science Policy in Germany," pp. 245–49; MPG-Archiv, Abt. I., Rep. 1A, Nr. 2435, p. 19. There were also two other divisions at the German Research Institute for Psychiatry at this time: the Chemical and Psychological Departments.

8. "IX Bericht über die Deutsche Forschungsanstalt für Psychiatrie (Kaiser-Wilhelm-Institut in München) zur Stiftungsratssitzung von 11. März 1929," Sonderabdruck der *Zeitschrift für die gesamte Neurologie und Psychiatrie* 122 (1929): 55–73. I thank Matthias Weber for making this article available to me.

9. Robert A. Lambert's officer's diary, February 4, 1932, RAC RG 1.1.

10. In a letter to Loeb, Spielmeyer relates a meeting with Glum in which the General Secretary told Spielmeyer that it would be an affront to him and Plaut if another Research Institute Department head—who was not scientifically more renowned—were to receive a significantly higher salary. Spielmeyer to Loeb, January 2, 1928, MPG-Archiv, Abt. I, Rep. 1A, Nr. 2434, p. 254. In a letter to the head of the Foundation Council, Glum mentions that Spielmeyer and Plaut were not aware of the extent of Rüdin's financial demands although his original conditions were apparently higher. Glum to Hauptmann, MPG-Archiv, Abt. I., Rep. 1A, Nr. 2434, pp. 310–11.

11. Rüdin to Plaut, May 14, 1925, MPG-Archiv, Abt. I, Rep. 1A, Nr. 2429, pp. 210–11.

12. Rüdin to Spielmeyer and Plaut, October 28, 1927, MPG-Archiv, Abt. I, Rep. 1A, Nr. 2433, p. 147.

13. Quote taken from Glum's letter to Hauptmann, MPG-Archiv, Abt. I., Rep. 1A, Nr. 2434, pp. 310–11; for the salary figures, see Weber, *Ernst Rüdin*, p. 153; Rüdin to Spielmeyer and Plaut, October 28, 1927, MPG-Archiv, Abt. I, Rep. 1A, Nr. 2433, p. 151; Volker Roelcke, "Psychiatrische Wissenschaft im Kontext nationalsozialistischer Politik und 'Euthanasie'. Zur Rolle von Ernst Rüdin und der Deutschen Forschungsanstalt für Psychiatrie/Kaiser-Wilhelm-Institut," in *Geschichte der Kaiser-Wilhelm-Gesellschaft im Nationalsozialismus. Bestandsaufnahme und Perspektiven der Forschung*, Bd. 1, ed. Doris Kaufmann (Göttingen: Wallstein, 2000), p. 117.

14. Spielmeyer to Loeb, January 2, 1928, MPG-Archiv, Abt. I, Rep. 1A, Nr. 2434, p. 254.

15. Volker Roelcke, "Programm und Praxis der psychiatrischen Genetik an der Deutschen Forschungsanstalt für Psychiatrie unter Ernst Rüdin. Zum Verhältnis von Wissenschaft, Politik und Rasse-Begriff vor und nach 1933," in *Rassenforschung an Kaiser-Wilhelm-Instituten vor und nach 1933,* ed. Hans-Walter Schmuhl (Göttingen: Wallstein, 2003), p. 43.

16. Plaut, "Einige Grundgedanken Kraepelins," p. 192.

17. Abschrift von der Genealogischen Abteilung der Deutschen Forschungsanstalt für Psychiatrie (Kaiser-Wilhelm-Institut), September 20, 1927, MPG-Archiv, Abt. I, Rep. 1A, Nr. 2435, p. 167.

18. Rüdin to Glum, September 15, 1931, MPG-Archiv, Abt. I, Rep. 1A, Nr. 2436, pp. 209–10.

19. Weber, *Ernst Rüdin,* pp. 160–61. Weber has correctly stressed that the Weimar government had already begun to loosen the protection of personal data in the interest of eugenic research—a tendency that indicated the relative value of the individual and the collective in the eyes of a democratic state.

20. Hans Luxenburger, "Leistungen der Zwillingsforschung für die Medizin," *Forschungen und Fortschritt* 8 (1932): 211–12.

21. Weber, *Ernst Rüdin,* pp. 142–43, 165–66; Luxenburger, "Die Bedeutung der Statistik für die psychiatrische Erblichkeitsforschung," *Psychiatrisch-Neurologische Wochenschrift* 31 (1929): 146–47. First quote on p. 146; Luxenburger, "Die wichtigsten neueren Fortschritte der psychiatrischen Erblichkeitsforschung," *Fortschritte der Neurologie, Psychiatrie und ihre Grenzgebiete* 1 (1929): 88. Second quote on p. 88.

22. Weber, *Ernst Rüdin,* p. 144.

23. Notgemeinschaft der Deutschen Wissenschaft: Anthropologische Erhebung der Deutschen Bevölkerung, RAC folder 187, box 20, series 717 S, RG 1.1.

24. Weber, *Ernst Rüdin,* p. 163.

25. The quotes are taken from Luxenburger, "Psychiatrische Heilkunde und Eugenik," *Das Kommende Geschlecht* 8 (1932): 32–33. Other publications devoted to eugenics include Luxenburger, "Über die Zunahme der Fruchtabtreibung vom Standpunkt der Volksgesundheit und Rassenhygiene," *Monatsschrift für Kriminalpsychologie und Strafrechtsreform* 18 (1927): 326–28; "Welche Folgerungen hat die Eugenik aus den Ergebnissen der psychiatrischen Erblichkeitsforschung zu ziehen?" *Münchener Medizinische Wochenschrift* 77 (1930): 2020.

26. Weber, *Ernst Rüdin,* p. 176; Although Rüdin did not openly support mandatory sterilization at this time, from his earliest writings he discussed the use of force in cases where the "unfit" could not be brought to see the necessity of voluntary sterilization or in particularly egregious cases of degeneration. Ibid., p. 183.

27. Ibid., pp. 175–78; quote on p. 178.

28. Forschungsanstalt für Psychiatrie, Munich Research 1930–31, RAC folder 55, box 9, series 717 A, RG 1.1.

29. Weber, *Ernst Rüdin,* pp. 163–64,

30. Gregg to Lambert, January 15, 1933, RAC folder 56, box 9, series 717 A, RG 1.1.

31. Rüdin to Loeb, November 26, 1932, RAC folder 56, box 9, series 717 A, RG 1.1; Weber, *Ernst Rüdin,* p. 164.

32. Gütt mentions Rüdin's international reputation in his field as a reason for selecting him to write the commentary to the Nazi Sterilization Law. Weber, *Ernst Rüdin,* p. 183.

33. The letters between the two men use the word *"Lieber"* instead of the more formal (and usual) *"Sehr geehrter"* as part of the opening salutation. This is a sign that there was more than a merely official relationship between them. The letters demonstrate a degree of warmth and trust totally absent in the correspondence between Fischer and Gütt. In a letter to Rüdin dated September 22, 1939, Gütt states the following: "I am certain that [our] friendship will last for the duration of our lives." Nachlaß Ernst Rüdin, MPG-Archiv, ZA 131 Rüdin, Box 1.

34. BAB R1501 126243/290–92.

35. BAB R1501 126244/297; 126245/177–78.

36. BAB R1501 126245/157.

37. Weber, *Ernst Rüdin,* p. 224.

38. For a discussion of the role of the Expert Advisory Council, see Heidrun Kaupen-Haas, "Die Bevölkerungsplaner im Sachverständigenbeirat für Bevölkerungs- und Rassenpolitik," in *Der Griff nach der Bevölkerung: Aktualität und Kontinuität nazistischer Bevölkerungspolitik,* ed. Heidrun Kaupen-Haas, Schriften der Hamburger Stiftung für Sozialgeschichte des 20. Jahrhunderts, Bd. 1 (Nördlingen: Greno, 1986), pp. 103–20; Peter Weingart et al., *Rasse, Blut und Gene: Geschichte der Eugenik und Rassenhygiene in Deutschland* (Frankfurt am Main: Suhrkamp, 1988), pp. 460–64.

39. Hans-Walter Schmuhl, *Rassenhygiene, Nationalsozialismus, Euthanasie: Von der Verhütung zur Vernichtung "lebensunwerten Lebens," 1890–1945* (Göttingen: Vandenhoeck & Ruprecht, 1987), pp. 154–56. The Frick quotes are taken from Schmuhl; Weber, *Ernst Rüdin,* pp. 181–83.

40. Heinz Müller, "Der Sachverständigenbeirat für Bevölkerungs-und Rassenpolitik des Reichsministers des Innern," in *Rassekurs in Egendorf: Ein rassehygienischer Lehrgang des Thüringischen Landesamts für Rassewesen,* ed. Karl Astel (Munich: Lehmann Verlag, 1935), pp. 185–86.

41. Weber, *Ernst Rüdin,* p. 183; Schmuhl, *Rassenhygiene,* p. 154.

42. Weber argues that Rüdin was not directly involved in writing the Law. Weber, *Ernst Rüdin,* p. 183. In a letter to a British colleague who spent time in the Genealogy Department, Rüdin states that he was "one of the creators of the Law." Naturally, it is difficult to know what he meant by "creators." Rüdin to Slater, April 20, 1936, Archiv des Max-Planck-Instituts für Psychiatrie (MPIP-HA), GDA 132.

43. BAB R1501 126248/148.

44. Rüdin's medical commentary of the Law delivered to psychiatrists can be found in Rüdin, "Das deutsche Sterilisationsgesetz (Medizinischer Kommentar)," in *Erblehre und Rassenhygiene im völkischen Staat,* ed. Ernst Rüdin (Munich: Lehmann, 1934), pp. 150–73. The list of diseases included in Section 1 of the Law and the definition of "genetic illness" are found on pp. 151–52.

45. Gisela Bock, "Nazi Sterilization and Reproductive Politics," in *Deadly Medicine: Creating the Master Race,* ed Susan Bachrach and Dieter Kuntz (Chapel Hill: University of North Carolina Press), p. 68.

46. This point is stressed by Roelcke in "Programm and Praxis," p. 56; Luxenburger, "Spezielle empirische Erbprognosse in der Psychiatrie," in *Erblehre und Rassenhygiene im völkischen Staat,* ed. Ernst Rüdin (Munich: Lehmann, 1934), pp. 143–49.

47. Rüdin, "Das deutsche Sterilisationsgesetz," pp. 152 and 154.

48. BAB 1501 126252/1, pp. 485–96; quote on p. 493.

49. Roelcke, "Programm und Praxis," pp. 46–47, n. 22.

50. BAB R1501 126789/259–63. Quote on p. 259; MPG-Archiv, Abt. I, Rep. 1A, Nr. 2451, p. 39.

51. MPG-Archiv, Abt. I, Rep. 1A, Nr. 2451, pp. 45–46.

52. MPG-Archiv, Abt. I, Rep. 1A, Nr. 2451, p. 83.

53. Spielmeyer to O'Brien, March 26, 1934, RAC folder 56, box 9, series 717A, RG 1.1.

54. Weber, *Enrst Rüdin,* p. 205.

55. "Die Deutsche Organisation für psychische Hygiene in den Jahren 1930–1936," MPIH-HA, GDA 127.

56. Roelcke, "Programm und Praxis," p. 58.

57. "Bericht über den erbbiologisch-rassenhygienischen Schulungskurs für Psychiater," MPIP-HA, GDA 19; Rüdin, *Erblehre und Rassenhygiene.*

58. "Rassenhygiene in ihren Auswirkungen auf Kriegs-und Militärdienst," MPIP-HA, GDA 85.

59. Glum to Rüdin, January 29, 1936, MPG-Archiv, Abt. I, Rep. 1A, Nr. 2438, p. 92.

60. "Mein Führer," November 22, 1935, BAB R43 II/1229.

61. "Vermerk," December 3, 1935, BAB R43 II/1229.

62. "Mein Führer," November 16, 1936, MPIP-HA, GDA 8.

63. "An den Herrn Reichsminister des Innern," August 26, 1938, and "An den Direktor des Kaiser-Wilhelm-Instituts für Genealogie und Demographie der Deutschen Forschungsanstalt für Psychiatrie . . .," January 26, 1939, MPIP-HA, GDA 8.

64. "Beweise für meine Einstellung zur Rassenfrage des deutschen Volkes," BAB 1501 126245/227.

65. Weber, *Ernst Rüdin,* pp. 193–94.

66. Ibid. For a discussion of the impact of the Law for the Reestablishment of the Professional Civil Service on Jews, see Avraham Barkai, *From Boycott*

to Annihilation: The Economic Struggle of German Jews, 1933–1945 (Hanover and London: University Press of New England, 1989).

67. For a detailed discussion of the Kaiser Wilhelm Society's policy toward its Jewish scientists, see Michael Schüring, *Minervas verstoßene Kinder: Vertriebene Wissenschaftler und die Vergangenheitspolitik der Max-Planck-Gesellschaft* (Göttingen: Wallstein, 2006); Kristie Macrakis, *Surviving the Swastika* (New York: Oxford University Press, 1993); Weber, *Ernst Rüdin*, p. 194.

68. Robert Lambert's officer's diary, September 6, 1933, RAC folder 56, box 9, series 717A, RG 1.1.

69. Alan Gregg's officer's diary, June 19, 1934, RAC folder 56, box 9, series 7171A, RG 1.1; Daniel O'Brien's officer's diary, December 6, 1935, RAC RG 12.1.

70. Spielmeyer to Gregg, October 22, 1934, November 14, 1934, RAC folder 56, box 9, series 717A, RG 1.1; Spielmeyer to O'Brien, April 3, 34, RAC folder 55, box 9, series 717A, RG 1.1.

71. O'Brien's officer's diary, October 27–28, 1933, RAC RG 12.1.

72. Alan Gregg to Andrew Woods, February 15, 1935, RAC folder 57, box 10, series 717A, RG 1.1; Weber, *Ernst Rüdin*, p. 197.

73. John van Sickle to Sydnor Walker, April 28, 1933; Sydnor Walker to John van Sickle, June 27, 1933, RAC folder 187, box 20, series 717S, RG 1.1.

74. Alan Gregg's officer's diary, June 19, 1934, RAC folder 56, box 9, series 717A, RG 1.1; Daniel O'Brien's officer's diary, March 23–24, 1933, RAC RG 12.1.

75. "Kraepelin Institute," RAC folder 58, box 10, series 717A, RG 1.1; Bruce Bliven to the RF, December 20, 1933; Thomas Appleget to Raymond Fosdick, February 2, 1934, RAC folder 56, box 9, series 717A, RG 1.1.

76. Appleget to George Strode, February 23, 1934, RAC folder 56, box 9, series 717A, RG 1.1.

77. Rüdin to Plaut, October 25, 1935, RAC folder 57, box 10, series 717A, RG 1.1.

78. Robert Lambert to D. O'Brien, December 2, 1935, RAC folder 57, box 10, series 717 A, RG 1.1.

79. von Stengle to Glum, MPG-Archiv, Abt. I, Rep. 1A, Nr. 2438, p. 88.

80. Cited from Weber, *Ernst Rüdin*, pp. 198–99,

81. D. O'Brien to R. H. Curtis, September 6, 1939, RAC folder 42, box 4, series 1, RG 6.1; Weber, "Psychiatric Research and Science Policy," p. 253.

82. Weber, *Ernst Rüdin*, pp. 199–200.

83. Glum to Rüdin, November 16, 1935, MPG-Archiv, Abt. I, Rep. 1A, Nr. 2452, p. 55.

84. Rüdin to Telschow, MPG-Archiv, Abt. I, Rep. 1A, Nr. 2453, p. 201.

85. Roelcke, "Programm und Praxis," pp. 48–53.

86. Ibid.

87. "Rassenhygiene in ihren Auswirkungen auf Kriegs-und Militärdienst," MPIP-HA, GDA 85; "Mein Führer," November 22, 1935, BAB R43 II/1229; "Mein Führer," November 16, 1936, MPIP-HA, GDA 8.

88. Rüdin allowed his Jewish colleague Neubürger to use the facilities of the Research Institute, although he was no longer drawing a salary from it; he also helped Fritz Kallmann, a Jewish colleague (but not member of the Genealogy Department) who studied schizophrenia. Rüdin enabled him to use the facilities of the Institute, he invited him to give a paper at an international congress in 1935, and he allowed a paper of his to be read at a professional conference in Germany. Rüdin also wrote a very positive letter of recommendation for Kallmann when it was clear that he desired to pursue his career in the United States. Rüdin's Testimonial, May 30, 1936, APS Franz Boas Papers; Kallmann immigrated to New York in 1936 and made an important name for himself in psychiatric genetics. MPG-Archiv, Abt. I, Rep. 1A, Nr. 2452, p. 93; Franz Kallmann's character witness statement of Rüdin, June 30, 1947, MPIP-HA: NLR 8.

89. MPG-Archiv, Abt. I, Rep. 1A, Nr. 2438, pp. 116–18.

90. Rüdin to Herrn X, June 19, 1933, MPIP-HA: GDA 63.

91. Rüdin to Carl Schneider, May 19, 1935, MPIP-HA: GDA 132.

92. Müller-Hill, *Murderous Science,* p. 133.

93. Weber, *Ernst Rüdin,* p. 250.

94. Rüdin to P. F. Waardenburg, May 13, 1937; Rüdin to Slater, April 20, 1936, MPIP-HA: GDA 132.

95. Gütt to Rüdin, August 8, 1933; Rüdin to Gütt, August 9, 1933; Luxenburger, "Die Sterilisierung Minderwertiger," BAB R 1501 126248/388–95. Quotes on pp. 389 and 395.

96. This quote is taken from Weber, *Ernst Rüdin,* p. 191.

97. Von Verschuer to Luxenburger, February 11, 1936; Luxenburger to von Verschuer, February 13, 1936; Von Verschuer to Luxenburger, February 15, 1936, MPIP-HA: GDA 148/II.

98. The Rüdin-von Verschuer conflict over use of twin subjects can be gleaned in the correspondence between the two men: von Verschuer to Rüdin, July 12, 1935; Rüdin to von Verschuer, July 24, 1935; von Verschuer to Rüdin, August 20, 1935; von Verschuer to Rüdin, March 15, 1941, MPIP-HA: GDA 132.

99. Fischer to Rüdin, June 17, 1937, MPIP-HA: GDA 131; Rüdin to von Verschuer, July 24, 1935, MPIP-HA: GDA 132.

100. Rüdin to Fischer, June 24, 1937, MPIP-HA: GDA 131.

101. James Watson, *The Double Helix. A Personal Account of the Discovery of the Structure of DNA,* Norton Critical Edition (New York: Norton, 1980).

102. For a discussion of the conflict, see Weber, *Ernst Rüdin,* pp. 253–56. Quotes are cited from pp. 254–55.

103. Gütt to the Reichsführer, February 7, 1938, BAB NS 19 3434/2, pp. 79–81.

104. BAB R22 4483/39–40.

105. R.R. to the Erbgesundheitsgericht München, April 6, 1942, Nachlaß Ernst Rüdin, MPG-Archiv, ZA 131 Rüdin, Box 3.

106. Weber, *Ernst Rüdin,* p. 256.

107. Ibid., p. 257.
108. BAB (formerly Berlin Documents Center), PK, Rüdin, Ernst, April 19, 1874.
109. Rüdin to Frick, April 29, 1939, Nachlaß Ernst Rüdin, MPG-Archiv, ZA 131 Rüdin, Box 1.
110. Professor Ernst Rüdin zum 65. Geburtstag von Dr. med. Heinz Riedel, Nachlaß Ernst Rüdin, MPG-Archiv, ZA 131 Rüdin, Box 1.
111. Roelcke, "Programm und Praxis," p. 63; Glum to Rüdin, January 29, 1936, MPG-Archiv, Abt. I, Rep. 1A, Nr. 2438, p. 92; Nr. 2453, p. 167.
112. Rüdin to Gütt, June 24, 1939, MPIP-HA: GDA 126.
113. For a full discussion of the Ahnenerbe, see Michael H. Kater, Das "Ahnenerbe" der SS 1935–1945, Ein Beitrag zur Kulturpolitik des Dritten Reiches, 2nd ed. (Munich: R. Oldenbourg Verlag, 1997).
114. Luxenburger and Streicher clashed over the so-called impregnation theory. Streicher believed that once an "Aryan" woman was defiled by a Jewish man, her children would harbor Jewish blood; Luxenburger insisted on the unscientific nature of this position. Weber, Ernst Rüdin, p. 240.
115. Rüdin to Wüst, June 24, 1939, BAB NS 21 352. I am grateful to Matthias Weber for making this key document known to me. He cites it and quotes from it in his book, Weber, Ernst Rüdin, pp. 259–60. However, I have gone into more detail in laying bare its contents.
116. Weber, Ernst Rüdin, p. 259; Professor Ernst Rüdin zum 65. Geburtstag von Dr. med. Heinz Riedel, Nachlaß Ernst Rüdin, MPG-Archiv, ZA 131 Rüdin, Box 1; "Bericht über Vorträge in Alt Rehse," MPG-Archiv, Abt. II, Rep. 1A, Rüdin Personalia, Nr. I.
117. Rüdin to Wüst, February 12, 1940, BAB NS 21 352.
118. "Aktenvermerk," September 19, 1939, BAB NS 21 352.
119. Ibid.
120. Brandt to Wüst, November 3, 1941; Wüst to Wolff, December 17, 1940; "Vermerk," June 13, 1941. The entire history of the Ahnenerbe's relationship to the Rüdin Institute is summarized in a report dated May 18, 1944, BAB NS 21 352.
121. "Aktenvermerk," April 1, 1940, MPG-Archiv, Abt. I, Rep. 1A, Nr. 2447, p. 168a.
122. Telschow to Wüst, April 1, 1940, MPG-Archiv, Abt. I, Rep. 1A, Nr. 2447, p. 168.
123. "Vermerk," May 18, 1944, BAB NS 21 352.
124. Weber, Ernst Rüdin, pp. 262–63.
125. Ibid., pp. 264–65; "Grobig an die Gaudozentenbundesführung," November 13, 1940, Nachlaß Ernst Rüdin, MPG-Archiv, ZA 131 Rüdin, Box 5, p. 11. Here the term "Staatsfeind" is used with reference to Luxenburger.
126. "Grobig an die Gaudozentenbundesführung," November 13, 1940, Nachlaß Ernst Rüdin, MPG-Archiv, ZA 131 Rüdin, Box 5.
127. Ibid., Grobig to Ströder, November 26, 1940.
128. Ibid., Deussen to Grobig, November 11, 1940.
129. Weber, Ernst Rüdin, p. 265.

130. Ibid., pp. 265–66; Thums to Grobig, February 4, 1944, Nachlaß Ernst Rüdin, MPG-Archiv, ZA 131 Rüdin, Box 3.

131. Rüdin to Wüst, April 13, 1937, Nachlaß Ernst Rüdin, MPG-Archiv, ZA 131 Rüdin, Box 3. In his letter of reference regarding Luxenburger, Rüdin also claims that he was an anticommunist. The only known anti-Semitic comment on record was a statement made by Luxenburger in a letter to Muckermann in 1932. Here he compared the frequent debates in the Reichstag to a "Nigger kraal" or a "Talmudic school." Weber, *Ernst Rüdin,* p. 179.

132. Luxenburger to Rüdin, June 3, 1940, Nachlaß Ernst Rüdin, MPG-Archiv, ZA 131 Rüdin, Box 3.

133. "Vermerk," October 5, 1944; Rüdin to Telschow, July 23, 1942; "Vermerk," May 5, 1942, MPG-Archiv, Abt. I, Rep. 1A, Nr. 2440; Weber, *Ernst Rüdin,* p. 282.

134. Ernst Rüdin "Zehn Jahre nationalsozialistischer Staat," *ARGB* B (1943), pp. 321–22. A copy of the article can be found in Nachlaß Ernst Rüdin, MPG-Archiv, ZA 131 Rüdin, Box 3.

135. This claim is mentioned by a high-ranking medical official in a testimonial in Rüdin's defense during his de-Nazification proceeding. Dr. Ast, October 8, 1949, Nachlaß Ernst Rüdin, MPG-Archiv, ZA 131 Rüdin, Box 3.

136. The term "political hornet's nest" was used by Rüdin's former colleague Karl Thums, see Thums to Grobig. February 4, 1944, Nachlaß Ernst Rüdin, MPG-Archiv, ZA 131 Rüdin, Box 3.

137. Bock, "Nazi Sterilization," p. 81.

138. Michael Burleigh, "Nazi 'Euthanasia' Programs," in *Deadly Medicine: Creating the Master Race,* ed. Susan Bachrach and Dieter Kuntz (Chapel Hill: University of North Carolina Press, 2004), pp. 127–30; for a fuller discussion of the professional situation of psychiatry and the "cure or kill" mentality among physicians, see Michael Burleigh, *Death and Deliverance: "Euthanasia in Germany c. 1900–1945* (Cambridge: Cambridge University Press, 1994). Also important is Götz Aly's pathbreaking work "Pure and Tainted Progress," in *Cleansing the Fatherland: Nazi Medicine and Racial Hygiene,* ed. Götz Aly, Peter Chroust, and Christian Pross; trans. Belinda Cooper (Baltimore: Johns Hopkins, 1994), pp. 156–235.

139. For a detailed discussion of the connection between "euthanasia" and the Holocaust, see Henry Friedlander, "From 'Euthanasia' to the 'Final Solution'" in *Deadly Medicine: Creating the Master Race,* ed. Susan Bachrach and Dieter Kuntz (Chapel Hill: University of North Carolina Press, 2004), pp. 155–83, 212–13. Friedlander's *The Origins of Nazi Genocide: From Euthanasia to the Final Solution* (Chapel Hill: University of North Carolina Press, 1995). The estimated total number of victims was taken from http://www .lebensunwert.at/ns-euthanasie.html.

140. This new research on the killing of asylum patients before September 1939 is outlined by Roelcke, in "Psychiatrische Wissenschaft," p. 128.

141. Burleigh, "Nazi 'Euthanasia' Programs," pp. 127–53, 211–12; http://www
.lebensunwert.at/ns-euthanasie.html.

142. Rüdin to de Crinis, July 21, 1941, MPIP-HA: NLR 3; Rüdin to de Crinis, May
9, 1943, Nachlaß Ernst Rüdin, MPG-Archiv, ZA 131 Rüdin, Box 3.

143. In addition to his talk at the Berlin Military Academy (see note 87 above),
Rüdin was asked to participate in a teach-in sponsored by the National
Socialist Professors' Association on racial biology. March 17, 1942, MPIP-
HA: GDA 8; he also participated in a racial hygiene seminar sponsored by
the Academy for Medical Training in Bavaria, September 22, 1934, Nachlaß
Ernst Rüdin, MPG-Archiv, ZA 131 Rüdin, Box 3.

144. Roelcke, "Psychiatrische Wissenschaft," p. 119.

145. Ibid., pp. 122–26.

146. Roelcke raises the question of Rüdin's involvement with the "euthanasia" action in this sense. He views the Munich director as someone who legitimized,
supported, and profited from it. Ibid., p. 129; for a discussion of the active
participants in the "euthanasia" action see the books listed in notes 138 and
139 above.

147. "Das Ethos der Rassenhygiene: Eine Unterredung mit Prof. Rüdin," July 25,
1935, *Sonderdienst der Wohlfahrts-Korrespondenz* in MPIP-HA: NLR 9; "Was
ist Rassenhygiene Eine Unterredung mit Prof. Dr. Rüdin," *Kongress-
Korrespondenz*, August 3, 1935, MPIP-HA: NLR 3.

148. Karl Binding and Alfred Hoche, *Die Freigabe der Vernichtung lebensunwerten
Lebens: Ihr Maß und ihre Form* (Leipzig: Meiner, 1920). Rüdin uses the term
"ballast" in a public talk sponsored by the Kaiser Wilhelm Society in 1935,
MPG-Archiv, Abt. I, Rep. 1A, Nr. 836, pp. 418–19. It was employed in Binding and Hoche's publication.

149. "Das Ethos der Rassenhygiene: Eine Unterredung mit Prof. Rüdin," July 25,
1935, *Sonderdienst der Wohlfahrts-Korrespondenz* in MPIP-HA: NLR 9.

150. Von Braunmühl to Rüdin, January 24, 1938, Nachlaß Ernst Rüdin, MPG-
Archiv, ZA 131 Rüdin, Box 3. For a fuller discussion of von Braunmühl and
insulin shock therapy, see Burleigh, *Death and Deliverance,* pp. 84–90.

151. Roelcke, "Psychiatrische Wissenschaft," pp. 131–32; Weber, *Ernst
Rüdin,* pp. 274–75; Burleigh, *Death and Deliverance,* pp. 84–85; Roemer to
Rüdin, September 16, 1941, MPIP-HA: NLR 9; Luxenburger to Holl,
October 31, 1945, MPG-Archiv, Abt. II, Rep. 1A, Rüdin Personalia,
Nr. 22.

152. Several postwar written and oral testimonies allege that Rüdin was against
the killings. Several state that the Munich director explicitly called them
"murder." Weber, *Ernst Rüdin,* p. 274; Luxenburger to Holl, October 31,
1945; Barlen to Rüdin, October 9, 1945, MPG-Archiv, Abt. II, Rep. 1A,
Rüdin Personalia, Nr. 22; Schulz to Forest, April 15, 1947; Willibald Scholz,
"Gutachterliche Äusserung über Professor Dr. med. Rüdin, April 24, 1947,
MPIP-HA: NLR 5.

153. For a discussion of Schneider's research program and Deussen's involvement in it, see Volker Roelcke, Gerrit Hohendorf, and Mike Rotzoll, "Psychiatric Research and 'Euthanasia': The Case of the Psychiatric Department at the University of Heidelberg, 1941–1945," *History of Psychiatry* 5 (1994): 517–32.

154. "Lebenslauf," Personal Akt Deussen, Universitätsarchiv Heidelberg (UAHD). I would like to thank Volker Roelcke for making these key Deussen documents available to me.

155. Rüdin to Achelis, August 28, 1944; Rüdin to Achelis, February 14, 1945, UAHD; Roelcke, "Psychiatrische Wissenschaft," pp. 138–41.

156. Rüdin to den Reichsgesundheitsführer, October 23, 1942, MPIP-HA: 129. I would like to thank Matthias Weber for making this document available to me. See, also, Roelcke, pp. 131 and 136; Weber, *Ernst Rüdin*, p. 279.

157. Quoted from Roelcke, Hohendorf, and Rotzoll, "Psychiatric Research," p. 522.

158. Ibid.

159. Rüdin to Nitsche, January 8, 1944, MPIP-HA: GDA 131.

160. Jürgen Pfeiffer, "Neuropathologische Forschung an 'Euthanasie'-Opfern in zwei Kaiser-Wilhelm-Instituten," in *Geschichte der Kaiser-Wilhelm-Gesellschaft im Nationalsozialismus. Bestandsaufnahme und Perspektiven der Forschung,* Bd. 1, ed. Doris Kaufmann (Göttingen: Wallstein, 2000), pp. 150–73.

161. For a discussion of this movement in psychiatry to compare clinical observations with histopathological data derived from dissections, see Roelcke, "Psychiatrische Wissenschaft," pp. 132–38.

CHAPTER FOUR

1. Otmar Freiherr von Verschuer, "Rassenhygiene als Wissenschaft und Staatsaufgabe," *Frankfurt: Akademische Reden Nr. 7* (Frankfurt: H. Bechold, 1936), pp. 8–9.

2. Bernhard vom Brocke, "Die Kaiser-Wilhelm-Gesellschaft im Kaiserreich: Vorgeschichte, Gründung und Entwicklung bis zum Ausbruch des Ersten Weltkriegs," in *Forschung im Spannungsfeld von Politik und Gesellschaft,* ed. Rudolf Vierhaus und Bernhard vom Brocke (Stuttgart: Deutsche Verlagsanstalt, 1990), pp. 160–61.

3. Bernhard vom Brocke, "Die Kaiser-Wilhelm-Gesellschaft in der Weimarer Republik: Ausbau zu einer gesamtdeutschen Forschungsorganisation," in *Forschung im Spannungsfeld von Politik und Gesellschaft,* pp. 318–29; "A Short History of the Harnack-Haus," Visualization and Mathematics 2002 (accessed November 2, 2004). Available at http://www-sfb288.math.tu-berlin.de/vismath/location/Harnackhaus_eng.html.

4. Vom Brocke, "Die Kaiser-Wilhelm-Gesellschaft im Kaiserreich.," p. 161; Archiv-MPG, Abt. I, Rep. 1A, Nr. 832, pp. 178, 230–33, 243.

5. For a discussion of the most recent evaluation of the KWS during the Third Reich, see Rüdiger Hachtmann, *Wissenschaftsmanagement im "Dritten Reich": Geschichte der Generalverwaltung der Kaiser-Wilhelm-Gesellschaft*, 2 Bde. (Göttingen: Wallstein, 2007).

6. Archiv-MPG, Abt. I, Rep. 1A, Nr. 836, pp. 322–23.

7. Hachtmann, *Wissenschaftsmanagement*.

8. Archiv-MPG, Abt. I, Rep. 1A, Nr. 808, p. 145.

9. Archiv-MPG, Abt. I, Rep. 1A, Nr. 833, p. 393; Nr. 834, pp. 20–21. Quote on Nr. 833, p. 393.

10. Archiv-MPG, Abt. I, Rep. 1A, Nr. 834, pp. 33–38. First quote on Nr. 834, p. 36; second quote on Nr. 834, p. 38.

11. Archiv-MPG, Abt. I, Rep. 1A, Nr. 834, p. 21.

12. Archiv-MPG, Abt. I, Rep. 1A, Nr. 834, p. 118; Nr. 833, p. 395.

13. Archiv-MPG, Abt. I, Rep. 1A, Nr. 834, p. 118.

14. Archiv-MPG, Abt. I, Rep. 1A, Nr. 834, p. 90. These quotes are taken from an advance press release of von Verschuer's talk. It almost certainly reflects an accurate summary of the actual talk, since it was written by von Verschuer himself.

15. Archiv-MPG, Abt. I, Rep. 1A, Nr. 834, p. 78.

16. Archiv-MPG, Abt. I, Rep. 1A, Nr. 836, pp. 402–3.

17. Archiv-MPG, Abt. I, Rep. 1A, Nr. 836, p. 417.

18. Archiv-MPG, Abt. I, Rep. 1A, Nr. 836, pp. 418–19.

19. Archiv-MPG, Abt. I, Rep. 1A, Nr. 809, p. 228.

20. Archiv-MPG, Abt. I, Rep. 1A, Nr. 810, pp. 408–12.

21. Archiv-MPG, Abt. I, Rep. 1A, Nr. 810, p. 413.

22. Lenz to Planck, July 29, 1935; Planck to Lenz, July 30, 1935. Archiv-MPG, Abt. I, Rep. 1A, Nr. 837, n.p.

23. Archiv-MPG, Abt. I, Rep. 1A, Nr. 838, p. 157.

24. Archiv-MPG, Abt. I, Rep. 1A, Nr. 838, pp. 163–66.

25. Archiv-MPG, Abt. I, Rep. 1A, Nr. 838, pp. 161–62.

26. Archiv-MPG, Abt. I, Rep 1A, Nr. 838, pp. 165–66, 169, 173.

27. Archiv-MPG, Abt. I, Rep 1A, Nr. 838, pp. 166, 185. There is an actual drawing of the seating arrangements for the dinner in the file.

28. Archiv-MPG, Abt. I, Rep. 1A, Nr. 838, pp. 197–98, 181, 200. Quotes on p. 198.

29. Archiv-MPG, Abt. I, Rep. 1A, Nr. 838, p. 198.

30. Ibid.

31. Archiv-MPG, Abt. I, Rep. 1A, Nr. 838, p. 58.

32. Archiv-MPG, Abt. I, Rep. 1A, Nr. 838, pp. 156, 160, 169, 269.

33. Archiv-MPG, Abt. I, Rep. 1A, Nr. 838, pp. 241–42.

34. Archiv-MPG, Abt. I, Rep. 1A, Nr. 838, p. 275.

35. Archiv-MPG, Abt. I, Rep. 1A, Nr. 838, pp. 202–3, 206, 231, 233, 234.

36. Memo from the Reichserziehungsministerium to the KWS, April 9, 1941, Archiv-MPG, Abt. I, Rep. 1A, Nr. 822, n.p.

37. Memo from the Office of the Reichsführer SS, October 22, 1941, Archiv-MPG, Abt. I, Rep. 1A, Nr. 822, n.p.
38. Eugen Fischer to Ernst Telschow, July 12, 1941, Archiv-MPG, Abt. 1, Rep. 1A, Nr. 822, n.p.
39. MPIP-HA, GDA 32.
40. Niels C. Lösch, *Rasse als Konstrukt: Leben und Werk Eugen Fischers* (Frankfurt am Main: Peter Lang, 1997), pp. 517–21.
41. Josef Mengele, "Tagung der Deutschen Gesellschaft für physische Anthropologie," *Der Erbarzt*, Nr. 10 (1937), pp. 140–41.
42. Der Reichs- und Preußische Minister für Wissenschaft, Erziehung und Volksbildung an den Herrn Präsidenten der Kaiser-Wilhelm-Gesellschaft, May 22, 1935, Archiv-MPG, Abt. I, Rep. 1A, Nr. 1052, p. no. illegible; Reichs- und Preußischer Minister des Innern an den Herrn Reichs- und Preußischen Minister für Wissenschaft, Erziehung und Volksbildung, April 4, 1937; BAB R 4901 2760/46.
43. "Richtlinien für die Leiter Deutscher Abordnungen zu Kongressen im Ausland," Politisches Archiv des Auswärtigen Amts (PAAA), Budapest 178/2; Jahresbericht der Deutschen Kongress-Zentrale, 1938, PAAA, Budapest 178/1, 40.
44. "Richtlinien für die Leiter Deutscher Abordnungen," PAAA, Budapest 178/1; Ibid., pp. 2, 3 and 6 (emphasis in the original on p. 3).
45. Ibid., pp. 6, 12–13.
46. Ibid., pp. 26–29.
47. "Ratschläge für Auslandsreisende, Anlage 2," PAAA, Budapest 178.
48. Eugen Fischer to Bernhard Rust, May 13, 1934, BAB R 4901 12384/2.
49. Archiv-MPG, Abt. I, Rep. 1A, Nr. 1051, p. 142; Walter Gross to the Reichsministerium des Innern, June 13, 1934, BAB R 1501 126245/244.
50. Ibid., p. 266. For a discussion of the denunciation campaign, see chapter 2.
51. Bericht über die Internationale Tagung der Anthropologischen und Ethnologischen Wissenschaften, August 5, 1934, Archiv-MPG, Abt. I, Rep. 1A, Nr. 1051, pp. 184–85.
52. Ibid., pp. 184–85.
53. Quote taken from Lösch, *Rasse als Konstrukt*, p. 267.
54. Quoted from Stefan Kühl, *Die Internationale der Rassisten: Aufstieg und Niedergang der internationalen Bewegung für Eugenik und Rassenhygiene im 20. Jahrhundert* (Frankfurt am Main: Campus, 1997), p. 131.
55. Hans Harmsen und Franz Lohse, eds., *Bevölkerungsfragen: Bericht des Internationalen Kongresses für Bevölkerungswissenschaft, Berlin, 26. August–1.September 1935* (Munich: J. F. Lehmann, 1936), p. 43. Emphasis in the original.
56. Eröffnungsansprache des Ehrenpräsidenten des Kongresses, Reichsminister Dr. Frick, *Bevölkerungsfragen*, pp. 6–12. First quote on p. 7; second quote on p. 12.
57. Ernst Rüdin, "Was ist Rassenhygiene," August 3, 1935, MPIP-HA, NLR 3; "Die Rassenhygieniker gegen die Kriegshetze," August 2, 1934, MPIP-HA, NLR 3.

58. Eugen Fischer to Herr Reichserziehungsminister Rust, Internationaler Bev-
ölkerungskongress in Paris, February 13, 1937, BAB R 4901 2760/5.

59. Ibid.; "Richtlinien," PAAA 178/1 Budapest, p. 10.

60. Fischer to von Verschuer, February 6, 1937, Nachlass von Verschuer,
Archiv-MPG, Abt. III, Rep. 86A, Nr. 291–93.

61. Eugen Fischer to Herr Reichserziehungsminister Rust, Bevölkerungskon-
gress in Paris, June 15, 1937, BAB R 4901 2750/99. Handwritten note
regarding Rüdin's nomination as substitute delegation leader was probably
written by Rust.

62. Kühl, *Die Internationale der Rassisten,* pp. 145–52; MPI-HA GDA 41. For
those who can read German and want a fuller discussion of Boas's position, see Doris Kaufmann, "'Rasse und Kultur': Die amerikanische Kul-
turanthropologie um Franz Boas (1858–1942) in der ersten Hälfte des 20.
Jahrhunderts—ein Gegenentwurf zur Rassenforschung in Deutschland," in
Rassenforschung an Kaiser-Wilhelm-Instituten vor und nach 1933, ed. Hans-
Walter Schmuhl (Göttingen: Wallstein, 2003), pp. 309–27.

63. Bericht über die Reise nach Paris zur Teilnahme an dem Internationalen
Kongreß für Bevölkerungswissenschaft, Universitätsarchiv Frankfurt, Akten
des Rektors (UFAR), Abt. 1, Nr. 47 (von Verschuer), p. 20; Rodenwaldt
quote from Kühl, *Die Internationale der Rassisten,* p. 150.

64. Fischer and von Verschuer often use the German term *"andersartig"* rather
than *"minderwertig"* to describe Jews, as the latter would not be considered
"scientific." See, for example, Fischer's use of the term *"andersartig"* in his
talk given in the Harnack House on February 1, 1933, where he describes
the Jews in this manner. MPG-Archiv, Abt. I, Rep. 1A, Nr. 808, pp. 93–96.

65. Bericht über die Reise nach Paris zur Teilnahme an dem Internationalen
Kongreß für Bevölkerungswissenschaft. UFAR, Abt. 1, Nr. 47 (von Verschuer), p. 20.

66. Bericht über die Pariser Kongresse 1937, MPIP-HA: GD 41/8; "Eugenik der
Geistesstörung" Vortrag zum Bevölkerungswissenschaftlichen Kongress,
Paris 1937, MPIP-HA: GDA 42/11–12 (first quote), 9 (second quote).

67. Bericht über die Pariser Kongresse 1937, MPIP-HA: GDA 41/8–9; Bericht
über die Pariser Kongresse 1937 von Professor Rüdin, November 1, 1937,
BAB R 4901 2750/198–99. Quote on p. 198: Rüdin an den Generaldirek-
tor der Kaiser-Wilhelm-Gesellschaft, February 2, 1939, Archiv-MPG, Abt. I,
Rep. 1A, Nr. 1062, p. 77a.

68. Rüdin to the Leitung der Auslandsorganisation der NSDAP, September
1937, MPIP-HA: GDA 40; quote found in Rüdin to die Deutsche Kon-
gresszentrale, January 3, 1937, MPIP-HA: GDA 40.

69. Bericht über die Pariser Kongresse 1937, MPIP-HA: GDA 41/1–5. Quote on
p. 4.

70. "Bedingungen und Rolle der Eugenik in der Prophylaxe der Geistesstörun-
gen," MPIP-HA: GDA 40; quote on p. 10.

71. To Herr Minister für Wissenschaft, Erziehung und Volksbildung durch den Herrn Präsidenten der Kaiser-Wilhelm-Gesellschaft, September 9, 1937, Archiv-MPG, Abt. I, Rep. 1A, Nr. 1057, p. 953; Ute Deichmann, *Biologen unter Hitler: Vertreibung, Karrieren, Forschung* (Frankfurt am Main: Campus, 1992), p. 79. Von Wettstein to the Reichsministerium für Wissenschaft, Erziehung und Volksbildung, July 27, 1938, MPG-Archiv, Abt. I, Rep. 1A, Nr. 1059, pp. 9–12; quote on pp. 9–10. Emphasis in the original.

72. Von Wettstein to the Reichsministerium für Wissenschaft, Erziehung und Volksbildung, July 27, 1938, MPG-Archiv, Abt. I, Rep. 1A, Nr. 1059, pp. 9–12; quote on pp. 9–10. Emphasis in the original; second quote on pp. 10–11. Emphasis in the original.

73. Ibid., pp. 11–12.

74. Deichmann, *Biologen unter Hitler,* pp. 172–73; Bernd Gausemeier, "Mit Netzwerk und doppeltem Boden: Die Botanische Forschung am Kaiser-Wilhelm-Institut für Biologie und die nationalsozialische Wissenschaftspolitik," in *Autarkie und Ostexpansion. Pflanzenzucht und Agrarforschung im Nationalsozialismus,* ed. Susanne Heim (Göttingen: Wallstein, 2002), pp. 180–205. The term *"patriotische Angelegenheit"* (translated here as "a matter of patriotic pride") is taken from Gausemeier, p. 183.

75. To the Ministerium für Wissenschaft, Erziehung und Volksbildung, July 27, 1938, MPG-Archiv, Abt. I, Rep. 1A, Nr. 1059, pp. 12–13. First quote on p. 12; second quote on p. 13

76. Ibid.

77. One can deduce von Wettstein's national-conservative outlook from some of his speeches. See, for example, the talk he held at his colleague Carl Correns's death in Gausemeier, "Mit Netzwerk und doppeltem Boden," p. 183.

78. To the Ministerium für Wissenschaft, Erziehung und Volksbildung, July 27, 1938, MPG-Archiv, Abt I, Rep. 1A, Nr. 1059, p. 13.

79. Ibid.

80. Ibid., pp. 13–14; second and third quotes on p. 14.

81. For a discussion on this point, see Susanne Heim, *Kalorien, Kautschuk, Karrieren: Pflanzenzüchtung und landwirtschaftliche Forschung in Kaiser-Wilhelm-Instituten, 1933–1945* (Göttingen: Wallstein, 2003); Susanne Heim, *Research for Autarky: The Contribution of Scientists to Nazi Rule in Germany* (Preprint no. 4 from the Research Program "History of the Kaiser Wilhelm Society in the National Socialist Era," 2001).

82. The lecture was published in *Proceedings of the Royal Society of London* 128 (1939): 62–68. Bericht über die vom 7.-14.6.39, durchgeführte Reise nach London, UFAR, Abt. 1, Nr. 47 (von Verschuer), p. 37; Fischer to von Verschuer, March 9, 1939, Archiv-MPG, Abt. III, Rep. 86A, Nr. 291–93.

83. For a discussion of these changes, see Hans-Jürgen Döscher, *SS und Auswärtiges Amt im Dritten Reich: Diplomatie im Schatten der "Endlösung"* (Berlin: Wolf Jobst Siedler, 1991).

84. Deutsche Kulturpropaganda im Ausland, September, 1939, Archiv-MPG, Abt. I, Rep. 1A, Nr. 1065, n.p.

85. Internationale Wissenschaftliche Verbände, Institute, Vereinigungen, usw., November 12, 191940, AB R 4901 3190/131–34. (It is not clear who authored this document).

86. Fühlungsnahme deutscher Wissenschaftler und Hochschullehrer mit politischen Wissenschaftlern, July 2, 1941, Archiv-MPG, Abt. I, Rep. 1A, Nr. 1067, n.p.

87. Archiv der Stiftung für Sozialgeschichte des 20. Jahrhunderts (SfS), Bremen, Bestand von Verschuer, Nr. 1; Korrespondenz zu Vorträgen, 1939–42, photocopy of the poster announcing von Verschuer's Vortrag, n.p. I am grateful to Karl-Heinz Roth for making this material available to me. I would also like to thank Mark Walker for pointing out the comparison to Werner Heisenberg.

88. UFAR, Abt. 1, Nr. 47 (Von Verschuer), p. 41. On the origins of the Bibliotheca Hertziana, see Bernhard vom Brocke and Hubert Laitko, eds., *Die Kaiser-Wilhelm-/Max-Planck-Gesellschaft und ihre Institute: Das Harnack-Prinzip* (Berlin and New York: Walter de Gruyter, 1996) p. 633.

89. Denkschrift: Die Gründung der Kulturwissenschaftlichen Abteilung der Bibliotheca Hertziana in Rom, PAAA, Rom 1322b. Quote on p. 4.

90. Bericht über die Vortragsreise nach Rom vom 10.- 21. 2. 40, UFAR, Abt. 1, Nr. 47 (von Verschuer), p. 43.

91. Reisebericht von Eugen Fischer, n.d. Archiv-MPG, Abt. I, Rep. 1A, Nr. 1067, n.p.

92. For a discussion of Heisenberg as a "goodwill ambassador," see Mark Walker, *Nazi Science: Myth, Truth and the German Atomic Bomb* (New York: Plenum Press, 1995), chap. 7.

93. For a discussion of the importance of the *Deutsches Institut* in Paris, see Frank-Rutger Hausmann, *"Auch im Krieg schweigen die Musen nicht": Die Deutschen Wissenschaftlichen Institute im Zweiten Weltkrieg,* 2nd ed. (Göttingen, Vandenhoeck & Ruprecht, 2002), pp. 100–30.

94. UFAR, Abt. 1, Nr. 47 (von Verschuer), p. 46.

95. Vortragsreise nach Paris betreffend, December 10, 1941, Archiv-MPG, Abt. I, Rep. 1A, Nr. 1067, n.p.

96. For a short yet useful discussion of the Wannsee Conference, see Doris L. Bergen, *War & Genocide: A Concise History of the Holocaust* (New York: Rowman & Littlefield, 2003), p. 159. Martin Gilbert gives 165,000 as the figure for the number of Jews in occupied France slated for extermination at the Wannsee Conference, in *The Routledge Atlas of the Holocaust,* 3rd ed. (London: Routledge, 2002), p. 85.

97. Eugen Fischer and Gerhard Kittel, *Das antike Weltjudentum. Tatsachen, Texte, Bilder,* Schriften des Reichsinstituts für Geschichte des neuen Deutschlands, Forschungen zur Judenfrage, Bd. 7 (Hamburg: Hanseatische Verlagsanstalt Aktiengesellschaft, 1943).

98. Etat et Santé, eds., *Cahiers de L'Institut Allemand* (Paris: Karl Epting, n.d.).
99. Ibid., pp. 85–110. Quote on p. 106.
100. Vortragsreise nach Paris betreffend, 10.12.41, Archiv-MPG, Abt. I, Rep. 1A, Nr. 1067, n.p. It is not clear who Fischer originally addressed this report to, although it was probably Rust.

CHAPTER FIVE

1. An exact copy of the decree was reported in Ernst Dobers and Kurt Higelke, eds., *Rassenpolitische Unterrichtspraxis: Der Rassengedanke in der Unterrichtsgestaltung der Volksschulfächer,* 3rd ed. (Leipzig: Klinkhardt, 1940), p. 354.
2. *Deutschabituraufsätze* of F.G. and P. K., Bismarck-Gymnasium, Berlin-Wilmersdorf, January 26, 1934, "Biologische Grundlagen der völkischen Rassenpflege," Bismarck-Gymnasium, 1934–35; Archiv und Gutachterstelle für Deutsches Schul- und Studienwesen (ADS).
3. Gertrud Scherf, "Vom deutschen Wald zum deutschen Volk: Biologieunterricht in der Volksschule im Dienste nationalsozialistischer Weltanschauung und Politik," in *Schule und Unterricht im Dritten Reich,* ed. Reinhard Dithmar (Neuwied: Hermann Luchterhand, 1989), p. 217. Scherf maintains that 85–90 percent of all school children only attended an eight-year elementary school; Peter Lundgreen, *Sozialgeschichte der deutschen Schule im Überblick, Teil II: 1918–1980* (Göttingen: Vandenhoeck & Ruprecht, 1981), pp. 110–11. Lundgreen offers statistics that suggest that the number of children attending secondary schools remained fairly constant during the years 1828–1937. According to Lundgreen, between 7 and 8 percent of school age children attended a secondary school; more boys attended than did girls. Not all, however, went to a college-preparatory higher school. Some went to a so-called middle school (*Mittelschule)* that ended in the tenth grade. The higher secondary school education required thirteen grades until the reform during the Third Reich in 1938. At that time it was reduced to twelve grades.
4. "Biologisches Schulungslager des NSLB, Gau Württemberg in Verbindung mit der Universität Tübingen," *Der Biologe* 4 (1935): 359; Ernst Lehmann, biology professor at the University of Tübingen and Nazi sympathizer, quotes Schemm; Fritz Donath and Karl Zimmermann, *Biologie, Nationalsozialismus und neue Erziehung,* 2nd ed. (Leipzig: Quelle and Meyer, 1933), p. 11. Examples of Hitler's proclamations on these subjects in *Mein Kampf* are located in a source book, "Hitler's *Mein Kampf:* Nation and Race," in *Inside Hitler's Germany: A Documentary History of Life in the Third Reich,* ed. Benjamin Sax and Dieter Kuntz (Lexington, MA: D.C. Heath and Company, 1992), pp. 189–95.
5. Hermann Merkle, "Rassen- und Bevölkerungskunde in UII und OI der höheren Lehranstalten," *Unterrichtsblätter für Mathematik und Naturwissenschaften* 39 (1933): 352.

6. There is a very large secondary literature as well as oral testimonies on the impact of Nazi education on the youth and the function of such education for the National Socialist state. Among the more important general secondary works are Dithmar, ed., *Schule und Unterricht im Dritten Reich*; Kurt-Ingo Flessau, *Schule der Diktatur: Lehrpläne und Schulbücher des Nationalsozialismus* (Frankfurt am Main: Fischer, 1979); Reiner Lehberger and Hans-Peter de Lorent, eds., *"Die Fahne hoch" Schulpolitik und Schulalltag in Hamburg unterm Hakenkreuz* (Hamburg: Ergebnisse, 1986); Karl-Christoph Lingelbach, *Erziehung und Erziehungstheorien im nationalsozialistischen Deutschland* (Frankfurt am Main: Dipa, 1987); Hans-Jochen Gamm, *Führung und Verführung: Pädagogik des Nationalsozialismus*, 3rd ed. (Munich: Paul List, 1990); Harald Scholtz, *Erziehung und Unterricht unterm Hakenkreuz* (Göttingen: Vandenhoeck & Ruprecht, 1985); Barbara Schneider, *Die Höhere Schule im Nationalsozialismus: Zur Ideologisierung von Bildung und Erziehung* (Cologne: Böhlau, 2000); Gregory Paul Wegner, *Anti-Semitism and Schooling under the Third Reich* (New York: Routledge Falmer, 2002). For biology education specifically, see Änne Bäumer-Schleinkofer, *NS-Biologie und Schule* (Frankfurt am Main: Peter Lang, 1992); Detlev Franz, "Biologismus von oben: Das Menschenbild in Biologiebüchern" (Diss., Duisburg, n.d.); Sheila Faith Weiss, "Pedagogy, Professionalism and Politics: Biology Instruction during the Third Reich," in *Science, Technology and National Socialism*, ed. Monika Renneberg and Mark Walker (Cambridge: Cambridge University Press, 1994), pp. 184–96 and 377–85. For oral testimonials regarding the impact of Nazi education on those who attended schools at that time, see Geert Platner, ed., *Schule im Dritten Reich: Erziehung zum Tod? Eine Dokumentation* (Nördlingen: Deutscher Taschenbuch Verlag, 1983); Harald Focke and Uwe Reimer, eds., *Alltag unterm Hakenkreuz: Wie die Nazis das Leben der Deutschen veränderten*, Bd. 1 (Hamburg: Rowohlt, 1979); Johannes Leeb, *"Wir waren Hitlers Eliteschüler" Ehemalige Zöglinge der NS-Ausleseschulen brechen ihr Schweigen*, 2nd ed. (Munich: Wilhelm Heyne, 2000); Arbeitsgruppe Pädagogisches Museum, ed., *Heil Hitler, Herr Lehrer: Volksschule, 1933–1945* (Hamburg: Rowohlt, 1983).

7. Hans-Walter Schmuhl, *Grenzüberschreitungen: Das Kaiser-Wilhelm-Institut für Anthropology, menschliche Erblehre und Eugenik, 1927–1945* (Göttingen: Wallstein, 2005), p. 224.

8. Hermann Eilers, *Die nationalsozialistische Schulpolitik: Eine Studie zur Funktion der Erziehung im totalitären Staat* (Opladen: Westdeutscher Verlag), p. 128.

9. The Cabinet Order and the Proceedings of the School Conference where Wilhelm II made these remarks are reprinted in *Deutsche Schulkonferenzen*, Bd. 1, Verhandlungen über Fragen des höheren Unterrichts, Berlin, December 4–17, 1890 (Glashütten im Taunus: Auvermann, 1972), pp. 4–13 and 72–73.

10. Lynn K. Nyhardt, "Teaching Community via Biology in Late-Nineteenth Century Germany," *Osiris 17* (2002): –70. The notion of *Lebensgemeinschaft*

was extremely popular among Nazi biology pedagogues. It was pushed as a paradigm for elementary school biology and even biology instruction in the lower grades of the secondary schools. For its use in elementary schools, see Paul Brohmer, *Biologieunterricht und völkische Erziehung* (Frankfurt am Main: Diesterweg, 1933); the paradigm of *Lebensgemeinschaft* is used in the middle school biology textbook by Ernst Kruse and Paul Wiedow, *Lebenskunde für Mittelschulen*, Bd.1 (Leipzig: Teubner, 1940). The concept is also employed by Erich Meyer and Karl Zimmermann for the third and fourth classes of the higher secondary schools in *Lebenskunde: Lehrbuch der Biologie für höhere Schulen*, Bd. 2 (Erfurt: Kurt Stenger, 1943). This book was one of only four approved for use in the biology classrooms of the higher secondary schools after the 1938 school reforms.

11. Wolfgang Geiger, "Staatsbürgerliche Erziehung und Bildung in der Endphase der Weimarer Republik," in *Schule und Unterricht in der Endphase der Weimarer Republik*, ed. Reinhard Dithmar (Neuwied: Luchterhand, 1993), p. 2. S. Weiss, "Pedagogy, Professionalism and Politics," p. 186.

12. Staatsarchiv Hamburg (STA-HH) 361 2II Oberschulbehörde (OSB) II: A27 Nr. 27.

13. Jakob Graf, *Vererbungslehre und Erbgesundheitspflege* (Munich: Lehmann, 1930), p. 255. Information on Graf's Nazi Party membership was obtained from the Berlin Document Center. He became member number 1,203,598 on August 1, 1932.

14. Philipp Depdolla, "Vererbungslehre und naturwissenschaftlicher Unterricht," in *Vererbung und Erziehung*, ed. Günther Just (Berlin: Springer, 1930), p. 292: Zentralinstitut für Erziehung und Unterricht, ed., *Erblehre- Erbpflege* (Berlin: Mittler & Sohn, 1933), preface and table of contents.

15. Cäsar Schäffer, *Leitfaden der Biologie I* (Leipzig: Teubner, 1930), pp. 117–23; Cäsar Schäffer, *Einführung in die Biologie: Zum Gebrauch an Höheren Schulen und zum Selbstunterricht (Grosse Ausgabe)*, 7th ed. (Leipzig: Teubner, 1929), pp. 300–66.

16. ADS, *Biologieabituraufsätze* of W.L., G.H., A.H., G.N., B.S., and J.W., Leibniz-Oberrealschule, Berlin-Charlottenburg, January 29, 1932, "Eugenik: Ein Überblick über die Bestrebungen zur Vergütung des Erbgefüges," Leibniz-Oberrealschule, 1932.

17. Karl Oberkirsch, "Die Stellung der Biologie in der Mittelschule," *Der Biologe* 1 (1931–32): 69–70; "Mitteilungen," *Der Biologe* 1 (1931–32): 58.

18. Wilhelm Quitzow, "Das Menschenbild im Biologieunterricht- von der Evolutionstheorie zum Sozialdarwinismus," in *Schule und Unterricht in der Endphase der Weimarer Republik*, p. 237; Weiss, "Pedagogy, Professionalism and Politics," pp. 187–88; Hans-Christoph Laubach, *Die Politik des Philologenverbandes im Deutschen Reich und in Preußen während der Weimarer Republik: Die Lehrer an höheren Schulen mit Universitätsausbildung im politschen Spannungsfeld der Schulpolitik von 1918–1933* (Frankfurt am Main: Peter Lang, 1986).

19. The Hitler quote is taken from Gilmer W. Blackburn, *Education in the Third Reich: Race and History in Nazi Textbooks* (Albany: SUNY Press, 1985), p. 13.

20. Regarding the 1933 decree, see Eilers's classic study, *Die nationalsozialistische Schulpolitik*, p. 15. For Frick's views, see Wilhelm Frick, *Kampfziel der deutschen Schulen* (Langensalza: Beyer & Söhne, 1933), pp. 12–14. On *völkisch* academics see, for example, Fritz Lenz, *Über die biologischen Grundlagen der Erziehung* (Munich: Lehmann, 1925); Paul Brohmer, *Biologie* (Frankfurt am Main, Diesterweg, 1932). Some of the plans for biology school education elicited from officials can be found in BAB, R 4901, 26852. Proposals for biology education in the more overtly political "Adolf Hitler Schools" can be found in STA-HH 361 2II Oberschulbehörde (OSB) VI: F VI d 9/10.

21. Renate Fricke-Finkelnburg, *Nationalsozialismus und Schule: Amtliche Erlasse und Richtlinien, 1933–1945* (Opladen: Leske & Budrich, 1989), p. 214.

22. Rudolf Benze, *Rasse und Schule* (Braunschweig: E. Appelhaus & Co., 1934), p. 28; Ernst Anrich, *Neue Schulgestaltung aus nationalsozialistischem Denken* (Stuttgart: W. Kohlhammer, 1933), p. 43.

23. Eugen Fischer, "Volkstum und Rasse als Frage der Nation," *Monatsschrift für die höhere Schule* 32 (1933): 263; BAB NS 5 VI, 1738/104, Eugen Fischer, "*Wozu Erb- und Rassenforschung?*"; BAB NS 5 VI, 1738/106, Fritz Lenz, "*Rassenlehre ist Erblehre*"; Günther Just, "Eugenik und Erziehung," *Die Mittelschule. Zeitschrift für das gesamte mittlere Schulwesen* 47 (1933): 431–34; Nikolai Timofeeff-Ressovsky, "Drosophila im Schulversuch," *Der Biologe* 3 (1933/34): 141; Jakob Graf, "Die Familienkunde im Unterricht der höheren Schule von der Lebenskunde (Biologie) aus gesehen," *Deutsche höhere Schule* 2 (1935): 199–208.

24. Biology instructors employed in college-preparatory higher schools often used the academic literature on various facets of racial science written by researchers like Fischer, Just, and other geneticists/eugenicists. This material, of course, was not meant to be used in a classroom setting. The number of new textbooks for the graduating classes of all schools and the so-called *Ergänzungshefte* (supplementary pamphlets) to older biology texts was enormous. The following is but a small selection of the available titles: Max Schwartz and Hans Wolff, *Kurzgefaßter Lehrgang der Biologie für die Abschlußklassen* (Frankfurt am Main: Diesterweg, 1936); Arthur Hoffmann, *Vom Erbgut und von der Erbgesundheit unseres Volkes* (Erfurt: Kurt Stenger, 1936); Otto Steche, *Leitfaden der Rassenkunde und Vererbungslehre der Erbgesundheitspflege und Familienkunde für die Mittelstufe* (Leipzig: Quelle & Meyer, 1934); Otto Hermann and Werner Stachowitz, *Einführung in die Vererbungslehre, Rassenkunde und Erbgesundheitspflege für die Mittelstufe* (Frankfurt am Main: Diesterweg, 1934); Hans Feldkamp, *Vererbungslehre, Rassenkunde, Volkspflege* (Münster: Aschendorffsche Verlagsbuchhandlung, 1935); Emil Jörns and Julius Schwab, *Rassenhygienische Fibel* (Berlin: Alfred Metzner, 1933). For the *Ergänzungshefte*, see Cäsar Schäffer and Heinrich Eddelbüttel,

Erbbiologische Arbeiten: Ergänzungsheft zum Biologischen Arbeitsbuch (Leipzig: Teubner, 1934); Leopold Trinkwalter, *Einführung in die Vererbungslehre, Familienkunde, Rassenkunde und Bevölkerungspolitik: Eine Ergänzung zur 80. Auflage von Schmeil "Der Mensch" und zu allen anderen menschenkundlichen Teilen des Unterrichtswerkes* (Leipzig: Quelle & Meyer, 1934). Other text supplements are listed in Bäumer-Schleinkofer, *NS-Biologie und Schule*, pp. 147–62.

25. Hoffmann, *Vom Erbgut*, p. 31; diagram (to be filled out by pupils), on p. 6 of the supplement, *Ein erbkundliches Arbeitsheft*.

26. Feldkamp, *Vererbungslehre*, pp. 19, 22, 25, 27, 34; Otto and Stachowitz, *Einführung in die Vererbungslehre*, pp. 22–24; quotation on p. 21; Feldkamp even offers a glossary where the internationally accepted genetic terms are translated into German.

27. Rudolf Benze and Alfred Pudelko, *Rassische Erziehung als Unterrichtsgrundsatz der Fachgebiete* (Frankfurt am Main: Diesterweg, 1937), p. 12. Quotes are found in Hoffmann, *Vom Erbgut*, p. 6; Albert Bauer, *Lebenskunde für die Abschlußklassen der höheren Lehranstalten* (Berlin and Leipzig: G. Freytag, 1937), p. 179.

28. Schäffer and Eddelbüttel, *Erbbiologische Arbeiten*, pp. 26–28; the author has an example of this *Ahnentafel* from the biology textbook of Erich Thieme, *Vererbung, Rasse und Volk* (Leipzig: Teubner, 1934).

29. Dieter Rossmeissl, *"Ganz Deutschland wird zum Führer halten . . ." Zur politischen Erziehung in den Schulen des Dritten Reiches* (Frankfurt am Main: Fischer, 1985), p. 115; Michael Burleigh, *The Third Reich: A New History* (New York: Hill and Wang, 2000), p. 355.

30. Steche, *Leitfaden*, pp. 37–38.

31. Ibid., p. 38; Feldkamp, *Vererbunglehre*, pp. 75–76.

32. The four mandated sets of biology textbooks designed for use in the higher secondary schools include Jakob Graf, *Biologie für Oberschule und Gymnasium* (Munich: Lehmann, 1940); Otto Steche, Erich Stengel, and Maximilian Wagner, *Lehrbuch der Biologie für Oberschulen und Gymnasien* (Leipzig: Quelle & Meyer, 1942); Erich Meyer und Karl Zimmermann, *Lebenskunde: Lehrbuch der Biologie für höhere Schulen* (Erfurt: Kurt Stenger, 1943); Karl Kraepelin, Cäsar Schäffer, and Gustav Franke, *Das Leben* (Leipzig: Teubner, 1942). These degrading pictures of the mentally handicapped appeared in volume 4 of Graf's work, p. 428, and volume 3 of Meyer and Zimmermann's book, p. 182. The quote is taken from Steche, Stengel, and Wagner, *Lehrbuch der Biologie*, volume 3, p. 182. For a discussion of the impact of German films denigrating the handicapped, see Michael Burleigh, *Death and Deliverance: "Euthanasia in Germany c. 1900–1945* (Cambridge: Cambridge University Press, 1994), chap. 6. The "euthanasia project" is referred to as the "opening act" of the "final solution" by Henry Friedlander, *The Origins of Nazi Genocide: From Euthanasia to the Final Solution* (Chapel Hill: University of North Carolina Press, 1995), p. 22.

33. Michael Burleigh," Psychiatry, German Society and the Nazi 'Euthanasia' Program," in *The Holocaust. Origins, Implementation, Aftermath*, ed. Omer Bartov (London and New York: Routledge, 2000), p. 16.

34. For a discussion of Hans F. K. Günther's work, see Wegner, *Anti-Semitism and Schooling*, pp. 14–15; Robert N. Proctor, *Racial Hygiene: Medicine under the Nazis* (Cambridge, MA: Harvard University Press, 1988), pp. 26–27; Patrik von zur Mühlen, *Rassenideologien: Geschichte und Hintergründe* (Berlin: Dietz, 1977), pp. 143–45 and 161–64.

35. Trinkwalter, *Einführung*, pp. 30–32; Otto and Stachowitz, *Einführung in die Vererbungslehre*, 38–42; E. Meyer and W. Dittrich, *Erb und Rassenkunde*, pp. 56–85; Feldkamp, *Vererbungslehre*, pp. 50–61; Graf, *Biologie*, vol. 4, pp. 33–36.

36. See, for example, Rust's discussion of *Rassenkunde* in his 1935 decree extending the teaching of racial science to the entire Reich in Dobers und Higelke, *Rassenpolitische Unterrichtspraxis*, p. 360. He reiterated this point again in the 1938 curriculum reform for secondary school biology in *Erziehung und Unterricht in der Höheren Schule* (Berlin: Weidmannsche Verlagsbuchhandlung, 1938), p. 147. August Hagermann, Hamburg District Expert for Biology in the NLSB, warned against alienating German children in his appraisal of a curriculum proposal for the teaching of biology in an Adolf Hitler school in STA-HH 361 2II Oberschulbehörde (OSB) VI: F VI d 9/10.

37. Bauer, *Lebenskunde für die Abschlußklassen*, p. 144.

38. For a discussion of Lenz's views on race, see Sheila Faith Weiss, "Race and Class in Fritz Lenz's Eugenics," *Medizinhistorisches Journal* 27 (1992): 5–25; Feldkamp, *Vererbungslehre*, pp. 54–55.

39. Karl Bareth and Alfred Vogel, *Erblehre und Rassenkunde für die Grund- und Hauptschule* (Bühl-Baden: Konkordia, 1937), p. 74. *"Unser Führer hat es schlicht und einfach ausgesprochen: Die Rasse erkennt man an der Leistung."* Emphasis in the original.

40. Feldkamp, *Vererbungslehre*, 5th ed. (1937), p. 80.

41. Bauer, *Lebenskunde für die Abschlußklassen*, p. 155; Schwartz and Wolff, *Kurzgefaßter Lehrgang*, p. 51.

42. Steche argued that the Jews were merely "racially different" from Germans in his text, *Leitfaden der Rassenkunde*, p. 76. He declined to say they were inferior. Schwartz and Wolff were harsher in their rhetoric and included the wording of the Nuremberg Laws. See *Kurzgefaßter Lehrgang*, p. 54. Meyer and Zimmermann included a pictorial image of the Nuremberg Laws in *Lebenskunde*, p. 187.

43. On the importance of anti-Semitic education in the elementary schools, see Wegner, *Anti-Semitism and Schooling;* Graf, *Biologie*, vol. 3, p. 152; Graf, *Biologie*, vol. 4, pp. 434 and 384.

44. *Geheimes Staatsarchiv Preußischer Kulturbesitz* (GSPK), Rep. 76 VI, Sekt. 1, Gen. Z, 369/31–33 and 369/15–18; other reports of such *Lehrgänge* for biology teachers can be found in BAB, R 4901/4607.

45. BAB NS 12, 293.
46. The information and quotes from Brecht are taken from Delia Nixdorf and Gerd Nixdorf, "Politisierung und Neutralisierung der Schule in der NS-Zeit," in *Herrschaftsalltag im Dritten Reich: Studien und Texte,* ed. Hans Mommsen (Düsseldorf: Patmos Verlag, 1988), p. 232. The German title of Brecht's play is *Furcht und Elend des Dritten Reiches.*
47. Weiss, "Pedagogy, Professionalism and Politics," pp. 193–94.
48. Nixdorf and Nixdorf, "Politisierung," p. 228.
49. Ibid., p. 226.
50. For a discussion of the conflict between teachers and the Hitler Youth, see ibid., pp. 250–58, 275.
51. Weiss, "Pedagogy, Professionalism and Politics," p. 194.
52. BAB NS 12, 845, *Vierteljahresberichte des Sachgebietes Biologie der Gaue/Tätigkeitsberichte der Gausachbearbeiter für Biologie und Rassenfragen.*
53. These testimonials of prominent West Germans are collected in Platner, ed., *Schule im Dritten Reich,* pp. 34–35, 44, 88, 89.
54. Gerd Radde et al., eds., *Schulreform, Kontinuitäten und Brüche: Das Versuchsfeld Berlin-Neukölln,* vol. 1, *1912–1945* (Opladen: Leske & Budrich, 1993), pp. 376–78.
55. For evidence regarding oral exams in racial science in the Bodensee area, see Rossmeissl, *"Ganz Deutschland wird zum Führer halten . . .,"* pp. 115, 191.
56. ADS, *Reifeprüfungsprotokolle* der Cecilienschule Berlin-Wilmersdorf, 1934/1937/1943; *Reifeprüfungsprotokolle* der Viktoria-Luise-Schule Berlin-Wilmersdorf, 1934/1936, In 1939/40 one of the topics for the oral racial science exam for girls stressing home economics at the Königin-Luise-Schule in Berlin-Friedenau reads as follows: "The Responsibility of the Woman in National Socialist Germany for the Purity of the Biological Substrate [of the *Volk*]."
57. ADS, Topics for the *Reifeprüfung* in Biology for Berlin Oberrealschulen for Boys and Girls/1935–35.
58. In the Vikoria-Luise-Schule, for example, the biology exit exam topic for the Easter, 1938 was entitled "The Microscopic Organization of the Skin." ADS, *Biologieabituraufsätze,* Viktoria-Luise-Schule, February 3, 1938, *"Der mikroskopische Aufbau der Haut."*
59. ADS, *Biologieabituraufsätze,* Leibniz-Oberrealschule, February 1, 1934, "Die Mendelschen Vererbungsregeln (Mono-und Dihybrid Kreuzung)," Easter, 1934.
60. ADS, *Biologieabituraufsätze,* Viktoria-Luise-Schule, January 31, 1934, "Die Bedeutung der Drosophilaversuche für die Erblehre," Easter, 1934.
61. ADS, Topics for the *Reifeprüfung,* Staatliche Augusta-Schule, 1936/37.
62. ADS, *Biologieabituraufsätze,* Leibniz-Oberrealschule, February 17, 1938, "Der Geburtensturz in Deutschland, die Tatsachen und ihre Bedeutung für die Gegenwart und Zukunft des deutschen Volkes, die Ursachen und die ersten Schritte der Regierung zur Behebung der drohenden Gefahr," Easter, 1938; February 3, 1938, "Der rassische Aufbau des deutschen Volkes, die

Gefahren der Rassenmischung und die Maßnahmen des nationalsozialistischen Staates zum Schutz des deutschen Blutes," Easter, 1938.

63. ADS, *Deutschabituraufsätze* of F.G. and P.K, Bismarck-Gymnasium, Berlin-
Wilmersdorf, January 26, 1934, "Biologische Grundlagen der völkischen
Rassenpflege," Bismarck-Gymnasium, 1934–35.

64. Fricke-Finkelnburg, *Nationalsozialismus und Schule*, p. 214.

CHAPTER SIX

1. Daniel J. Kevles, *In the Name of Eugenics: Genetics and the Uses of Human
Heredity* (New York: Knopf, 1985), p. 112.

2. Ibid., p. 114.

3. Robert Proctor, *Racial Hygiene: Medicine under the Nazis* (Cambridge, MA,
and London: Harvard University Press, 1988), pp. 119–30.

4. Gerhard Baader et al., "Pathways to Human Experimentation, 1933–1945:
Germany, Japan and the United States," *Osiris* 20 (2005): 112 and 230–31.

5. Kevles, *In the Name of Eugenics*, pp. 171–72; Diane B. Paul, *Controlling Human Heredity, 1865 to the Present* (Atlantic Highlands, NJ: Humanities Press,
1995), p. 120.

6. H. S. Jennings, *The Biological Basis of Human Nature* (New York: Norton,
1930), pp. 249–50.

7. Elof Axel Carlson, *The Unfit: A History of a Bad Idea* (Cold Spring Harbor,
NY: Cold Spring Harbor Press, 2001). pp. 347–48; the quote was taken from
the abstract of his speech, http://adsabs.harvard.edu/abs/1933SciMo. .37 . . .
40M.

8. Carlson, *The Unfit*, p. 349. All quotes cited from Carlson.

9. Kevles, *In the Name of Eugenics*, pp. 122–23.

10. Lancelot Hogben, *The Nature of Living Matter* (New York: Alfred A. Knopf,
1931), pp. 207–8.

11. Ibid., p. 215.

12. Quote taken from Diane Paul, "Eugenics and the Left," *Journal of the History
of Ideas* 45 (1984): 571.

13. Kevles, *In the Name of Eugenics*, pp. 181–84.

14. Nils Roll-Hansen, "Norwegian Eugenics: Sterilization as Social Reform," in
*Eugenics and the Welfare State: Sterilization Policy in Denmark, Sweden, Norway
and Finland*, ed. Gunnar Broberg and Nils Roll-Hansen (East Lansing: Michigan State University Press, 1996), pp. 158–67.

15. Gunnar Broberg, "Eugenics in Sweden: Efficient Care," in *Eugenics and the
Welfare State*, pp. 91–95.

16. A copy of L. C. Dunn's letter to John Merriam, July 3, 1935, can be found
in C-2-3:7, the Laughlin Papers. See, also, Kevles, *In the Name of Eugenics*, p.
355n18.

17. H. J. Muller to Robert A. Lambert, June 7, 1933, RAC folder 46, box 4, series
1.1, RG, 6.1.

18. Julian S. Huxley and A. C. Haddon, *We Europeans: A Survey of "Racial" Problems* (London: Harper and Brothers, 1936), pp. vii and 49

19. J. B. S. Haldane, *Heredity and Politics* (New York: W. W. Norton, 1938), p. 97.

20. Ibid., p. 154.

21. Gunnar Dahlberg, *Race, Reason and Rubbish,* trans. Lancelot Hogben (London: George Allen and Unwin, 1942). The first quote appears on the inside title page.

22. Ibid., p. 225; for his general treatment of the Jews, see chap. XII.

23. Ibid., first quote on p. 232; second quote on p. 230.

24. Victoria F. Nourse, *In Reckless Hands: Skinner v. Oklahoma and the Near Triumph of American Eugenics* (New York: W. W. Norton and Company, 2008), pp. 129–30. Quote taken from p. 130.

25. Frederick Osborn, *Preface to Eugenics* (New York: Harper and Brothers, 1940), pp. 295, 297, and 299.

26. Ibid., pp. 22, 169.

27. Frederick Osborn to Harry H. Laughlin, June 17, 1936, C-2-2:6, the Laughlin Papers.

28. Ibid., June 14, 1936.

29. Otmar von Verschuer to Paul Popenoe, August 28, 1946, Archiv-MPG, III, Rep. 86A, Nr. 551.

30. http://www.pioneerfund.org/; Stefan Kühl, *The Nazi Connection: Eugenics, American Racism and German National Socialism* (New York: Oxford University Press, 1994), pp. 5–10; quote on p. 6.

31. Harry H. Laughlin, "Answer to Jennings in Reference to the Futility of Eugenical Sterilization," D-2-2:24; "Race Conditions: The United States, 1939," B:5:1B:4; "Proposed Plans for a Study of the Family Life of Army Aviators," D-2-3:14, Laughlin Papers.

32. Quote taken from Kühl, *The Nazi Connection,* p. 37.

33. C. M. Goethe, January 12, 1936, http//www.eugenicsarchive.org/ID1038.

34. C. M. Goethe, October 27, 1937, http//www.eugenicsarchive.org/ID1035.

35. Harry H. Laughlin to Eugen Fischer, July 31, 1935, C-4-4:7, Laughlin Papers.

36. Harry H. Laughlin to Eugen Fischer, July 31, 1935; Harry H. Laughlin to Colonel W. P. Draper, W. P. Draper to Harry H. Laughlin, August 15, 1935; August 28, 1935, D-2-3:14, Laughlin Papers.

37. Kühl, *The Nazi Connection,* pp. 34–35.

38. "U.S. Eugenicist Hails Nazi Racial Policy," *New York Times,* August 29, 1935; Waldemar Kaempffert to Harry H. Laughlin, October 15, 1935; J. H. Landman to Harry. H. Laughlin, September 13, 1935, D-2-3:5, Laughlin Papers.

39. See, for example, "Eugenical Sterilization in Germany," *Eugenical News* 18 (1933): 93; Clarence G. Campbell, "The German Racial Policy," *Eugenicial News* 21 (1936): 25–29; "Seeking 'Race Purity' in Germany," and "Verschuer's Institute," *Eugenicial News* 21 (1936): 58–59; Marie E. Kopp, "The German Program of Marriage Promotion through State Loan," *Eugenical News* 21 (1936): 121–29.

40. C. Thomalia "The Sterilization Law in Germany," trans. Alice Hellmer, *Eugenical News* 19 (1934): 137.
41. Robert Cook, "A Year of German Sterilization," *Journal of Heredity* 26 (1935): 489.
42. Hilda von Hellmer Wullen, "Eugenics in Other Lands: A Survey of Recent Developments," *Journal of Heredity* 28 (1937): 269–70.
43. L. H. Snyder, "A New Edition of Baur-Fischer-Lenz," *Journal of Heredity* 27 (1936): 310.
44. "A Letter from Dr. Ploetz," *Eugenical News* 19 (1934): 129.
45. Franz Kallmann, "Hereditary Reproduction and the Eugenic Procedure in the Field of Schizophrenia," *Eugenical News* 23 (1938): 105.
46. "The Hereditary Aspect of Pathology, *Eugenical News* 21 (1936): 21–22; the *Journal of Heredity* reviewed von Verschuer's first coauthored book, *Zwillingstuberkulose,* published in 1933. Lawrence H. Snyder, "The Inheritance of Tuberculosis," *Journal of Heredity* 25 (1934): 26–27.
47. Harry H. Laughlin to Colonel W. P. Draper, March 15, 1937, D-2-3:14, Laughlin Papers.
48. Harry H. Laughlin, "Eugenics in Germany," *Eugenicial News* 22 (1937): 65–66; see also Kühl's assessment in *The Nazi Connection,* pp. 48–50.
49. "Eugenics in Germany: Motion Picture Showing How Germany Is Presenting and Attacking Her Problems in Applied Eugenics," C-2-3:2, Laughlin Papers.
50. " Ehrenpromotionen" Medizinische Fakultät, Universitätsarchiv Heidelberg (UAHD), B-1523/7b.
51. "Bericht des vorläufigen Arbeitsausschusses an den Rektor über die grundsätzlichen Fragen der 550 Jahrfeier," UAHD, B-1523/4.
52. "Das Ausland und die Heidelberger Universität," Sonderbeilage der *Heidelberger Neuesten Nachrichten,* n.d., UAHD, B-1812/119.
53. The Generalkonsulat New York to the Deutsche Botschaft, Washington, DC, December 15, 1936, UAHD, B-1523/7b; http//www.eugenicsarchive .org ID# 1221.
54. Otmar von Verschuer to Harry Laughlin, September 2, 1936, Archiv-MPG, III, Rep. 86A, Nr. 457.
55. Kühl, *The Nazi Connection,* p. 85.
56. Madison Grant, *Die Eroberung eines Kontinents: Die Verbreitung der Rassen in Amerika,* trans. Else Mez (Berlin: Alfred Metzner, 1937), p. vii; Osborn's quote is on p. xi. Kühl states that it was Grant's 1916 book, the 1925 translated version of *The Passing of the Great Race,* that contained Fischer's foreword. See a detailed discussion of the Fischer-Grant connection in Jonathan Peter Spiro, *Defending the Master Race: Conservation, Eugenics and the Legacy of Madison Grant* (Lebanon, NH: University Press of New England, 2009), pp. 357–60.
57. Spiro, *Defending the Master Race,* pp. 370–71.
58. Mrs. George H. Webb to Adolf Hitler, July 26, 1933, BAB, R1501/126248.

59. There are several dozen newspaper articles dealing with aspects of Nazi anti-Semitism and eugenics included in the Laughlin Papers. C-2-7:2 and E-1-4:2 are examples of such collections. As a representative example for the infiltration of "race" into American electoral politics, see "The Kansas Isms: Racial Issues Is Interjected Into Gubernatorial Campaign," September 26, 1938, C-2-7:2, Laughlin Papers. The paper from which this article was taken is not given.

60. Eleonore von Trott zu Solz, "In Praise of Hitlerism," C-2-6:3, Laughlin Papers.

61. Ibid.

62. Conversation with Helmut von Verschuer, October 21, 2007. For an account of Eleonore von Trott zu Solz and her son Adam, see Benigna von Krusenstjern, *"daß es Sinn hat zu sterben - gelebt zu haben": Adam von Trott zu Solz 1904–1944; Biographie* (Göttingen: Wallstein, 2009).

63. Hyman Achinstein to the Director of the Department of Eugenics of the Carnegie Institution of Washington, DC, May 10, 1934, C-4-7:1, Laughlin Papers.

64. Quote taken from Kühl, *The Nazi Connection,* p. 76.

65. Garland E. Allen, "The Eugenics Record Office at Cold Spring Harbor: An Essay in Institutional History," *Osiris,* 2nd ser., 2 (1986): 254.

66. Falk Ruttke, "Zwei-Jahres-Versammlung der Internationalen Federation Eugenischer Organisationen in Zürich vom 18. bis 21. Juli 1934," February 13, 1935, MPIP-HA: GDA 33.

67. Kühl, *Die Internationale der Rassisten,* p. 128

68. Ludovic Naudeau, "La stérilisation des 'inaptes' en Allemagne," *Le Phare,* August 14, 1933, BAB R1501/126248/444.

69. William H. Schneider, *Quality and Quantity: The Quest for Biological Regeneration in Twentieth-Century France* (New York: Cambridge University Press, 1990), p. 187; quotes taken from "Die Internationale Föderation der Gesellschaften für Eugenik," excerpted from the *Revue Anthropologique* 43 (1933): 388–89, MPIP-HA: GDA 34.

70. Alfred Mjøen to Ernst Rüdin, January 7, 1934, MPIP-HA: GDA 33.

71. Kühl, *Die Internationale der Rassisten,* pp. 128–29. All quotes taken from Kühl.

72. Ibid.

73. Georges Schreiber "Der Nationalsozialismus und die eugenische Sterilization in Deutschland," excerpted from *Le Siècle Médical,* December 1, 1934, MPIP-HA: GDA 35.

74. Ibid.

75. Falk Ruttke, "Zwei-Jahres-Versammlung der Internationalen Föderation Eugenischer Organisationen in Zürich vom 18. bis 21. Juli 1934," February, 13, 1935, MPIP-HA: GDA 33.

76. "Die Rassenhygieniker gegen die Kriegshetze," *Völkischer Beobachter,* August 2, 1934, MPIP-HA: NLR 3.

77. Kühl, *Die Internationale der Rassisten,* p. 130.

78. "Was ist Rassenhygiene," Eine Unterredung mit Prof. Dr. Rüdin," August 1935, MPIP-HA: NLR 3.
79. Kühl, *Die Internationale der Rassisten,* p. 138.
80. Ibid.
81. Gunnar Dahlberg, "To the King," MPIP-HA: GDA 113; all quotes taken from this document. Kühl, *Die Internationale der Rassisten,* p. 138.
82. Alfred Mjøen to Ernst Rüdin, January 7, 1934, MPIP-HA: GDA 33; all quotes taken from this document. Roll-Hansen, "Norwegian Eugenics: Sterilization as Social Reform," p. 166; Kühl, *Die Internationale der Rassisten,* p. 138.
83. Kühl, *Die Internationale der Rassisten,* p. 139.
84. Ernst Rüdin to Karl Astel, June 12, 1936, MPIP-HA: GDA 36.
85. Kühl, *Die Internationale der Rassisten,* p. 139; Ernst Rüdin to Dr. Kapp, May 8, 1936, MPIP-HA: GDA 36. All quotes are taken from this document.
86. "Organizations: The Twelfth Meeting of the International Federation of Eugenics Organizations," *Eugenical News* 21 (1936): 106.
87. Kühl, *Die Internationale der Rassisten,* pp. 140–41; Eugen Fischer to Ernst Rüdin, July, 13, 1936, MPIP-HA: GDA 36.
88. Torsten Sjögren to Ernst Rüdin, September 18, 1936, MPIP-HA: NLR 3; Kühl, *Die Internationale der Rassisten,* p. 141.
89. "IV Internationaler Kongress für Rassenhygiene (Eugenik) Wien, 26. Bis 28. August 1940," MPG-Archiv, I, Rep. 1A, Nr. 95; Kühl, *Die Internationale der Rassisten,* pp. 142–43.
90. Kühl, *Die Internationale der Rassisten,* pp. 153–54; Ernst Rüdin to Dr. H. Nilsson-Ehle, March 16, 1936, MPIP-HA: GDA 36. Quotes taken from this letter.
91. "Seventh Genetics Congress to Have General Eugenics Section," *Journal of Heredity* 27 (1936): 344; "Seventh International Congress of Genetics, C-2-5:3, Laughlin Papers.
92. Kühl, *Die Internationale der Rassisten,* p. 155. All quotes taken from Kühl.
93. Ibid.; www.eugenicsarchiv.org/ID#1765; Ernst Rüdin to Dr. H. Nilsson-Ehle, March 16, 1936, MPIP-HA: GDA 36. Rüdin also sent this letter to Mjøen, Lundborg, Sjögren, and other like-minded Swedish colleagues.
94. Mark B. Adams, Garland E. Allen, and Sheila Faith Weiss, "Human Heredity and Politics: A Comparative Institutional Study of the Eugenics Record Office at Cold Spring Harbor (United States), the Kaiser Wilhelm Institute for Anthropology, Human Heredity and Eugenics (Germany), and the Maxim Gorky Medical Genetics Institute (USSR)," *Osiris,* 2nd ser., 20 (2005): 253; H. J. Muller, *Out of the Night* (New York: Vanguard Press, 1935).
95. Otto Mohr to Milislav Demerec, September 9, 1937, October 25, 1937; Otto Mohr to A. J. Muralov, January 7, 1937; Otto Mohr to A. J. Muralov and N. I. Vavilov, March 13, 1937, APL Genetics Society of America Papers: 575.06:G.
96. "Russians Abandon Genetics Meeting," C-2-5:3, Laughlin Papers; quotes taken from this article. Kühl, *Die Internationale der Rassisten,* p. 156.

97. "Social Biology and Population Improvement," *Nature* 144 (1939): 521–22, quote taken from this article; "Plan for Improving Population," *Eugenical News* 24 (1939): 63–64; Kühl, *Die Internationale der Rassisten*, pp. 156–57.

98. *Ier Congrès Latin d' Eugénique: Rapport* (Paris: Masson et Cie, Editeurs, 1937); Kühl, *Die Internationale der Rassisten*, pp. 144; Nancy Leys Stepan, *"The Hour of Eugenics": Race, Gender and Nation in Latin America* (Ithaca, NY: Cornell University Press, 1991), p. 19.

99. See, for example, Henry Friedlander, *The Origins of Nazi Genocide: From Euthanasia to the Final Solution* (Chapel Hill: University of North Carolina Press, 1995).

100. Dr. Lemme to Ernst Rüdin, January 7, 1937, and January 25, 1937; Ernst Rüdin to Dr. Lemme, January 29, 1937, MPIP-HA: GDA 50.

101. Herbert Linden to Ernst Rüdin, February 7, 1940, MPIP-HA: NLR 3.

102. "Protokoll der Sitzung am 12. November 1940: Die Internationalen Verbände," BAB, 4901/3191, 3190, 3058.

103. "Protokoll," BAB, 4901/3191/9.

104. Ibid., 3191/40.

105. Ibid., 3191/43.

106. Kühl, *Die Internationale der Rassisten*, p. 169; "Vermerk: Anthropologie," BAB, 4901/3058/5. Quotes taken from Kühl.

107. BAB, 4901/3190/131–32.

108. Ibid., 3190/135.

109. Sheila Faith Weiss, *Race Hygiene and National Efficiency: The Eugenics of Wilhelm Schallmayer* (Berkeley: University of California Press, 1987), chap. 1.

CONCLUSION

1. Max Weinrich, *Hitler's Professors: The Part of Scholarship in Germany's Crimes against the Jewish People*, 2nd ed. (New Haven, CT: Yale University Press, 1999), p. 240; I found a portion of this very appropriate quote for my thesis in Jonathan Peter Spiro, *Defending the Master Race: Conservation, Eugenics, and the Legacy of Madison Grant* (Lebanon, NH: University Press of New England, 2009), p. 376

2. For a discussion of these theses, see Mark B. Adams, Garland E. Allen, and Sheila Faith Weiss, "Human Heredity and Politics: A Comparative Institutional Study of the Eugenics Record Office at Cold Spring Harbor (United States), the Kaiser Wilhelm Institute for Anthropology, Human Heredity and Eugenics (Germany) and the Maxim Gorky Medical Genetics Institute (USSR)," *Osiris*, 2nd ser., 20 (2005): 259n70.

3. Ibid., p. 259nn71–72.

4. Detlev J. K. Peukert, *The Weimar Republic: The Crisis of Classical Modernity*, trans. Richard Devson (New York: Hill and Wang, 1989), pp. 275–81.

5. Adams, Allen, and Weiss, "Human Heredity and Politics," p. 259.

6. Gerhard Baader, Susan Lederer, Morris Low, Florian Schmaltz, and Alexander von Schwerin, "Pathways to Human Experimentation, 1933–1945: Germany, Japan, and the United States," *Osiris*, 2nd ser., 20 (2005): 205–31.

7. This thesis has been discussed in Adams, Allen, and Weiss, "Human Heredity and Politics," p. 260, and Sheila Faith Weiss, "Human Genetics and Politics as Mutually Beneficial Resources: The Case of the Kaiser Wilhelm Institute for Anthropology, Human Heredity, and Eugenics during the Third Reich," *Journal of the History of Biology* 39 (2006): 73–83.

8. Adams, Allen, and Weiss, "Human Heredity and Politics," p. 260.

9. Ibid., pp. 255, 257, and 260.

10. Ibid., pp. 260–61.

11. Ibid., p. 261.

12. Henry Friedlander, *The Origins of Nazi Genocide: From Euthanasia to the Final Solution* (Chapel Hill: University of North Carolina Press, 1995), and "Physicians as Killers in Nazi Germany: Hadamar, Treblinka and Auschwitz," in *Medicine and Medical Ethics in Nazi Germany: Origins, Practices, Legacies,* ed. Francis R. Nicosia and Jonathan Huener (New York: Berghahn Books, 2002), pp. 56–73.; Michael Kater, "Criminal Physicians in the Third Reich: Toward a Group Portrait," in *Medicine and Medical Ethics in Nazi Germany,* pp. 77–92.

13. These motivations have been discussed in Weiss, "Human Genetics and Politics as Mutually Beneficial Resources," pp. 80–81.

14. Robert Gellately, *Lenin, Stalin, and Hitler: The Age of Social Catastrophe* (New York: Alfred Knopf, 2007), pp. 108–9 and 446–47. For a discussion of the brutal nature of the war on the Eastern Front, see Omer Bartov, *Hitler's Army: Soldiers, Nazis, and War in the Third Reich* (New York: Oxford, 1991).

15. Eugen Fischer, "Fünfzig Jahre im Dienste der menschlichen Erbforschung und Anthropologie: Lebenserinnerung und Einblicke in die Entwicklung dieser Wissenschaft," (unpublished ms., Freiburg, ca. 1945–47); discussion with Fischer's grandson, Eberhard Fischer, 2003. Hermann Fischer ultimately fell on the Eastern Front in July 1942. Niels C. Lösch, *Rasse als Konstrukt: Leben und Werk Eugen Fischers* (Frankfurt am Main: Peter Lang, 1997), p. 391.

16. For a discussion of this factor, see Adams, Allen, and Weiss, "Human Heredity and Politics," p. 262.

17. http://skeptic.com/cognitivedissonance.html.

18. William Sheridan Allen, *The Nazi Seizure of Power: The Experience of a Single German Town, 1922–1945,* rev. ed. (New York: Franklin Watts, 1984), pp. 303, 191, and 287–89. The first edition of the book was published in 1965.

19. Adams, Allen, and Weiss, "Human Heredity and Politics," p. 262. Although this is a multiauthored article, this significant thesis was formulated and written by Garland Allen.

Index